Urban Aquaculture

This page intentionally left blank

Urban Aquaculture

Edited by

Barry Costa-Pierce

Rhode Island Sea Grant College Program
University of Rhode Island
USA

Alan Desbonnet

Rhode Island Sea Grant College Program
University of Rhode Island
USA

Peter Edwards

Agricultural and Aquatic Systems Program
Asian Institute of Technology
Thailand

and

Dale Baker

New York Sea Grant
Cornell University
USA

CABI Publishing

CABI is a trading name of CAB International

CABI Head Office
Nosworthy Way
Wallingford
Oxfordshire OX10 8DE
UK

Tel: +44 (0)1491 832111
Fax: +44 (0)1491 833508
Email: cabi@cabi.org
Web site: www.cabi.org

CABI North American Office
875 Massachusetts Avenue
7th Floor
Cambridge, MA 02139
USA

Tel: +1 617 395 4056
Fax: +1 617 354 6875
Email: cabi-nao@cabi.org

A catalogue record for this book is available from the British Library, London, UK.

Library of Congress Cataloging-in-Publication Data
Urban aquaculture / edited by Barry Costa-Pierce … [et al.].
p. cm.
Includes bibliographical references and index.
ISBN 0-85199-829-1 (alk. paper)
1. Urban aquaculture. 2. Water reuse. I. Costa-Pierce, Barry A. II. Title.

SH141.U73 2005
639.8′09173′2—dc22

2004017470

ISBN-13: 978-0-85199-829-9

First published 2005
Transferred to print on demand 2012

Printed and bound in the UK by CPI Group (UK) Ltd, Croydon CR0 4YY.

Contents

Contributors

Norman J. Blake, *College of Marine Sciences, University of South Florida, 140 Seventh Avenue South, St. Petersburg, FL 33701, USA.* (nblake@marine.usf.edu)

Craig L. Browdy, *Waddell Mariculture Center, Marine Resources Research Institute, South Carolina Department of Natural Resources, P.O. Box 12559, Charleston, SC 29487, USA.* (browdycl@musc.edu)

Stuart W. Bunting, *Institute of Aquaculture, University of Stirling, Stirling FK9 4LA, UK.* (s.w.bunting@stir.ac.uk)

Joseph K. Buttner, *Northeastern Massachusetts Aquaculture Center and Department of Biology, Salem State College, 352 Lafayette Street, Salem, MA 01970, USA.* (joe.buttner@salemstate.edu)

Barry Costa-Pierce, *Rhode Island Sea Grant College Program, University of Rhode Island, Graduate School of Oceanography, Narragansett, RI 02882, USA.* (bcp@gso.uri.edu)

Niels De Pauw, *Laboratory of Environmental Toxicology and Aquatic Sciences, Department of Applied Ecology and Environmental Biology, Ghent University, J. Plateaustraat 22, B-9000 Ghent, Belgium.* (niels.depauw@Ugent.be)

Alan Desbonnet, *Rhode Island Sea Grant College Program, University of Rhode Island, Graduate School of Oceanography, Narragansett, RI 02882, USA.* (aland@gso.uri.edu)

Peter Edwards, *Agricultural and Aquatic Systems Program, School of Environment, Resources and Development, Asian Institute of Technology, P.O. Box 4, Klong Luang, Pathumthani 12120, Thailand.* (pedwards@inet.co.th)

J. Adam Frederick, *Maryland Sea Grant Extension Program, Center of Marine Biotechnology, 701 East Pratt Street, Baltimore, MD 21202, USA.* (frederic@mdsg.umd.edu)

Clifford A. Goudey, *Massachusetts Institute of Technology, Sea Grant College Program, Building NE20–376, 3 Cambridge Center, Cambridge, MA 02139, USA.* (cgoudey@mit.edu)

Nitai Kundu, *Institute of Wetland Management and Ecological Design, Department of Environment, Government of West Bengal, Kolkata, India.* (npk1967@yahoo.co.in)

David C. Little, *Institute of Aquaculture, University of Stirling, Stirling FK9 4LA, UK.* (d.c.little@stir.ac.uk)

Greg Lutz, *Louisiana State University, Aquaculture Research Station, 2410 Ben Hur Road, Baton Rouge, LA 70820, USA.* (glutz@agctr.lsu.edu)

Brandy M. Moran, *Massachusetts Institute of Technology, Sea Grant College Program, Building NE20–376, 3 Cambridge Center, Cambridge, MA 02139, USA.* (bmmoran@mit.edu)

Shaun M. Moss, *The Oceanic Institute, Makapu'u Point, 41–202 Kalanianaole Highway, Waimanalo, HI 96795, USA.* (smoss@oceanicinstitute.org)

Madhumita Mukherjee, *Department of Fisheries, Government of West Bengal, Kolkata, India.* (madhukol@rediffmail.com)

Minh Phan Van, *Department of Science, Technology and Environment, 244 Dien Bien Phu Street, District 3, Ho Chi Minh City, Vietnam.* (minhuyen@hcm.vnn.vn) *Current address: Laboratory of Environmental Toxicology and Aquatic Ecology, Ghent University, J. P.lateaustraat 22–9000 Ghent, Belgium.*

Allen R. Place, *Center for Marine Biotechnology, University of Maryland Biotechnology Institute, Suite 236, Columbus Center, 710 East Pratt Street, Baltimore, MD 21202, USA.* (place@umbi.umd.edu)

Vo Quy Hoan, *Hanoi Agricultural University, Gia Lam District, Hanoi, Vietnam.* (huanhuy@hotmail.com)

Cathy A. Roheim, *Department of Environmental and Natural Resource Economics, University of Rhode Island, 112 Woodward Hall, Kingston, RI 02881, USA.* (crw@uri.edu)

John J. Roy, *The Sound School Regional Aquaculture Center, 60 South Water Street, New Haven, CT 06519, USA.* (hawk1959@mindspring.com)

Martin P. Schreibman, *Aquatic Research and Environmental Assessment Center, Brooklyn College/CUNY, 2900 Bedford Avenue, Brooklyn, NY 11210, USA.* (martins@brooklyn.cuny.edu)

Harold J. Schreier, *Center for Marine Biotechnology, University of Maryland Biotechnology Institute, Suite 236, Columbus Center, 710 East Pratt Street, Baltimore, MD 21202, USA.* (schreier@umbi.umd.edu)

Colin R. Steven, *Center for Marine Biotechnology, University of Maryland Biotechnology Institute, Suite 236, Columbus Center, 710 East Pratt Street, Baltimore, MD 21202, USA.* (StevenC@afip.osd.mil)

John Stubblefield, *Center for Marine Biotechnology, University of Maryland Biotechnology Institute, Suite 236, Columbus Center, 710 East Pratt Street, Baltimore, MD 21202, USA.* (stubble@umbi.umd.edu)

Yossi Tal, *Center for Marine Biotechnology, University of Maryland Biotechnology Institute, Suite 236, Columbus Center, 710 East Pratt Street, Baltimore, MD 21202, USA.* (tal@umbi.umd.edu)

Michael B. Timmons, *Department of Biology and Environmental Engineering, Cornell University, 302 Riley-Robb, Ithaca, NY 14850, USA.* (mbt3@cornell.edu)

Chester B. Zarnoch, *Aquatic Research and Environmental Assessment Center, Brooklyn College/CUNY, 2900 Bedford Avenue, Brooklyn, NY 11210, USA.* (czarnoch@brooklyn.cuny.edu)

Yonathan Zohar, *Center for Marine Biotechnology, University of Maryland Biotechnology Institute, Suite 236, Columbus Center, 710 East Pratt Street, Baltimore, MD 21202, USA.* (zohar@umbi.umd.edu)

Preface

The world is in the midst of the greatest human migration of all time, with millions of people moving from rural, inland areas to coastal cities. The United Nations estimates that by 2025 the world's urban population will increase to 5.1 billion people, equivalent to the entire human population on Earth in 1930. Meeting basic human needs for protein foods by urban consumers and disposal of human wastes concentrated in urban centres will be a major challenge for the future. As we cannot expect to catch more food from the sea, the world must turn to farming the waters – not just hunting them – and rapidly accelerate the 'blue revolution'. The expansion of aquaculture will have to occur not only in rural areas but also in the urban areas of the world that can be centres not only of marketing and distribution, but also of recycling and production. Traditional, often ancient aquaculture methods integrate wastewater disposal, retain and recycle nutrients, and also produce safe drinking water within peri-urban spaces next to urban centres. These ecosystem-conscious aquatic farms are models for modern ecological engineering. Marcel Proust said, 'The real voyage of discovery consists not in seeking new landscapes but in having new eyes.'

The late ecologist Eugene Odum defined cities as fabricated, heterotrophic parasites on the global landscape. Our challenge is to make cities more autotrophic, more self-feeding, and assist the evolution of urban agriculture and aquaculture and other 'green businesses' in the underutilized urban and peri-urban environments of the world. As rural lands disappear, brownfields, abandoned industrial buildings and warehouses of the 20th century could become integrated aquaculture/agriculture/energy parks of the 21st century. But urban aquaculture is not only the growing of aquatic plans and animals in the cities and the peri-urban neighbourhoods. Cities are the most important marketplaces for all aquaculture products. Many of these markets are now holding tonnes of live products for consumers. New aquaculture businesses have evolved to maintain these valuable, live commodities. And there lies the seed of potential for these supply-type urban businesses to evolve even further. Actually, when we think about it, this keeping of fish in markets is how aquaculture started in Asia some 2000 years ago (Costa-Pierce and Effendi, 1988). Fish were kept live in woven baskets and bamboo cages in ponds and canals outside markets. Fishmongers noticed how they became tame and began to feed them with market wastes, and the rest is history. Asia became the world's aquaculture leader through the development of market-driven aquaculture technologies that met human needs.

Urban areas are centres of change and innovation, and of rapid developments in education. As urban areas become completely human-dominated ecosystems with people increasingly separated from nature, integrated aquaculture systems put in our schools can not only help reconnect people to the natural world, distant times and their ethnic roots, but also offer many future possibilities as invaluable tools to increase mathematics and science capabilities in our urban educational systems.

Lewis Mumford, in his 1938 *Introduction to The Culture of Cities*, said that 'The city is a fact in nature, like a cave, a run of mackerel, or an ant-heap. But it is also a conscious work of art, and it holds within its communal framework many simpler and more personal forms of art.' It is our challenge and our duty to encourage the art of aquaculture in urban areas and plan creatively for its beauty and utility in revitalized cities.

Barry Costa-Pierce and Alan Desbonnet

Reference

Costa-Pierce, B.A. and Effendi, P. (1988) Sewage fish cages of Kota Cianjur. Naga. *The ICLARM Quarterly* 11(2), 7–9.

Mumford, L. (1938) *An Introduction to the Culture of Cities.* Harcourt, Brace & Co, New York.

1 A Future Urban Ecosystem Incorporating Urban Aquaculture for Wastewater Treatment and Food Production

Barry Costa-Pierce and Alan Desbonnet
Rhode Island Sea Grant College Program, University of Rhode Island, Graduate School of Oceanography, Narragansett, USA

Abstract

Predicted increases in wastewater flows from seven megacities by 2050 show that even with large capital investments in sewage treatment to treat an ever-increasing volume of wastewater, nitrogen loadings to coastal oceans will continue to increase dramatically. Furthermore, it is uncertain if nations will be able to afford the huge capital investments required to gain nitrogen reduction via tertiary or reverse osmosis treatments, or if these nitrogen reductions will be enough to fall below a threshold that would prevent the eutrophication of coastal waters. Carrying capacity models for use in coastal waters to guide the management of wastewater nutrient loadings are not widely available, making cost–benefit analyses difficult, if not impossible.

In order to avoid continuing damage to coastal ecosystems from increased wastewater nutrient loadings, future urban ecosystems must plan for: (i) diverting a greater percentage of wastewaters inland, and more aggressively developing water markets for wastewaters; (ii) increasing research investments into understanding the ecological roles and carrying capacities of wastewater inputs to protect (and even possibly enhance) nearshore ecosystems; and (iii) developing new, innovative urban ecosystems that make productive use of wastewaters for aquaculture and agriculture.

A pilot scale aquaculture–wetland ecosystem (AWE) using wastewaters was developed in Los Angeles County, California (USA) that grew luxuriant crops of Chinese water spinach and tilapia for food, and water hyacinths as soil conditioners (mulch and composts). The AWE removed over 97% of the nutrients from tertiary-treated wastewaters.

Introduction

By 2015, the United Nations Population Division (1994) estimates there will be 27 mega-cities of 10 million or more people – 23 of them in economically developing countries – and most of these will be located on the coast. By 2025, the world's urban population will increase to 5.1 billion people, equivalent to the entire human population on planet Earth in 1930 (United Nations Population Division, 1994). This massive trend towards urbanization of the landscape will have major influences on ecosystems at regional and global scales (Alberti *et al.*, 2003). To sustain these people the United Nations World Commission on Environment and Development projects a 5–10-fold expansion in industrial activities (WCED, 1987). Matthews *et al.* (2000)

(eds B. Costa-Pierce, A. Desbonnet, P. Edwards and D. Baker)

project a 50% increase in world population, a 500% increase in global economic activity and a 300% increase in global energy consumption and manufacturing activity.

The huge human demand on the Earth's freshwater and marine resources could lead to massive losses of biodiversity and the complete dismantling of the remaining intact coastal ecosystems if we defer the necessary planning and innovative natural and social engineering needed to ensure the sustainability of water resources for both nature and millions of people. Given these challenges, Vitousek *et al.* (1997) recommend: a slowing of the rate of population growth, since humanity's growth drives all resource use and waste generation; a reduction in the rate of human impacts, since ecosystems can react to lower rates of impacts and stabilize with moderate levels of human-induced changes; and accelerating ecosystem-level research and management understanding.

Coastal oceans are among the most heavily used and modified areas of the planet, suffering a disproportionate amount of habitat destruction and pollution. Human activities in the coastal zone deliver sewage, solid wastes, refuse (marine debris), sediments, dust, pesticides and oil hydrocarbons to rivers, estuaries and coastal oceans. It is estimated that about 80% of marine pollution originates from land-based sources and activities (Vitousek *et al.*, 1997). Assessing the range, magnitude and delivery of land-based sources of pollution to coastal oceans is a major global effort.

There is a growing chorus of voices saying the planet is entering into a time of water crisis (Falkenmark, 1989). Humanity is now one of the major driving forces of the Earth's hydrologic cycle, using more than half of the world's freshwater runoff (Vitousek *et al.*, 1997). Water is increasingly scarce in regions of limited rainfall such as in the Middle East, Africa and portions of Asia, to name a few (Gleick, 1998; Speidel *et al.*, 1988). Most of the world's rivers are dammed – there are 36,000 dams – and the number of dams is increasing (Costa-Pierce, 1997; UNEP/GRID, 2002). Many major rivers (Colorado, Nile and Ganges) are so heavily used that little or no water reaches their deltas and the sea. Only 2% of rivers in the USA run free, and many major inland water areas (Aral Sea, Lake Chad, etc.) have nearly dried up or been reduced to small, saline lakes. It is estimated that two out of every three people will live in water-stressed areas by the year 2025, and that about 450 million people in 29 countries currently suffer from water shortages (UNEP/GRID, 2002). In Africa, it is estimated that 25 countries will be experiencing water stress (below 1700 m^3/head/year) by 2025. Recycling of wastewater is one of the main options when looking for new sources of water in water-scarce regions.

While availability of water is a major issue in water-scarce regions, water quality issues are of concern at a global scale. Clean water supplies and sanitation are major problems in many parts of the world, with 20% of the global population lacking access to safe drinking water. Water-borne diseases from faecal pollution of surface waters are a major cause of illness in developing countries (Blumenthal *et al.*, 2000). Polluted water is estimated to affect the health of 1.2 billion people, and contributes to the death of 15 million children annually (UNEP/GRID, 2002).

Inputs of nutrients to the coastal zone from wastewaters, development and agriculture pose a major threat to the sustainability of coastal marine ecosystems. Increased wastewater discharges result in dramatic ecosystem changes to nearshore environments, have human health impacts, and huge economic consequences for tourism, recreation, fishing and aquaculture interests (CGER, 1993). Wastewaters are, however, highly valuable sources of freshwaters that are enriched in nutrients – resources that are much too valuable to be wasted by simple disposal and dilution in global oceans. Further treatment of wastewaters creates a product of sufficient quality that it can, and should, be utilized (Asano and Levine, 1998). The reuse of treated wastewaters for agricultural and industrial purposes should be part of a comprehensive urban planning process to reduce water consumption, improve the quality of receiving waters and reduce overall environmental impacts.

Current Solutions

Sewage often accounts for 50% or more of the nitrogen inputs to various estuaries and may be a more controllable input than those from agriculture and other non-point sources (Nixon and Pilson, 1983; Table 1.1).

Currently there are only two solutions to the disposal of wastewaters in coastal regions: (i) discharge to coastal oceans; and (ii) enormous capital investments to treat wastewaters using secondary or tertiary treatment, or expensive reverse osmosis technologies. The major problems with these alternatives are that wastewaters cause environmental/ocean contamination leading to eutrophication, toxic pollution, harmful algal blooms and other contamination (CGER, 1993) and the fact that the large capital investments for sewage treatment to tertiary treatment (and beyond) are not affordable in poorer countries. Indeed, even many of the rich countries struggle to obtain the fantastic amounts of capital required for wastewater treatment to tertiary levels and beyond.

It is a common perception that tertiary sewage treatment plants (TSTPs) remove nutrients completely and are 'environmentally friendly' (Table 1.2). However, TSTPs are not a panacea and will not solve the

Table 1.1. Dissolved inorganic nitrogen inputs (DIN in mmol/m^2/year) and percent sewage to selected bays and estuaries (CGER, 1993).

Bays/estuaries	DIN inputs	Estimated % from sewage
Kaneohe Bay, Hawaii	230	78
Long Island Sound, New York	400	67
Chesapeake Bay, Maryland–Virginia	510	33
Apalachicola Bay, Florida	560	2
Narragansett Bay, Rhode Island	950	41
Raritan Bay, New Jersey	1,460	86
North San Francisco Bay, California	2,010	45
New York Bight, New York–New Jersey	31,900	82

Table 1.2. Engineering solutions alone cannot solve the wastewater crisis. Typical percentage removal capabilities for a range of wastewater treatment processes (CGER, 1993) indicate that complete nitrogen and phosphorus removal from wastewaters is impossible.

Treatment levels	Conventional primary	Chemically enhanced primary (CEP)	Conventional biological secondary preceded by conventional primary	Biological secondary preceded by CEP	Nutrient removal preceded by conventional biological secondary and conventional primary	Reverse osmosis
Nitrogen removal (as mg/l TN)	2–28	26–48	0–63	N/A	80–88	97
Phosphorus removal (as mg/l TP)	19–57	44–96	10–66	83–91	95–99	99

problem alone. While large capital investments in TSTPs will certainly assist in reducing nutrient inputs to coastal oceans, the facts are that: (i) the sheer magnitude of the predicted increased volumes of urban wastewater flows has the potential to increase, in some cases dramatically, the discharges of total nutrients given continued secondary treatment (Table 1.3); and (ii) while TSTPs will reduce overall nitrogen loadings, they do not reduce nutrient levels below those that stimulate primary productivity in aquatic environments. For example, total inorganic nitrogen concentrations (ammonia-N, nitrite-N, nitrate-N) in wastewater effluents from the Pomona, California TSTP averaged 14.1 mg/l (range 8.8–20.6 mg/l) over a year-long monitoring period (Costa-Pierce, 1998). So, while nutrient reduction will be a good course to pursue, it is unclear if TSTPs can provide the degree of treatment required to sustain healthy coastal ecosystems.

With the huge projected increases in wastewater discharges in the future (Table 1.3) – even with treatment levels obtainable by employing reverse osmosis technologies (Table 1.4) – nutrient reductions via engineering solutions alone may not provide a complete solution to the nutrients problem as long as the wastewater discharge pipes continue to be 'pointed

Table 1.3. Estimated increases in population growth, wastewater discharges, and nitrogen loadings from seven megacities up until 2050.

	Population (millions)[a]	Water use (10^3 l per year per person)[b]	Discharge (10^9 l per year)[c]	Nitrogen input (10^6 kg per year)[d]				
				Raw[e]	Primary[f]	Secondary[g]	Tertiary[h]	Reverse osmosis[i]
2000								
New York City	16.7	259	4334	173.3	138.7	86.7	26.0	8.7
Tokyo	26.4	157	4152	166.1	132.9	83.0	24.9	8.3
Calcutta	13.1	18	235	9.4	7.5	4.7	1.4	0.5
Los Angeles	13.2	259	3422	136.9	109.5	68.4	20.5	6.8
Mumbai	16.1	18	290	11.6	9.3	5.8	1.7	0.6
São Paulo	18.0	91	1635	65.4	52.3	32.7	9.8	3.3
Jakarta	11.0	12	132	5.3	4.2	2.6	0.8	0.3
2025								
New York City	18.7	259	4843	193.7	155.0	96.9	29.1	9.7
Tokyo	27.2	157	4277	171.1	136.9	85.5	25.7	8.6
Calcutta	19.9	18	357	14.3	11.4	7.1	2.1	0.7
Los Angeles	15.1	259	3898	155.9	124.7	78.0	23.4	7.8
Mumbai	26.8	18	481	19.3	15.4	9.6	2.9	1.0
São Paulo	22.2	91	2020	80.8	64.4	40.4	12.1	4.0
Jakarta	20.3	12	243	9.7	7.8	4.9	1.5	0.5
2050								
New York City	20.4	259	5279	211.1	168.9	105.6	31.7	10.6
Tokyo	27.3	157	4279	171.1	136.9	85.6	25.7	8.6
Calcutta	32.2	18	580	23.2	18.6	11.6	3.5	1.2
Los Angeles	15.9	259	4118	164.7	131.8	82.4	24.7	8.2
Mumbai	36.0	18	648	25.9	20.7	13.0	3.9	1.3
São Paulo	23.1	91	2106	84.2	67.4	42.1	12.6	4.2
Jakarta	23.9	12	287	11.5	9.2	5.7	1.7	0.6

[a]From United Nations Population Division (2002).
[b]From Gleick (1998). Water use from 2000 to 2050 is assumed to be constant.
[c]Derived from Population × Water Use estimates.
[d]Based on a Total N concentration of 40 mg/l in untreated domestic sewage (Corbitt, 1989).
[e]Assumes 0% N removal.
[f]Assumes 20% N removal.
[g]Assumes 50% N removal.
[h]Assumes 85% N removal (CGER, 1993).
[i]Assumes 95% N removal (Hocking, 1998).

Table 1.4. Estimated nitrogen loadings from seven megacities will still increase even while employing expensive, highly sophisticated reverse osmosis technologies.

	Nitrogen load[a] (10^6 kg per year)	
	2000	2050
New York City	8.7	10.6
Tokyo	8.3	8.6
Calcutta	0.5	1.1
Los Angeles	6.8	8.2
Mumbai	0.6	1.3
São Paulo	3.3	4.2
Jakarta	0.3	0.6

[a]Assumes 95% N removal (Hocking, 1998) for 2000 and 2050.

towards the coast'. While investments in sewage treatment will undoubtedly increase worldwide in the next century, the quantities of nutrients discharged to the ocean will continue to cause degradation of coastal ecosystems. It is important that new, more comprehensive strategies be developed and that more productive uses be found for the increasing volumes of nutrient-rich wastewaters being generated by urban areas.

Wastewater Strategies for a Future Urban Ecosystem

More comprehensive strategies are needed to develop future urban ecosystems, and research will be needed to provide information by which to develop the strategies. Research will have to be conducted from a perspective that incorporates human activities as a part of ecosystem functions, not just as a perturbation to them, and with a focus on maintaining life support systems while meeting human needs (Palmer *et al.*, 2004). These strategies will need to involve the following:

- Greater knowledge is needed on the impacts of wastewater discharges to coastal oceans. Science-based information will allow more accurate determination of the carrying capacities of recipient coastal waters so that the natural treatment capacity for urban wastewaters that cause no environmental harm can be determined more effectively (CGER, 1993).
- Inland water markets for wastewaters need to be developed, and then expanded.
- Development of new, innovative urban ecosystems to treat wastewaters need to be encouraged. These systems will use the principles of ecological and social design and engineering (Lyle, 1994; Costa-Pierce, 2002; Mitsch and Jorgensen, 2004).

Wastewater Impacts of Discharges on Coastal Oceans

Nitrogen is the primary limiting factor for marine plankton production in marine ecosystems (Ryther and Dunstan, 1971). Significant increases in nitrogen and solids in coastal waters create symptoms referred to as 'eutrophication'. The symptoms of eutrophication are many: food chain shifts from phytoplankton and diatom-based to flagellate-based food webs (Doering *et al.*, 1989; Rosenberg *et al.*, 1990; Smayda, 1989); anoxic conditions causing catastrophic impacts to populations of fishes and invertebrates (Mearns and Word, 1982; Hansson and Rudstam, 1990; Rosenberg *et al.*, 1990); increased abundances of nuisance species such as filamentous algae, macrophytes and jellyfish (Baden *et al.*, 1990; Rosenberg *et al.*, 1990); and losses of desirable vascular species, such as eelgrass, and of economically important species such as oysters and other shellfish species (Bricelj and Kuenstner, 1989; Ryther, 1989; Tracey *et al.*, 1989).

There are many examples of adverse impacts on coastal ecosystems from the accelerating inputs of wastewaters. Increased nitrogen loadings have been implicated in stimulating nuisance algal blooms and in degrading the structure and function of coastal marine ecosystems of the Southern California Bight off San Pedro Bay (Eppley *et al.*, 1972; Thomas *et al.*,

1974; MacIssac *et al.*, 1979; Mearns, 1981; Reish, 1984). Similar results have been reported for the Swedish coast of the Baltic Sea (Rosenberg *et al.*, 1990).

In some estuaries, naturally occurring turbidity limits light penetration, which in turn limits phytoplankton response to nutrient increases, while in other estuaries the phytoplankton blooms themselves limit light penetration (CGER, 1993). Malone (1982) reports on light limitation effects on eutrophication impacts in the New York Bight. In this case, nutrients are transported offshore from the Hudson River and New York Harbor – where light penetration is limited and nutrient input high – creating impacts to offshore ecosystems rather than impacting estuaries and adjacent nearshore ecosystems that receive wastewater discharges directly. In Chesapeake Bay, it has been found that light limitation resulting from phytoplankton blooms in response to nutrient inputs has created shading associated with the loss of eelgrass, an important bay macrophyte (Kemp *et al.*, 1983; D'Elia, 1987). Similar impacts have been recorded in the Dutch Wadden Sea (Gieson *et al.*, 1990) and in Australia (Cambridge *et al.*, 1986).

However, despite the 'bad reputation' of nutrients as agents of ecosystem change, nutrients are necessary for the maintenance of sustainable aquatic ecosystems. In moderation, nutrient inputs can stimulate primary productivity in coastal waters, providing for increased food availability to shellfish and other filter-feeding organisms, leading to productive marine food webs. Stimulation of primary productivity can be harnessed to provide food for juvenile fishes, which in turn can be cultured to food products, or fed to target culture species. The difference between beneficial and deleterious is aptly stated in a report published by the National Academy of Sciences: 'Whether or not nutrient inputs should be considered excessive depends in part upon the physics and ecological sensitivity of the receiving water body' (CGER, 1993). The challenge ahead lies in better defining the carrying capacities of receiving waters such that wastewater nutrient loadings perform ecosystem functions that are beneficial rather than detrimental. Good, practically applied science that informs management will have to be fostered to achieve that end.

Development of Inland Water Markets for Wastewaters

As human populations increase globally, the need for potable water will increase accordingly and, as water is a limited, finite resource, planning for water recycling and reuse will have to become 'mainstreamed' as a societal norm. Already wastewaters are used for irrigation of crops in various countries: an estimated 60% of the total effluent is used in Israel, 25% in India and 24% in South Africa (Blumenthal *et al.*, 2000). Use of wastewaters for irrigation of crops has been reported to increase yields by as much as 30% (Blumenthal *et al.*, 2000). In France, golf courses are becoming major consumers of wastewaters for use in greens-keeping and grounds maintenance (Bontoux and Courtois, 1996). The World Health Organization (WHO) has developed a set of guidelines for the use and treatment of wastewaters for reuse (World Health Organization, 1989).

However, while water markets have developed, they certainly have not developed rapidly enough to stem the increased disposal of wastewaters in the ocean. For example, in Los Angeles County, California (USA), ten wastewater reclamation plants discharge approximately 1966 km^3 per day of wastewater effluents from treatment systems ranging from non-disinfected secondary, to coagulated, filtered and chlorinated tertiary wastewaters. While there has been an increased reuse of reclaimed wastewaters in inland areas of southern California since 1970, water markets have been too slow to develop. Only an estimated 5% (91 km^3 per day) of the wastewater generated by the ten treatment plants is sold, and mostly to non-agricultural users (of the 322 users of reclaimed water only nine were agricultural; SDLA, 1995). Over 90% of wastewaters is discharged to the San Gabriel River and then to the ocean, or directly into the ocean at San Pedro Bay. This situation occurs even given the fact that reclaimed water in Los Angeles County is available at a substantial discount in comparison with potable water. In 1995, the costs of reclaimed water ranged from US\$61.78 to US\$411.24/km^3, a 15–72% discount from purchased water from the Colorado River (SDLA, 1995).

Wastewater storage reservoirs are a practical alternative in arid and semi-arid regions where agricultural production is limited by access to natural water resources. In Mexico, wastewater is pumped into storage reservoirs prior to use for crop irrigation. Raw wastewater effluent is pumped into a storage reservoir, where it is retained for a period of up to 7 months (Blumenthal *et al.*, 2000). Effluent then moves into a second reservoir where it is retained for an additional 2–6 months, attaining an even higher quality that has been deemed safe by WHO for irrigation, provided that the edible portion of the crop is not in direct contact with irrigation water. Wastewater from the second reservoir applied on crops that are consumed directly, such as salad greens, was found to increase diarrhoea, especially in children (Blumenthal *et al.*, 2000). Blumenthal *et al.* (2000) note similar results from wastewater storage reservoirs used in Israel, where wastewater is suitable for restricted irrigation. Blumenthal *et al.* (2000), however, document the use of sequential batch-fed storage reservoirs (3-week retention time) in Brazil that produced water quality with faecal coliform concentrations of less than 1000 colonies/ 100 ml, suitable for direct use in the irrigation of crops for human consumption. In the Netherlands, artificial lakes have been constructed to capture river water, which was then piped to subsurface aquifers that had been overdrawn by urban populations (Martinez and Psuty, 2004). While the Netherlands example does not incorporate wastewater treatment, similar concepts could be applied to treated wastewaters such that they could be recycled into aquifers for reuse in urban ecosystems.

While the possibilities for the use of wastewater as a commodity on the open market has many opportunities, present efforts in this regard are meagre, at best, and very little information is available at both national and global scales (Postel, personal communication). The future is wide open in this regard for innovative thinkers and strategic planners who can perhaps create an economically viable solution to integrating wastewater reuse into the design and functions of future urban ecosystems.

Using Ecological and Social Engineering to Develop New, Urban Aquaculture Ecosystems

Odum urged society to consider the fundamental ecological characteristics of urban centres (Odum, 1993). As pointed out in Chapters 3–7 of this volume, there is a fascinating wisdom based upon indigenous knowledge of the farming elders in India, Vietnam and other Asian countries on wastewater use. This knowledge serves as the basis for developing a futuristic approach emphasizing technological engagement and education on wastewater issues which will help to reverse the mentality of ‘flush it and forget it’. The wisdom in Odum’s thinking in regards to the current problem is that there needs to be greater attention to the use of wastewaters as ‘misplaced resources’.

Growing fish and vegetables using wastewaters produced by urban populations has been practised throughout much of recorded history, as is noted by Bunting and Little in Chapter 8. Various authors of this book, as noted in Chapters 3–7, show that fish production using wastewater effluents is a critical form of income and livelihood maintenance in less developed countries, and has evolved over time in response to rapid urbanization. As urban populations grow, and the demand for aquatic foods increases proportionately, the income-producing potential of integrated wastewater aquaculture/agriculture ecosystems will increase similarly. While integrating wastewater aquaculture into urban ecosystems is desirable for economic and ecosystem benefits, there are concerns for public health that must be addressed, as noted in Chapters 3–7 of this book. WHO has developed guidelines for the use of wastewaters in aquaculture (World Health Organization, 1989), based on the studies of Strauss (1985). The WHO guidance sets limits on faecal coliform concentrations so that the flesh of cultured fish will not be contaminated; skin and viscera must be handled carefully and discarded to avoid cross-contamination.

In November of 2002, a workshop in Hyderabad, India on the use of wastewaters for agricultural irrigation resulted in the development of The Hyderabad Declaration (2002), which outlines best practices for the use of wastewaters in agriculture. The declaration, as provided below, while focused on agricultural use, is directly applicable to aquaculture as well.

The Hyderabad Declaration on Wastewater Use

Rapid urbanization places immense pressure on the world's fragile and dwindling fresh water resources and over-burdened sanitation systems, leading to environmental degradation. We as water, health, environment, agriculture, and aquaculture researchers and practitioners from 27 international and national institutions, representing experiences in wastewater management from 18 countries, recognize that:

1.1 Wastewater (raw, diluted or treated) is a resource of increasing global importance, particularly in urban and peri-urban agriculture.
1.2 With proper management, wastewater use contributes significantly to sustaining livelihoods, food security and the quality of the environment.
1.3 Without proper management, wastewater use poses serious risks to human health and the environment.
We declare that in order to enhance the positive outcomes while minimizing the risks of wastewater use, there exist feasible and sound measures that need to be applied. These measures include:
2.1 Cost-effective and appropriate treatment suited to the end use of wastewater, supplemented by guidelines and their application.
2.2 Where wastewater is insufficiently treated, until treatment becomes feasible:
(a) Development and application of guidelines for untreated wastewater use that safeguard livelihoods, public health and the environment.
(b) Application of appropriate irrigation, agricultural, post-harvest, and public health practices that limit risks to farming communities, vendors, and consumers.
(c) Education and awareness programs for all stakeholders, including the public at large, to disseminate these measures.
2.3 Health, agriculture and environmental quality guidelines that are linked and implemented in a step-wise approach.
2.4 Reduction of toxic contaminants in wastewater, at source and by improved management.
3. We also declare that:
3.1 Knowledge needs should be addressed through research to support the measures outlined above.
3.2 Institutional coordination and integration together with increased financial allocations are required.
4. Therefore, we strongly urge policy-makers and authorities in the fields of water, agriculture, aquaculture, health, environment and urban planning, as well as donors and the private sector to safeguard and strengthen livelihoods and food security, mitigate health and environmental risks and conserve water resources by confronting the realities of wastewater use in agriculture through the adoption of appropriate policies and the commitment of financial resources for policy implementation.

An Experimental Urban Wastewater Ecosystem

In a pilot study in California, an experimental aquaculture–wetland ecosystem (AWE) and management protocols were tested to see if the available tertiary-treated wastewater could be used inland to simultaneously accomplish irrigation, aquatic food production and inorganic nitrogen removal from wastewaters received from the Pomona, California Tertiary Sewage Treatment Plant (TSTP) (Costa-Pierce, 1998). The experimental ecosystem comprised a wastewater supply tank, aquaculture ponds and an artificial wetland. The pilot aquaculture–wetland ecosystem was able to utilize wastewaters to:

- Grow luxuriant crops of Chinese water spinach (*Ipomea aquatica*): an edible vegetable with a high food value, which has a proven ability to control water quality (Abe *et al.*, 1992) (Fig. 1.1).
- Grow tonnes of water hyacinths (*Eichhornia crassipes*) that were harvested regularly and made into compost for soil conditioning. Tissue nitrogen concentrations of the water hyacinths were high enough in order for the pond plants to be 'self-composting' by heaping them together in windrows (Fig. 1.2).
- Grow healthy tilapia (*Oreochromis* spp.). The AWE produced over 1000 kg of fish per season.

The AWE accomplished significant protein production while simultaneously removing 97% of the inorganic N in the original tertiary-treated wastewater. Pond effluents were treated effectively by the in-pond plants and wetland components of the ecosystem (Fig. 1.3).

This aquatic agroecosystem using wastewater was a research and demonstration site in not only the productive uses of wastewaters, but also in the ecosystem restoration of a damaged peri-urban environment. The AWE was not only important for its production of aquatic foods but also important for its 'ecosystem services' as artificial wetlands attracting wildlife (Lyle, 1994). In the arid and stark urban environment of Los Angeles County, California (USA), the ponds and wetlands were observed to be vital habitats for a variety of threatened bird species, notably black-crowned night herons, blue and green herons and rails. Further technical details of the pilot-scale system, and the challenging and unique methods found to manage and control water quality with the high nutrient loadings in the arid climate, are described in Costa-Pierce (1998). A future AWE would include a treatment reservoir

Fig. 1.1. Overview of the polyculture ponds with floating water hyacinths and Chinese water spinach mats contained by floating booms. Water quality control was achieved by initially establishing a bed of water hyacinths over 50% of the pond area and by flushing ponds with wastewater 50% a day. Chinese water spinach was then planted into the hyacinths bed, and as it established, water hyacinths were removed for composting, Chinese water spinach established to cover 50% of the pond surface and wastewater flushing rates decreased to 20% a day (Costa-Pierce, 1998).

Fig. 1.2. Water hyacinths were harvested weekly from the pond surfaces to maintain a pond cover of 50%, and plants were heaped into windrows. Plant tissue nitrogen concentrations were high enough that plants were 'self-composting', e.g. no additional nitrogen was needed to make rich composts as soil conditioners.

Fig. 1.3. Overview of the artificial wetland. All outflow discharges from the culture ponds that received wastewaters as inputs flowed into an approximate 0.1 ha wetland planted with *Typha latifolia* and seeded with duckweed. Duckweed was harvested monthly from the wetland and used as a soil conditioner.

and use solar energy to power a recycling loop to pumped storage (Fig. 1.4).

While development of such aquaculture–wetland ecosystems is potentially new productive uses for treated, nutrient-rich wastewaters in the peri-urban areas around major urban centres, the economic, regulatory and public perceptions about these systems, none of which were addressed in the Costa-Pierce (1998) study, need much more examination. Blumenthal *et al.* (2000) conclude that wastewater, provided it maintains a concentration of 4×10^4 faecal coliforms or less, poses no health threat, from domestic or recreational contact, to persons older than 5 years of age. Increasing the reuse and efficiency of wastewater recycling by developing wastewater food production ecosystems, especially in arid regions of the world like Los Angeles (Khalil and Hussein, 1997), and keeping these wastewaters inland and away from the coastal ocean, are priority issues for global sustainability (Vitousek *et al.*, 1997). Such aquatic agroecosystems could make cities less heterotrophic and damaging to coastal ecosystems (Odum, 1993), and be new, important, non-polluting contributors to world food production.

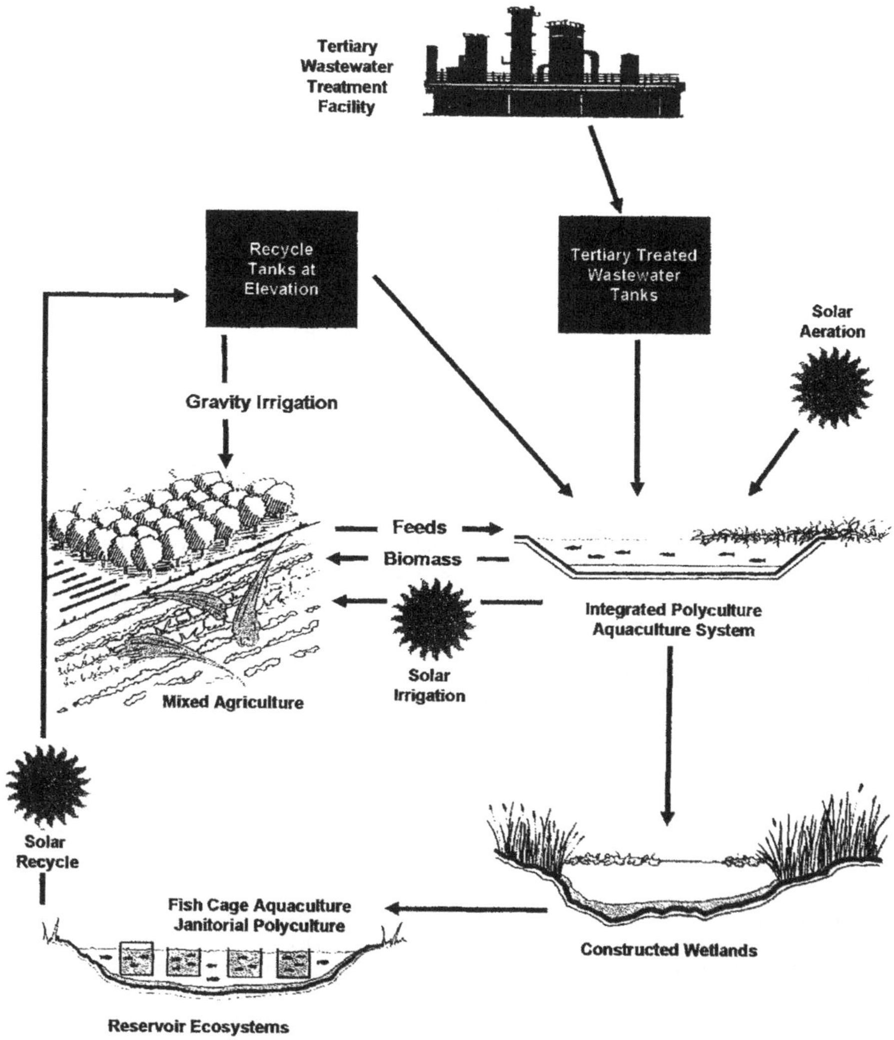

Fig. 1.4. Design of an integrated aquaculture–wetland ecosystem where aquatic protein production and nutrient removal from tertiary-treated wastewaters are accomplished by fish polyculture, in-pond aquatic plant culture, wetland and reservoir ecosystems integrated with solar arrays for aeration and water pumping. A reservoir below the wetland provides additional protein production, wastewater treatment and nutrient and water recovery, and tightens the integration with terrestrial food production ecosystems.

Conclusions

Marcel Proust said 'The real voyage of discovery consists not in seeking new landscapes but in having new eyes.' The expansion of aquaculture worldwide will have to occur not only in rural areas but also in the urban areas of the world that can be centres not only of marketing and distribution, but also of production, reuse and recycling. Our challenge is to make cities more autotrophic, more self-feeding and less ecologically damaging, and to assist the evolution of urban agriculture and aquaculture and other 'green businesses' in the urban and peri-urban environments of the world.

As urban areas become completely human-dominated ecosystems with people increasingly separated from nature, integrated aquatic ecosystems can help not only contribute to global food production and water sustainability while reducing impending environmental harm to coastal oceans, but also reconnect people to the natural world, distant times and their ethnic roots.

Acknowledgement

We would like to honour the recent passing of Professor John Lyle, founder of the Center for Regenerative Studies (CRS) at Cal Poly-Ponoma, whose vision made the CRS a reality.

References

Abe, K., Ozaki, Y. and Kihou, N. (1992) Use of higher plants and bed filter materials for domestic wastewater treatment in relation to resource recycling. *Soil Science and Plant Nutrition* 39, 257–267.

Alberti, M., Marzluff, J.M., Shulenberger, E., Bradley, G., Ryan, C. and Zumbrunnen, C. (2003) Integrating humans into ecology: opportunities and challenges for studying urban ecosystems. *Bioscience* 53(12), 1169–1179.

Asano, T. and Levine, A.D. (1998) Wastewater reclamation, recycling and reuse: an introduction. In: Asano, T. (ed.) *Wastewater Reclamation and Reuse.* Technomic Publishing Co., Lancaster, Pennsylvania, pp. 1–56.

Baden, S.P., Loo, L.O., Pihl, L. and Rosenberg, R. (1990) Effects of eutrophication on benthic communities including fish: Swedish west coast. *Ambio* 19, 112–133.

Blumenthal, U.J., Peasey, A., Ruiz-Palacios, G. and Mara, D.D. (2000) *WELL Study: Guidelines for Wastewater Reuse in Agriculture and Aquaculture: Recommended Revisions Based on New Research Evidence.* Task No. 68, Part 1. Water, Engineering and Development Centre, Loughborough University, UK.

Bontoux, J. and Courtois, G. (1996) Wastewater reuse for irrigation in France. *Water Science and Technology* 33(10/11), 45–49.

Bricelj, M. and Kuenstner, S. (1989) The feeding physiology and growth of bay scallops and mussels. In: Cosper, E.M., Carpenter, E.J. and Bricelj, V.M. (eds) *Novel Phytoplankton Blooms: Causes and Impacts of Recurrent Brown Tides and Other Unusual Blooms.* Springer, Berlin, pp. 491–509.

Cambridge, M., Chaffings, A., Brittan, C., Moore, L. and McComb, A. (1986) The loss of seagrasses in Cockburn Sound, Western Australia. II. Possible causes of seagrass decline. *Aquatic Botany* 24, 269–285.

CGER (Commission on Geosciences, Environment and Resources) (1993) *Managing Wastewater in Coastal Urban Areas.* National Academy Press, Washington, DC.

Corbitt, R.A. (ed.) (1989) *Standard Handbook of Environmental Engineering.* McGraw-Hill, New York.

Costa-Pierce, B.A. (1997) *From Farmers to Fishers: Developing Reservoir Aquaculture for People Displaced by Dams.* World Bank Technical Paper 369. The World Bank, Washington, DC.

Costa-Pierce, B.A. (1998) Preliminary investigation of an integrated aquaculture–wetland ecosystem using tertiary-treated municipal wastewater in Los Angeles County, California. *Ecological Engineering* 10, 341–354.

Costa-Pierce, B.A. (2002) Ecology as the paradigm for the future of aquaculture. In: Costa-Pierce, B.A. (ed.) *Ecological Aquaculture: The Evolution of the Blue Revolution.* Blackwell Science, Oxford, pp. 339–372.

D'Elia, C.F. (1987) Nutrient enrichment of the Chesapeake Bay – too much of a good thing. *Environment* 29, 6–33.

Doering, P.H., Oviatt, C.A., Beatty, L.L., Banzon, V.F., Rice, R., Kelly, S.P., Sullivan, B.K. and Frithsen, P.J. (1989) Structure and function in a model coastal ecosystem: silicon, the benthos and eutrophication. *Marine Ecology Progress Series* 62, 293–306.

Eppley, R., Carlucci, A., Holm-Hansen, O., Kiefer, D., McCarthy, J. and Williams, P. (1972) Evidence for eutrophication in the sea near southern California coastal sewage outfalls – July, 1970. *California Cooperative Oceanic Fisheries Investigation Report* 16, 74–83.

Eppley, R., Sapienza, C. and Renger, E. (1978) Gradients in phytoplankton stocks and nutrients off southern California in 1974–76. *Estuarine Coastal Marine Science* 7, 291–301.

Falkenmark, M. (1989) The massive water scarcity now threatening Africa: why isn't it being addressed? *Ambio* 18, 112–118.

Gieson, W., van Katwijk, M.M. and den Hartog, C. (1990) Eelgrass condition and turbidity in the Dutch Wadden Sea. *Aquatic Biology* 37, 71–85.

Gleick, P.H. (1998) *The World's Water 1998–1999. The Biennial Report on Freshwater Resources.* Island Press, Washington, DC.

Hansson, S. and Rudstam, L.G. (1990) Eutrophication and Baltic fish communities. *Ambio* 19, 123–125.

Hocking, M.B. (1998) *Handbook of Chemical Technology and Pollution Control.* Academic Press, San Diego, California.

Kemp, W.M., Twilley, R.R., Stevenson, J.C., Boynton, W.R. and Means, J.C. (1983) The decline of submerged vascular plants in upper Chesapeake Bay: summary of results concerning possible causes. *Journal Marine Technology Society* 17, 78–85.

Khalil, M.T. and Hussein, H.A. (1997) Use of waste water for aquaculture: an experimental field study at a sewage-treatment plant. *Egyptian Aquaculture Research* 28, 859–865.

Lyle, J. (1994) *Regenerative Design for Sustainable Development.* John Wiley & Sons, New York.

MacIssac, J., Dugdale, R., Huntsman, W. and Conway, H. (1979) The effect of sewage on uptake of inorganic nitrogen and carbon by natural populations of marine phytoplankton. *Journal Marine Research* 37, 51–66.

Malone, T.C. (1982) Factors influencing the fate of sewage-derived nutrients in the lower Hudson estuary and New York bight. In: Mayer, G.F. (ed.) *Ecological Stress and the New York Bight: Science and Management.* Estuarine Research Federation, Columbia, South Carolina, pp. 389–400.

Martinez, M.L. and Psuty, N.P. (eds) (2004) *Coastal Dunes: Ecology and Conservation.* Ecological Studies Series 171, Springer-Verlag, New York.

Matthews, E., Amann, C., Bringezu, S., Fischer-Kowalski, M., Huttler, W., Kleijn, R., Moriguchi, Y., Ottke, C., Rodenburg, E., Rogich, D., Schandl, H., Schutz, H., van der Voet, E. and Weisz, H. (2000) *The Weight of Nations: Material Outflows from Industrial Economies.* World Resources Institute, Washington, DC.

Mearns, A. (1981) Effects of municipal discharges on open coastal ecosystems. In: Geyer, R.A. (ed.) *Marine Environmental Pollution, 2, Dumping and Mining.* Elsevier, Amsterdam, pp. 25–66.

Mearns, A.J. and Word, J.Q. (1982) Forecasting the effects of sewage solids on marine benthic communities. In: Mayer, G.F. (ed.) *Ecological Stress in the New York Bight: Science and Management.* Estuarine Research Federation Symposium, Columbia, North Carolina.

Mitsch, W.J. and Jorgensen, S.E. (2004) *Ecological Engineering and Ecosystem Restoration.* John Wiley & Sons, New York.

Nixon, S.W. and Pilson, M.E.Q. (1983) Nitrogen in estuarine and coastal marine ecosystems. In: Carpenter, E.J. and Capone, D.G. (eds) *Nitrogen in the Marine Environment.* Academic Press, New York, pp. 565–648.

Odum, E. (1993) *Ecology and Our Endangered Life Support Systems.* Sinauer Pub., Sunderland, Massachusetts.

Palmer, M., Bernhardt, E., Chornesky, E., Collins, S., Dobson, A., Duke, C., Gold, B., Jacobson, R., Kingsland, S., Kranz, R., Mappin, M., Martinez, M., Micheli, F., Morse, J., Pace, M., Pascual, M., Palumbi, S., Reichman, O.J., Simons, A., Townsend, A. and Turner, M. (2004) Ecology for a crowded planet. *Science* 304, 1251–1252.

Reish, D. (1984) Domestic wastes. In: Kinne, O. (ed.) *Marine Ecology. Volume 5: Ocean Management.* John Wiley and Sons, New York, pp. 1711–1760.

Rosenberg, R., Elmgren, R., Fleischer, S., Jonsson, P., Persson, G. and Dahlin, H. (1990) Marine eutrophication case studies in Sweden. *Ambio* 19, 102–108.

Ryther, J.H. (1989) Historical perspective of phytoplankton blooms on Long Island and the green tides of the 1950s. In: Cosper, E.M., Carpenter, E.J. and Bricelj, V.M. (eds) *Novel Phytoplankton Blooms: Causes and Impacts of Recurrent Brown Tides and Other Unusual Blooms.* Springer, Berlin, pp. 375–382.

Ryther, J.H. and Dunstan, W.M. (1971) Nitrogen, phosphorus and eutrophication in the coastal marine environment. *Science* 171, 1009–1012.

SDLA (Sanitation Districts of Los Angeles County) (1995) Annual status report on reclaimed water use, fiscal year 1994–95. SDLA, Whittier, California.

Smayda, T.J. (1989) Primary production and the global epidemic of phytoplankton blooms in the sea: a linkage? In: Cosper, E.M., Carpenter, E.J. and Bricelj, V.M. (eds) *Novel Phytoplankton Blooms: Causes and Impacts of Recurrent Brown Tides and Other Unusual Blooms.* Springer, Berlin, pp. 449–483.

Speidel, D.H., Ruedisili, L.C. and Agnew, A.F. (eds) (1988) *Perspectives on Water: Uses and Abuses.* Oxford University Press, Oxford.

Strauss, M. (1985) Health aspects of nightsoil and sludge use in agriculture and aquaculture. Part II. Pathogen survival. International Reference Centre for Waste Disposal Report No. 04/85. SANDEC, Duebendorf.

The Hyderabad Declaration (2002) The Hyderabad Declaration on Wastewater Use in Agriculture. A workshop – 'Wastewater Use in Irrigated Agriculture: Confronting the Livelihood and Environmental Realities'. 11–14 November 2002, Hyderabad, India. Sponsored by the International Water Management Institute (IWMI, based in Colombo, Sri Lanka) and the International Development Research Centre (IDRC, based in Ottawa, Canada).

Thomas, W., Seibert, D. and Dodson, A. (1974) Phytoplankton enrichment experiment and bioassays in natural coastal seawater and in sewage outfall receiving waters off southern California. *Estuarine and Coastal Marine Science* 2, 191–206.

Tracey, G.A., Steele, R.L., Gatzke, J., Phelps, D.K., Nuzzi, R., Waters, M. and Anderson, D.M. (1989) Testing and application of biomonitoring methods for assessing environmental effects of noxious algal blooms. In: Cosper, E.M., Carpenter, E.J. and Bricelj, V.M. (eds) *Novel Phytoplankton Blooms: Causes and Impacts of Recurrent Brown Tides and Other Unusual Blooms.* Springer, Berlin, pp. 557–574.

UNEP/GRID. (2002) Vital water graphics. *An Overview of the State of the World's Vital Marine and Freshwater Resources.* http://www.unep.org/VitalWater/summary.htm

United Nations Population Division (1994) *World Urbanization Prospects: the 1994 Revision.* United Nations, New York.

United Nations Population Division (2002) *World Urbanization Prospects: the 2001 Revision.* United Nations, New York. (Table A.11. The 30 largest urban agglomerations ranked by population size, 1950–2015.)

Vitousek, P.M., Mooney, H.A., Lubchenco, J. and Milillo, J.M. (1997) Human domination of Earth's ecosystems. *Science* 277, 494–499.

WCED (World Commission on Environment and Development) (1987) *Our Common Future.* Oxford University Press, Oxford.

World Health Organization (1989) *Health Guidelines for the Use of Wastewater in Agriculture and Aquaculture.* Technical Report Series 778. World Health Organization, Geneva.

2 Viewing Urban Aquaculture as an Agroindustry

Greg Lutz
Louisiana State University, Baton Rouge, USA

Abstract

Although a number of aquaculture projects have been proposed or attempted in urban areas in the past several decades, many have faced serious questions related to sustainability. Truly sustainable urban aquaculture must be technically, financially and economically sound. Each of these areas of emphasis presents tremendous challenges for entrepreneurs and communities wishing to develop aquaculture facilities in an urban setting. In terms of planning and feasibility analysis, urban aquaculture can be viewed as an agroindustry: an enterprise that processes agricultural inputs into consumable products while adding value. This review presents the technical, regulatory, financial and marketing constraints currently facing urban aquaculture development, with examples of economic impacts and potential solutions.

Introduction

The term agroindustry relates to a variety of enterprises that process or fundamentally transform agricultural raw materials. Although characterizing urban aquaculture may be almost as difficult as defining aquaculture itself, many forms of urban aquaculture fit the concept of agroindustry. 'Raw' materials such as feed and fingerlings are transformed into value-added products, be they living organisms or fresh fillets. Traditional agroindustries vary in the complexity of the transformation processes they employ and, as a rule, the more complex the process, the greater the requirements for capital, technology and management. These same relationships are apparent in urban aquaculture enterprises.

Over the past several decades, an entire discipline has evolved to facilitate the planning, analysis and evaluation of agroindustries (Austin, 1981). Much of this work has been undertaken through international institutions such as the World Bank. The process of agroindustrial development involves a number of steps, including identification of opportunities, project design, project analysis and implementation.

Each year, institutions of higher education throughout the world produce countless professionals with industrial or agricultural expertise, but very few with both. This same division of focus is widely apparent in governmental institutions and multilateral agencies. These disciplines are rarely integrated to the degree required for sound agroindustrial project analysis.

Many urban aquaculture enterprises have failed as a result of inadequate planning and analysis, but emerging agroindustrial methods are well-suited to this type of endeavour.

Defining Agroindustry

Agroindustries are quite varied. Examples include facilities for cleaning and packaging fresh vegetables, pencil manufacturing, potato storage and packing houses, sugar mills, and poultry production and processing. In spite of their differences, all share a number of common characteristics. To varying degrees, the raw materials they transform exhibit seasonality, perishability and variability in quality. In contrast, the finished products they market are subject to more constant demand. This results in a fundamental difference between agroindustry and traditional manufacturing, and creates greater challenges in terms of inventory management, balancing supply and demand, scheduling production, and coordinating internal acquisition, process and marketing activities.

In the case of urban aquaculture, features typical of agroindustries are often obvious, especially with regard to seasonality of inputs. In many cases, fry and/or fingerlings are available for only a short period of time each year. Production processes, storage facilities and scheduling must be designed accordingly. One example in North America involves the production of hybrid striped bass in indoor recirculating systems. Fingerlings from this artificial cross are only available during a 2-month period each year, and several large producers have developed the concept of 'cold-banking', which relies on housing fingerlings *en masse* at low temperatures until production facilities and scheduling can accommodate their growout. Although this practice requires significant outlays for facilities, labour and energy, it is necessary to meet the constant demand required by target markets. Unfortunately, risk of losses to disease or technical malfunction is ever present.

Perishability issues typical of agroindustries are also encountered frequently in urban aquaculture. Storage space is often limited and, in reality, critical inputs such as feed and fingerlings are often highly perishable. Inputs are also often highly variable. Examples include feed quality and/or palatability, fingerling size and health status, chlorine/chloramine content of municipal water supplies, etc. Indeed, in many intensive aquaculture facilities even energy supplies must be considered somewhat variable, necessitating on-site generators and switching systems.

In the future, water will probably also figure as a major raw material cost in urban aquaculture processes. Herein lies one important divergence of urban aquaculture from the general agroindustry model: raw materials are typically the most important cost components of agroindustries, but they may be partially or wholly eclipsed in an urban aquaculture enterprise by cost factors such as labour, energy, liquid oxygen, interest on borrowed capital and insurance fees.

Compared to typical manufacturing, agroindustrial production, planning and management can be extremely challenging. Less-standardized inputs result in additional pressure on production scheduling, cash flow and quality control than in traditional manufacturing. These challenges, in turn, require a much higher level of integration and information flow within the enterprise. In an urban aquaculture enterprise, these variations are compounded as various batches of fish move through the production cycle, requiring segregation, grading and maintenance until they are harvestable. Space, feed and labour must all be appropriately allocated to avoid undue stress on production stocks. Agroindustries are also somewhat unique in their dependence on a number of distinct economic sectors, from agricultural inputs to service industries, utilities and distribution enterprises. Each of these linkages results in the potential for interruptions in day-to-day operations, especially in an urban aquaculture enterprise.

In spite of the uncertainties that typically characterize these businesses, few other investments can rival well-planned agroindustries in terms of creation of economic value. In most cases, this statement also applies to urban aquaculture. The key qualifier is planning. Effective planning and evaluation of agroindustrial projects requires the incorporation of elements from both agricultural and manufacturing analysis. As stated above, this combination of disciplines is often difficult to attain, in institutions or individual analysts. None the less, key aspects of both manufacturing and agricultural analysis have been successfully integrated on numerous occasions to allow for design and implementation of well-conceived agroindustrial enterprises.

Planning and Evaluating Agroindustries

While an agroindustrial enterprise may include production, transport, distribution and marketing activities to varying degrees, its primary focus must be the value-added transformation of agricultural products. Accordingly, traditional agroindustrial analysis focuses on the transformation process, and on associated procurement and marketing activities. Since the transformation process is generally dependent on and determined to a great extent by procurement systems, procurement planning is often a logical starting point for agroindustrial analysis.

Procurement

Key considerations in planning and analysis of procurement activities include quantity (availability), quality, timing (seasonality, perishability, lead time required to secure), cost (including cost determinants and pricing) and organization of the procurement system (composition and integration). In the case of an urban aquaculture enterprise, appropriate questions might include:

- Who are the key suppliers for any particular input? This is particularly important if fry/fingerlings must be purchased on the open market. This is also an important consideration for feed requirements.
- What have their production capacities been over the past 5–10 years? It is important to evaluate to what extent competition for inputs such as fingerlings and speciality feeds may impact their availability or prices. Annual supply variation must be characterized and major causes must be identified.
- What, if any, problems have the key producers experienced historically? In some instances, suppliers may not be able to meet their obligations due to unforeseen problems resulting from inadequate facilities or management. A fingerling supplier that has suffered from poor spawning success in three of the past five years may not be a reliable choice for a long-term contractual arrangement. Similarly, a feed supplier that has inconsistent access to ration components such as fish meal may be a liability in terms of long-term planning.
- What are their plans for the immediate future? Is retirement, a sale to new owners or a shift to an alternative species being planned? If a tilapia grower finds that his fingerling supplier intends to convert entirely to yellow perch and hybrid bluegill, alternatives will be somewhat limited.
- Who are the other major customers (competing users)? Are they planning to expand (and thereby present a more lucrative account for key inputs such as feed or fingerlings)?

In terms of quality of inputs, key considerations for fry or fingerlings might include size availability, health status, genetic background (performance), uniformity, coloration and adaptability to the production system (process) in question. For feeds, palatability, formulation (digestibility), consistency, fines content and storage features (perishability) are all important characteristics. In an urban aquaculture setting, perhaps even more emphasis could be placed on regular feed availability and rapid delivery to avoid the need for extensive on-site storage. To a certain extent, the quality of inputs such as feed and fingerlings must reflect market demands. Some examples from urban consumers in developed nations include use of hormone-free fingerlings, 'organic' feed formulations, or particular coloration in finished products, be they living fish or fillets.

Seasonality, while less of a consideration for feed inputs, can be especially important in the procurement of fingerlings. As mentioned above, for many aquatic species, fry or fingerlings are available only for a limited period of each year. Maintaining seedstock for growout and harvests throughout the year requires specialized holding and grading facilities, while greatly increasing the risk of catastrophic losses. Health status of fry and fingerlings upon delivery becomes even more important under conditions wherein they must be held for months prior to growout. Identification of suppliers willing to take on part of the management and risk associated with expanding the period of availability for appropriately sized fingerlings can be a high priority for many intensive aquaculture facilities.

Another important aspect of procurement analysis involves input costs and cost determinants. Underlying factors affecting costs of inputs will include supply and demand relationships (i.e. fingerling producers and competing users), opportunity costs for operations producing fingerlings (whether to sell to potential competitors, attempt to growout in-house or simply dispose of excess fingerlings), the procurement system structure (hatcheries, live-haulers, brokers) and logistics (who is where). Pricing, for operational and planning purposes, may involve spot pricing, wherein a going market price based on multiple suppliers is used at any point in time to determine an input cost. Alternatives involve contract pricing from designated suppliers, which may be based on spot prices, on a cost of production plus a certain agreed-to premium, or on a fixed base price plus an additional charge based on designated economic and market factors.

Such arrangements have been successfully used for both fingerlings and feed supplies in urban aquaculture enterprises. In many cases, favourable prices can be negotiated with sufficient lead time and commitment. Clearly, contract pricing allows for more long-term procurement relationships to be established, but it also increases dependency on outside entities. Procurement cost analysis should define and accommodate historical variation in input costs, while attempting to determine those factors that have contributed to historical variability.

Analysis of the fry/fingerling procurement system structure deserves special consideration in aquaculture growout operations. Numbers of producers, their capacities and their transport capabilities must all be determined, as well as their geographic locations in relation to the facility in question and to other, competing users. The product mix of each supplier also becomes an important consideration, since competitive pressures or emerging opportunities may result in individual operations ceasing to produce the particular species being sought. Ownership patterns are important indicators of the stability of key suppliers, and should not be overlooked. Flow of fry/fingerlings from producers to growout facilities, either directly or through live-haulers and brokers must be analysed to determine centres of power in the procurement system. Power among individual entities at each step of the procurement pathway is indicated by both volume (portion) of total production handled and margins realized between suppliers and purchasers (Austin, 1981).

In some instances, especially for aquatic species that can be reared entirely in recirculating systems or at various times throughout the year, it may make more sense from logistical and financial standpoints to adopt 'backward vertical integration' and develop in-house facilities for fingerling production. While this approach provides for more control and flexibility in the production process, it also requires substantial capital and a more-or-less permanent alteration of the enterprise. Backward integration should be evaluated during the procurement analysis for any inputs that could potentially be elaborated in-house. Questions that must be answered will relate to the efficiency of the proposed process, the scale required for economic efficiency, genetic impacts (inbreeding depression) and hidden costs in terms of diversion of management, production areas, labour and other resources. This approach may be best utilized by urban aquaculture enterprises when addressing new markets or in developing new products (species).

Production (transformation) technology

Once a clear picture begins to emerge of procurement systems for required inputs, analysis and planning activities can focus on the production (transformation) processes that will be employed. Key considerations will involve the technology to be utilized, facility location requirements, inventory management procedures, required supplies and inputs (other than raw materials), process control and programming, and potentially profitable by-products. A number of factors must be considered in planning and evaluating alternative production (transformation) technologies, including labour vs. capital requirements (especially for handling, sorting and packing activities), energy requirements, raw material needs and productive use of available capacity.

Evaluation of the production process must consider both financial and economic perspectives. While return on investment might be maximized through adoption of high levels of automation and process control, efforts to shift technology more toward the use of manual labour may make more sense if the overall goal is local or regional economic development (Austin, 1981). In most situations, these two needs must be balanced to some degree in order to secure both investment and local political support.

In terms of facility location requirements, access to inputs, markets and labour are often overlooked or not given sufficient emphasis in urban aquaculture development. Similarly, utilities and water availability (potable, cooling/heating and process water) are often overlooked. Far too often, undue importance is placed on site costs for urban aquaculture, especially when this outlay should be viewed as an annualized cost component for any agroindustrial enterprise. While neglected manufacturing or warehousing facilities often appear well-suited for conversion to urban aquaculture at first glance, most are fundamentally flawed structures for such purposes in terms of drainage, tolerance of exposure to high levels of humidity and salts, and biosecurity.

Availability of working capital and sufficient lines of credit should be key considerations when determining technology and site requirements. Using a series of bioeconomic models, Lutz and Roberts (1998) illustrated the impact of interest payments on profitability of a 46 t per year recirculating tilapia aquaculture system. Financing 60% of the initial outlay and all operating expenses at 10% interest extended the break-even horizon from 4 years to 11 years, and all but eliminated profitability for the model enterprise.

Markets and marketing

Of course, while intense analysis must be undertaken to estimate costs associated with the procurement and transformation processes to be used, any urban aquaculture enterprise must be based on a sound market analysis and marketing plan. Markets and marketing are far too often considered secondary characteristics of urban aquaculture enterprises. There is a fascination with production technology among many aquaculturists that blinds them to the more basic requirement for any agroindustry: to sell a value-added product at a price that allows both financial and economic sustainability.

Marketing analysis must take into consideration both consumer attributes and the competitive environment (Shaw, 1990). Consumers must be defined in terms of who they are, how they might decide to purchase the proposed product, be it live fish or processed fillets, and when, where and how often they would be most likely to make such a purchase. Potential consumers must be characterized in terms of socioeconomic status, income levels, language, ethnicity, urban or suburban location, and education levels.

Perhaps the most fundamental question in a marketing analysis would be: why? Why would a consumer purchase the product in question? How sensitive are the potential consumers to price variation; and how prevalent is price discounting? How do potential consumers define quality, and how important is it to them? Chapter 18 of this volume provides insight into some of the qualifiers that consumers consider when purchasing aquaculture products.

Similarly, the competitive environment must be defined in terms of the current market structure, including the current and previous market shares of competitors, the basis for competition among suppliers, barriers to entry for new producers and any existing institutional constraints. Key considerations include defining potential product substitutes, as well as the potential for raw material (fingerling) suppliers to forward integrate, or for distributors to backward integrate. Barriers to entry involve economies of scale, absolute cost advantages and vertical control of the procurement or marketing systems. Analysis should consider whether penetration pricing (i.e. below cost of production) will be required to gain initial market share.

Once these factors have been examined, a marketing plan can be developed to include product specifications, pricing guidelines, promotion, distribution channels and integration with procurement and production processes. These concepts, however, are beyond the scope of this discussion.

Financial and economic analysis

While transformation, procurement and marketing activities form the basis of systems analysis, financial and economic analyses are also required for sound planning and evaluation of agroindustries. Once analyses have been completed for the procurement, processing and marketing factors, more traditional approaches such as whole farm budgeting and partial budgeting can be applied (Jolly and Clonts, 1993).

Financial analyses are generally essential for access to capital, while economic analyses are typically of greater interest to local authorities and communities where proposed agroindustries will be located. Financial analyses focus primarily on factors such as return on investment, while economic analyses address costs and benefits to local, regional and national economies. Each has its own format and key considerations. O'Rourke (1996) provides illustrations of some economic analyses particularly useful for analysis of recirculating aquaculture systems. Financial and economic analyses, however, cannot be effectively developed until systems analysis has been completed, and often both must indicate favourable results to allow for the implementation phase to proceed.

An Illustrative, Fictional Case Study

The opportunity

An experienced fish culturist and several relatives and business associates decide to establish an urban aquaculture facility in a large metropolitan area in a temperate climate. The basic plan for the enterprise is to supply urban markets for live and/or fresh fish. The species proposed for start-up operations is the 'rainbow bass', a fast-growing carnivorous fish native to natural waters to the south of the region where the facility will be established. Several farms produce rainbow bass in ponds in the areas where the species occurs naturally.

Market characterization

Market demand for rainbow bass in the area includes both white table cloth restaurants and upscale seafood counters. Some flexibility in demand exists in the retail side of the market, but restaurants require regular supplies and consistent quality to commit a place on their menus for this product. Several established rainbow bass producers move their product through large seafood brokers in the city in question. These producers do not harvest and deliver on a regular basis, and extended transport times and related stress occasionally result in quality issues for both retailers and consumers. Market advantages for the proposed facility would involve the ability to deliver regularly, to respond rapidly to meet special customer demands from time to time, and to label the product as locally grown.

Transformation process and materials

Rainbow bass require high-protein/high-energy diets to achieve sufficient growth and maintain health. Additionally, recirculating system production requires feeds to be substantially fortified with vitamins and minerals. Rainbow bass have a low tolerance for ammonia, so bio-filtration must be highly efficient and reliable. Like many other carnivorous finfish, rainbow bass exhibit some degree of growth variability, with associated concerns for cannibalism and fish-on-fish aggression within culture systems.

Site considerations

In preliminary discussions, local government is pushing for one particular site – a long-unused warehousing facility. Authorities are proposing significant tax incentives and other assistance for this urban aquaculture venture, but these offers are linked to the development of this specific facility.

Procurement issues

Rainbow bass have a limited spawning season in the autumn of every year. Depending on weather and other factors, supplies of fingerlings are abundant some years and quite limited in others. Over the past several years, between 18 and 30 million fingerlings have been available for growout annually. These extremes of abundance or scarcity also affect market prices of cultured rainbow bass from year to year as a result of fluctuating supplies.

There are three large hatcheries doing business in rainbow bass fingerlings, along with one medium-sized hatchery. All are located to the south, in the region where rainbow bass occur naturally and pond production takes place. Two of the larger hatcheries are relative newcomers to the business, and both have had personnel changes within the past 6 months. The hatcheries typically prefer to sell advanced fingerlings, trained on feed, directly to growout operations. However, when nursery facilities are filled to capacity, hatcheries often sell some portion of their fry to four or five operations that purchase them when available and raise them to fingerlings for sale. One of these operations has large, diversified facilities for a variety of species, and normally controls roughly half of the fry-to-fingerling sector of the supply chain. Clearly, the trade-off for the hatchery facilities involves the additional revenue that can be earned by selling fry that would otherwise go to waste versus the additional availability of fingerlings on the market and resultant price reduction.

In either case, fry are available for only a limited period each year of approximately 3–4 weeks, and fingerlings are available for no more than 8 weeks. Hatcheries will accept deposits from persons wishing to reserve orders at a pre-set price prior to the spawning season on a first-come, first-served basis. Fry-to-fingerling operations are less inclined to make such arrangements, at least where pricing is concerned.

A local feed mill, immediately outside of the metropolitan area targeted for the enterprise, has the capability to produce a feed suitable for culturing rainbow bass. However, special arrangements are required to produce this feed as it is not part of the plant's regular product line. The plant cannot produce speciality orders in quantities of less than 5 t. Alternative feed sources, near existing pond production facilities, are at least 190 km distant and although they can provide suitable feed in smaller quantities, delivery costs from these sources are substantial.

Decisions and strategies

The decision is made to focus marketing through regularly scheduled deliveries to several restaurants, and to move additional product, as available, through local seafood markets. Integration of procurement, process and marketing goals results in a plan to use a combination of cold-banking, staggered stocking dates and densities, and grading to spread harvests over time for regular deliveries and to avoid periods when supplies are relatively abundant due to batch harvests from pond producers to the south.

The determination is made to configure production (transformation) technologies to rely on high levels of process control and monitoring, as well as high intensity measures such as liquid oxygen supplementation and an innovative bio-filtration approach. But plans also allow the facility to generate employment through labour-intensive animal husbandry (grading, harvesting, feeding, backwashing filters, etc.). Financial projections based on this technology suggest positive cash flow horizon after 21 months. Additionally, economic analysis by local authorities is very positive in terms of generation of employment and taxes.

The filtration system to be used relies on a novel yet very promising bio-filtration design which has proved highly efficient in university laboratory-scale systems. This type of filter requires high flow rates and liquid oxygen to operate efficiently.

Municipal water supplies available at the site are chlorinated, and would require treatment prior to use for fish culture. The decision is made to drill an independent well on-site to meet production water needs.

Unforeseen developments

In initial discussions to arrange for procurement of fingerlings, several hatchery operations cite fixed capacities as a limit on availability of advanced fingerlings. As a result, only one can commit in advance to supply fingerlings, and only 60% of the projected need for first year operations. A decision is made to try to contract with two smaller fry-to-fingerling operations in an effort to meet fingerling needs.

A number of permits are required from the regional and state groundwater authorities prior to attaining permission to drill the production water well. This delays development of other aspects of the facility. None the less, lease payments and monthly notes on capital expenditures and interest must be made in the interim.

The warehouse facility to be used for the project has limited areas suitable for feed storage. This complication is compounded by the need to purchase feed in bulk loads. Quality problems arise in stored feed due to condensation and humidity, frequently resulting in the need to discard feed before it can be used. Temporary arrangements must be made to purchase feed in smaller loads from alternate suppliers, while the feed storage area is retrofitted with appropriate heating, ventilating and air-conditioning (HVAC) equipment.

Bio-filtration problems result from difficulties in maintaining sufficient flow rates in scale-up of the prototype systems originally proposed. Ultimately, custom-fabricated sand filters must be installed to attain efficient nitrification.

Fingerlings are delivered on time from the large facility which agreed in advance to supply fingerlings, but shipments are delayed from one fry-to-fingerling operation and completely unavailable from the other due to a poor spawning season and difficulty obtaining fry from hatchery sources. Two established pond growout operations in turn drive spot prices up in efforts to meet their fingerling requirements. In the tight supply situation, reluctance is apparent on the part of most fry-to-fingerling operations to commit limited supplies to a new buyer with no established track record.

Cold banking and grading both result in unforeseen problems with diseases, especially in combination. Bio-filtration components operate at very low efficiencies at low temperatures, resulting in poor water quality and compounding disease problems. The resulting need to increase exchange rates on cold banking systems increases utility use associated with cooling water.

Initial growth projections are not applicable to later groups of fish. Greater variability in growth, in combination with production methods, results in an overall extension of average time to harvest. This results in problems correlating available tank space to accommodate the necessary harvest schedule. Difficulties meeting size and quantity needs of restaurant customers results in the loss of one of these accounts and the need to divert more variable-sized product through seafood markets. Penetration pricing is not an option due to the costs of transformation processes utilized and to cash flow constraints resulting from other problems (see above) during the first year of operation. The operation suffers from overall reductions in revenue and increased marketing costs.

The outcome

Ultimately, after re-negotiating lease obligations, re-financing certain expenses and bringing in new investors, the facility is eventually well-established and operates with marginal profitability. Local government and business leaders point to it as an example of a success story in urban economic development. Management is interested in developing production techniques for alternative species, as well as technologies or facilities for in-house fingerling production.

Conclusions

The typical agroindustrial project, including most urban aquaculture enterprises, has four phases of development: identification of the opportunity, analysis and design, implementation, and evaluation. There has never been a shortage of 'opportunity identification' where urban aquaculture is concerned; and there have been numerous examples of attempted implementation. Unfortunately, there have been far fewer opportunities for on-going evaluation of urban aquaculture enterprises due to a very low success rate. This, in turn, reflects fatal flaws in the analysis and design phases of most urban aquaculture enterprises. Procurement, transformation and marketing components of an urban aquaculture facility must be thoroughly evaluated and integrated *a priori*. Hopefully, increasing emphasis on these aspects of project development will boost the success rate for such enterprises in the years to come.

References

Austin, J.E. (1981) *Agroindustrial Project Analysis.* Economic Development Institute Series in Economic Development. The World Bank, The Johns Hopkins University Press, Baltimore, Maryland.

Jolly, C.M. and Clonts, H.A. (1993) *Economics of Aquaculture.* The Haworth Press, Binghamton, New York.

Lutz, C.G. and Roberts, K.J. (1998) Investment and management aspects of owner-operator scale greenhouse tilapia systems. In: Libey, G.S. (ed.) *Proceedings of the Second International Symposium on Recirculating Aquaculture.* Virginia Polytechnic Institute and State University, Blacksburg, Virginia, pp. 98–105.

O'Rourke, P.D. (1996) The economics of recirculating aquaculture systems. In: Libey, G.S. and Timmons, M.B. (eds) *Successes and Failures in Commercial Recirculating Aquaculture*, NRAES-98. Northeast Regional Agricultural Engineering Service, Ithaca, New York, pp. 61–78.

Shaw, S.A. (1990) *Marketing. A Practical Guide for Fish Farmers.* Blackwell Scientific Publications, Oxford, UK.

3 Opportunities and Constraints to Urban Aquaculture, with a Focus on South and Southeast Asia

David C. Little and Stuart W. Bunting
University of Stirling, Stirling, UK

Abstract

Any assessment of the opportunities and constraints to urban aquaculture requires that a clear idea of the concept is in hand. Definitions of urban aquaculture that reflect current thinking about urban development and urban food production, more specifically in South and Southeast Asia (SSEA), are initially discussed. The importance and specific characteristics of both aquaculture development and urban growth in this part of the world leads to the concepts of both 'extended metropolitan regions' (EMR) and the 'waste economy' model being used to explain the nature of urban aquaculture. Three main trajectories are advanced to explain the conditions under which urban aquaculture develops in the SSEA region: (i) opportunistic development of aquaculture within the changing physical aquatic landscape occurring through urbanization, (ii) continuation of existing aquaculture, often in a highly modified form, as urbanization occurs around it and (iii) strategic locating of new enterprises in urban areas to take advantage of specific resources such as nutrients, markets, information or some combination of these. The multiple roles that urban aquaculture performs at a household, local and municipal level are explained. Benefits extend far beyond the producer households and, because of urban–rural links, far beyond urban areas *per se*. In particular, the poor benefit through their role as service providers and consumers. Trading of wastes and other inputs, and processing and sale of aquaculture products, are important sources of employment. Migration for employment has affected market niches for products of urban aquaculture in urban areas and influenced consumption in rural areas. At a local community and municipal level, benefits accrue from the role of urban aquaculture in waste disposal and treatment, food production, flood control, employment, environmental quality and recreation. A variety of factors constrain the development of urban aquaculture, including loss of suitable physical locations and changes in waste disposal, that are often influenced by broader development trends. Changing attitudes and expectations in urban areas, especially as they become more affluent, are also related to changing demand. The 'openness' of aquatic systems make urban aquaculture particularly vulnerable to contamination and theft, both factors which discourage investment, although successful examples demonstrate how aquaculture may remain viable in urban areas. The dynamic role of entrepreneurs in urban aquaculture, reusing urban-derived wastes, is particularly important. Intensification strategies are also varied, and proven to maintain the competitiveness of aquaculture within the urban environment in the SSEA region. The proposed research and development agenda advocates holistic analysis of risks and benefits, and ensuring urban aquaculture is central to efforts towards improved public health. Improved education about the issues and enforcement of legislation among stakeholders is required. A better understanding of bioaccumulation of contaminants in aquatic farming systems is urgently required towards these ends.

Introduction

Rapidly increasing levels of urbanization occurring in many less developed countries are a major demographic and developmental phenomenon. Maintaining food security (Koc *et al.*, 1999), both in terms of quantity and quality, and ensuring sanitary waste disposal for dynamic towns and cities are major challenges, especially in the context of urbanization on seasonally flood-prone land in Asia. We explore what urban aquaculture is, where it is occurring and its roles in people's livelihoods before examining the conditions under which it has developed. We then assess the evidence for the sustainability of urban aquaculture as part of modern development, and review the major research and action needs to support it.

What is Urban Aquaculture and Where Does it Occur in Asia?

Aquaculture, as is the case with food production more generally, is mainly associated with rural areas, though typically aquaculture has strong linkages with urban centres. This is probably associated with the recent history of aquaculture in most of South and Southeast Asia; fish culture as a traditional activity is not widespread (Edwards *et al.*, 1996). Its rapid development in recent decades has been driven by access to information and markets that are more readily available in cities, but also due to availability of wastes. In contrast, fishing and collecting of wild stocks of indigenous aquatic plants and animals still typically meet most people's needs in rural Asia. Aquaculture, both rural and urban, is usually based on the stocking of exotic species produced in hatcheries. Around cities, higher yields of fish from resource-intensive culture systems are more likely to be achieved, compared to rural areas in which a paucity of nutrients often restricts farmers to polycultures of fish feeding low in the food chain. In rural areas, yields from stocking are often complemented by wild species entering, or self-recruiting in, culture systems.

Conventional definitions of urban and rural food production have relied heavily on location, and there has been little distinction between intra- and peri-urban locations. In general, both areas appear to occur in the cities of South and Southeast Asia to a greater or lesser extent, and an all-embracing urban aquaculture label seems more sensible. Whereas rural aquaculture has tended to become linked with 'traditional' and 'agricultural' practices, urban has become synonymous with 'industrial'. As rural aquaculture has been defined as systems 'meeting the needs of, and fitting with resources available to, small-scale farming households' (Edwards *et al.*, 2002), it might be assumed that urban aquaculture is supported by urban/industrial resources. This dichotomy is overly simplistic, however, as nutrient linkages between aquaculture and nearby food production are often weak in rural locations and, in contrast, urban systems can be well integrated with respect to water and nutrient reuse. A major characteristic of urban aquaculture systems in Asia is the link with waste and wastewater reuse. Nor can urban and rural aquaculture be defined solely in terms of the markets for their products, and the development of enterprises for export purposes fits uneasily in either urban or rural categories. Such export-oriented aquaculture is typically 'stand-alone', albeit with linkages to ports and supplies of industrially produced pelleted feed and other inputs.

Connection of urban aquaculture with the 'urban ecosystem' has been proposed to give urban food production a stronger conceptual basis (Mougeot, 1999). This definition stresses the importance of linkages with the urban nutrient cycle and broader urban food system (Smit, 1996). The various characteristics of urban agriculture may make this broad definition particularly relevant. Others might consider that urban aquaculture fits neatly into a 'waste economy' model as coined by Whitney and White (1992), for example, viewing the economic and social flows associated with wastes in a holistic approach.

Clearly aquaculture in and around cities such as Ho Chi Minh City, Vietnam does not fit comfortably within metropolitan boundaries, which often include rural areas (see Chapter 6 of this volume). In contrast, concentrations of aquaculture further away in neighbouring provinces have strong urban characteristics. Producers can commute to and from the city, and farms can supply products to urban consumers on a daily basis. Urban aquatic production systems may be relatively distant from urban centres, but strongly linked and interdependent through flows of water and nutrients. The definition of urban aquaculture cannot, therefore, be uncoupled from the nature of urbanization that typically occurs around cities. These patterns are determined by road development. Von Thünen (1875), Lösch (1964) and Mougeot (1999) have commented that the travel-time band around cities is typically more star-shaped than circular, and is a product of the characteristics of the road network. This contrasts with the 'rings' of productive horticulture (and aquaculture) that were located around traditional urban centres (von Thünen, 1875). The concept of the 'extended metropolis' (McGee, 1991), or more recently EMRs (Extended Metropolitan Regions; McGee and Robinson, 1995), encapsulate the idea of urban areas now rarely having single centres, but rather sprawling regions of mixed land use that have both rural and urban characteristics. These so-called *desakota* – areas of mixed agricultural and non-agricultural activities that stretch between large city cores – are embryonic in many Southeast Asian countries attempting to develop provincial centres away from their core cities, and these are typically areas of aquaculture development. McGee (1995) suggests that structural change, globalization and transactional revolution all affect how EMRs develop. The importance of these factors to urban aquaculture is considered later.

Urban aquaculture typically has many linkages with rural aquaculture, especially with regard to movements of products and knowledge, and there may be a tendency towards relocation from urban to rural as infrastructure and linkages improve. In terms of understanding livelihood strategies, strict distinctions between rural and urban are unhelpful because of these strong linkages (Farrington *et al.*, 2002).

Definitions of aquaculture, and the level of intensity in relation to the urban environment, are discussed elsewhere (Bunting and Little, 2003). Generally, extensive systems relate to aquaculture reliant on background levels of natural food; semi-intensive systems, which encompass a large range of systems and production levels, are based on managed nutrient inputs that supplement background levels; and intensive culture systems, usually of a high density of organisms, are solely dependent on imported nutrients, usually as commercial feeds.

There is an important interface between culture and the exploitation of unmanaged stocks. These stocks may be of plant or animal origin, and management practices may vary over time in relation to alternative sources of fish and livelihood opportunities. Management of aquatic organisms in this way has been coined 'proto aquaculture' (Beveridge and Little, 2002), suggesting a relatively small degree of control of the stock life cycle being imposed, but resulting in improved outcomes compared with exploitation alone. The relatively higher levels of background natural food levels in and around urban areas, compared to more nutrient-limiting rural areas, may make such systems productive and sustainable.

Under What Conditions Has Urban Aquaculture Developed?

Urban aquaculture appears to develop as the physical landscape changes in three ways. First, opportunistically as people take advantage of places where fish can be cultured, or aquatic plants grown, that are created by the process of urbanization such as borrow pits (holes left behind from the removal of soils for construction use elsewhere) and wastewater canals. Second, aquaculture can become urban as enterprises get overtaken by urban development but are sustained, often in a highly modified form. Third, operations that become established strategically in urban areas because of a need to be close to nutrients, markets, information or, as is most usual, some combination of these factors.

In Asia, a high proportion of aquaculture is urban based on this definition. The high perceived value of many aquatic foods as dietary items, and the pattern, pace and control of urbanization in much of Asia, probably explains this situation. Settlement close to water is normal for communities of people, and historically some of the longest established and densest populations in Asia have occurred close to river floodplains (Beveridge and Little, 2002). The importance of annual floods mediated by monsoon rains to such people is indicated by the primacy of aquatic foods in Asian diets – rice, an aquatic grass, and a variety of other aquatic plants and animals have been, and remain, staples. Urban lifestyles reduce opportunities for subsistence livelihood strategies based on the farming of wet rice and capture of fish, but demand for these products remains strong. Urban markets for aquatic products appear to have been a major stimulus to the development of aquaculture in South and Southeast Asia.

The highly perishable nature of aquatic food products also gave producers close to local markets a competitive edge. Rapid deterioration and decline in value occur in fish and vegetables even over short time periods in tropical conditions. Live marketing of fish is possible for local producers, and even when live haulage techniques and quality of infrastructure improve, risks remain higher. Thus, tilapias produced in ponds and cages in urban areas and marketed live in provincial cities of northeast Thailand have a premium over even much larger iced fish transported from lower cost systems in central Thailand. Cage culture of common carp in and around Ho Chi Minh City can be sustained because of the value advantage of marketing live or very fresh fish. The concept of rings of production around cities, first identified by von Thünen (1875) for perishable vegetables, persists in the region, especially where transportation infrastructure remains undeveloped.

A major feature of urban compared to rural areas is the high cost of land, making food production *per se* an unlikely first investment choice. However, agriculture in urban areas may be a niche function (de Zeeuw *et al.*, 2000) and is often carried out on land not owned by the user. Food production may also be a transitional use of land and water. Land and water bodies earmarked for construction by government or the private sector may be used informally for aquaculture on a temporary basis. Access may be controlled by informal rents, leases, inheritance or traditional rights. Typically, fear of eviction, and/or social conflicts, constrain investment, particularly if holdings are not home-based. Thus, households managing patches of aquatic vegetables around the periphery of a municipal urban lake will approach management decisions differently than a household producing the same crop in a small pond that they own within their homestead. Urban administrators often attempt some level of control on the type of development through zoning or other mechanism, which combined with speculative land ownership ensures cities and their environs are often rich in undeveloped patches. In flood-prone Asia, many of these are used for aquatic food production.

Urban development of floodplains also demands the use of large quantities of fill or 'borrow' which are most cost effectively removed from the periphery of cities giving rise for further opportunistic aquaculture. As urban areas expand, such borrow pits, which are often very deep, are typically bypassed, remaining as water resources throughout metropolitan areas. These water bodies, partly because of the expense involved in refilling, often become permanent resources, typically developing a range of functions regardless of ownership. In recent decades, the falling cost of mechanization has accelerated hole-digging in many places, especially during periods of construction boom. The now common development of extensive suburbs has encouraged a related practice of selling topsoil for both infilling and use in urban landscaping. This has resulted in large areas of shallow waterbodies around the periphery of Bangkok and other larger cities, often next to residential and industrial settlement. This type of land is of little value for rice production, but suitable for other aquatic crops, especially plants. Typically the production of such crops is unrecorded, as aquatic plants fall outside most national statistics, being considered neither horticulture nor aquaculture.

Change from rice monoculture to more diversified systems around urban areas is thus stimulated by rising demand for soil and products that can be profitably grown in deeper areas. Farmers that continue trying to grow rice often face problems as changing land use around them – ditch–dyking and pond construction – make it difficult or impossible to manage both water and pests. Increasing levels of unpredictable flood/run-off from urbanizing areas often exacerbate this problem. This situation, together with the higher returns to land of more intensive horticulture and aquaculture, tends to pull remaining rice farmers towards conversion of rice fields to ditch–dyke and/or pond systems. Alternatively, farmers abandon rice and let the land go idle; in one area of Hanoi in which aquaculture is well developed, 10% of the total area is fallow (Van den Berg *et al.*, 2003). Lack of necessity to work the land because of alternative income sources and a tendency for farmers to await impacts of changing government land policy before further investment probably explain this phenomenon.

The evidence for overtaking of conventional aquaculture by urbanization is circumstantial, but a typical course of events can be constructed. Intensification and shifts to higher value and/or air-breathing species (e.g. carps/tilapias to catfish/snakeheads) that improve returns to land, water and capital are common. Species of fish capable of air-breathing, rather than reliant on oxygen dissolved in the water, are more tolerant of poorer water quality, and can be raised at higher densities. Typically, they are also more valuable than other fish species. Food fish production often also evolves to a value-added, specialized activity, such as production of juveniles, ornamentals or fattening of fish. Many of the recreational 'fishing parks' where individuals pay a fee to catch fish for sport, and later to eat, in and around Asian cities, have been modified from food fish farms. Aquaculture enterprises that become surrounded by urban sprawl tend to be faced with water quality and other problems, but trade these off through a variety of technical and management innovations.

The opportunistic location of aquaculture close to urban centres is principally related to the availability of cheap nutrients, whether as fertilizers or feeds. Producing juvenile tilapias on wastewater around Ho Chi Minh City still appears to be more cost effective than at locations closer to customers within the Mekong Delta, despite transportation costs (Little and Hulata, 2000).

Urbanization and industrialization typically occur simultaneously as low cost labour is essential. Cafeteria wastes from factories, and by-products from slaughterhouses located on the outskirts of Bangkok, support local intensive catfish production. The high moisture levels and rapid deterioration in quality of such wastes encourage their reuse locally. Local availability of inputs to urban aquaculture undoubtedly explains the importance of 'clusters' of producers and traders. Industrial and trading clusters have long been identified (Marshall, 1919) as being critical for the availability of buyers and suppliers, allowing formation of specialized skilled labour, and facilitating the informal transfer of knowledge.

A major factor in the development of urban aquaculture is the well-developed entrepreneurial process that occurs in and around urban areas. Many urban fish farmers, especially in the early stages of development, either have careers in government or other salaried employment. Starting a farm business is therefore supported by relatively easy access to technical, market and financial knowledge. Knowledge of planning decisions with regard to urban land use and resource availability is also advantageous.

In summary, a combination of factors encourages urban aquaculture, including the process of urban-industrial development itself. Changing landforms lead to changing land use and access. Improved availability of nutrients, information and markets combine to stimulate and support development of urban aquaculture, mainly by entrepreneurs.

What Are the Roles of Urban Aquaculture and How Do People Benefit?

The production of aquatic foods has numerous roles in urban areas, and brings benefits in many ways. Viewed from the household, local community or municipality, different perspectives emerge; urban aquaculture can be an important source of nutrition or employment to a household, or an important component of urban waste disposal and flood control to local authorities.

The nature of benefits partly relates to the scale of urban aquaculture. Aquaculture production systems range from small, household-level operations, through informal exploitation of community water bodies for plants and animals for subsistence needs, to completely specialized commercial ventures that are capital intensive and employ large numbers of people.

Urban aquaculture also benefits people as consumers and intermediaries as well as producers. The importance of local food production in the community may relate more to such benefits, particularly as employment, than directly to producers on the farm. Neither is urban aquaculture always a dominant activity for urban households. Urban households, particularly comprising poorer people, are likely to be involved in aquaculture as a part-time activity within a complex portfolio of livelihood options as they attempt to reduce risks and their vulnerability (Moser, 1998).

Links between urban and rural areas also mean that rural people can and do benefit from urban aquaculture as movement of both people and products ensure that aquaculture in and around cities is having numerous impacts in rural areas.

Roles of urban aquaculture

Urban planners in the SSEA often underestimate the role of urban aquaculture in waste disposal and reuse (Table 3.1). This may partly be explained by the fact that the waste economy (Whitney and White, 1992)

Table 3.1. Roles of urban aquatic food production from different perspectives.

	Level of production system			
			Urban households	
Role	Municipal	Local community	Rich	Poor
Waste disposal and treatment	++	++++	+++	++
Food production	+	++	+	+++
Housing		+	+	++
Flood control	+++	++++	++	+
Employment	++	++	+	+++
Environmental quality	+++	+++	+++	+
Recreation	++	++	++	+

++++ Very important; ++ important; + unimportant.

requires all aspects to be analysed together. Urban administrations typically lack capacity for this type of analysis, but the role of urban aquaculture in reducing organic wastes that would otherwise impose greater costs and public health risks is rarely acknowledged, and sometimes denied. High profile exceptions to this, such as the East Kolkata wetlands, have become prominent because of the strength of civil society and non-government activists. At a local level, however, reuse of waste in urban aquaculture is likely to be more obvious, especially if municipal waste removal is lacking or costly. Perspectives among better off households to waste disposal and reuse are likely to be more related to the nature and impacts on local environmental (air, water) and amenity quality than their role in food production with which poorer households are more concerned. The value of periphery dyke areas of urban water bodies as sites for settlement of poor households, particularly as informal squatter communities, is clear. Such areas are increasingly favoured as sites for high-value housing in downtown areas of Asian cities, a trend that often leads to the exclusion of the poor. Enlightened municipalities may view urban wetlands for their flood control and/or amenity value, but are more likely to perceive them as areas ripe for development. Retaining or increasing water bodies within cities is also a strategy to ameliorate downstream impacts of storm water run-off from urban areas (Haughton and Hunter, 1996). A better understanding of the value of water bodies used to produce food at the local community and household levels could inform more strategic and pro-poor development.

Access and ownership

An important issue arises over the definitions of aquaculture being limited to strictly those owned by individuals, corporate bodies or the State (Rana, 1998). Clearly much production from aquatic systems occurs in common property or from a common pool resource. Beveridge and Little (2002) argue that the key criterion distinguishing farming from collection is any intervention that increases yields sustainably. This may be achieved either through ownership of stock, or controls on access to, and benefits from, such stocks.

Earlier we suggested that urban aquaculture may in many cases be encouraged by uncertain tenure, complex access rights, and ineffective planning and enforcement. It is likely that the opportunistic use of water and wastes, together with a place to stay in urban areas, are a magnet for the poor, especially migrants seeking low-cost accommodation. The tendency for informal housing, that is not always temporary, to develop around water bodies, is well established, and the linkage with water for sanitation and multipurpose use are well known. Thus, for migrants in many cities, whether Calcutta, Dhaka or Ho Chi Minh City, water bodies become a source of secondary income, nutrition and employment.

Urban–rural linkages

The flow of people, products and ideas between rural and urban areas results in the benefits of urban aquaculture being felt far outside urban areas. Small carps and tilapias produced on feed-lot livestock waste around Bangkok are iced and then marketed in poorer areas of the country, particularly north and northeastern regions where demand for cheap, freshwater fish is strong. Urban aquaculture can also lead to changes in rural areas through the flow of ideas and attitudes. Iaquinta and Drescher (2000) describe this as a social-psychological dimension of urbanization. Large-scale rural–urban migration may change the food culture in both directions. For example, the settlement of poor, fish-eating Bengali and Bihari migrants in Delhi has created a new and large demand for small cheap freshwater fish in Haryana State. Conversely, the long established and widespread practice of northeastern Thais migrating to Bangkok for construction and service employment may have led to small farmed fish becoming popular in their home communities (Little and Edwards, 2003). This in turn may have stimulated local production to compete with iced fish imported into the northeast region. The return of remittances of migrant workers, new consumption habits and even knowledge about production and marketing, may be powerful forces for change in rural areas of many countries of the SSEA region.

Trading networks

Recent definitions of urban agriculture include processing and trade to production (Mougeot, 1999). This commodity approach makes sense as value addition is an important concept for many aquatic products. Trading and distribution of urban wastes to rural aquaculture is a common and tangible linkage between urban centres and aquaculture in outlying areas that brings benefits to people, especially the poor, in a variety of ways. Thus, the trading of urban nightsoil for juvenile fish production in northern Vietnam (Prax *et al.*, 2002) and the contractual removal and sale of pig manure from feedlot units around Bangkok (Little and Edwards, 2003), lead to direct employment benefits for the poor and, indirectly, to the local community and wider municipality through environmental remediation (Fig. 3.1). Mongeot's observations about small and dispersed units making up an extensive and decentralized supply system within immediate reach of a massive consumption market for urban agriculture seems particularly true for aquatic products. Although returns to labour for producing and trading aquatic products may be low compared to other employment opportunities in urban areas, urban aquaculture probably increases the number and size of niches for marginal groups to gain employment, much of it of a petty trading nature. The women involved in processing and selling small, cheap tilapias in a ready-to-cook form in and around Bangkok's largest freshwater fish market (Sapan Pla) are an example of such niche filling (Fig. 3.2). This type of activity has also contributed to the ever-increasing acceptability of tilapias over the last three decades in rural northeast Thailand as mentioned above.

Fig. 3.1. Purchase of pig manure by an entrepreneur, who then transports it by pick-up truck to nearby fish farmers, is common practice in urbanizing areas in provinces around Bangkok, Thailand.

Fig. 3.2. The availability of small, cheap tilapias provides a livelihood opportunity to poorer people; these fish have been purchased cheaply at a major city market, de-scaled, gutted and are being sold to migrant construction workers in Bangkok.

Fish raised in urban water bodies may also be cheaper and more accessible to the poor than alternatives. In Coimbatore, Tamil Nadu, small fish from eutrophic urban water bodies are caught, sold, purchased and consumed by the poor. Value addition in the marketing chain may take other forms such as self-employed, low-caste fish cleaners who work in markets in West Bengal. The skilled cleaning of fish, especially if small-sized, is in high demand by more affluent customers, and has become an opportunity for assured, aseasonal employment for the very poor.

What Factors Constrain the Development of Urban Aquaculture?

A variety of factors combine to undermine current and potential urban aquaculture in the SSEA region. The rapid growth of Asian economies in recent decades has led to concomitant physical and social changes in urban settlements. Although some may provide short-term opportunities, many can result in longer-term atrophy of the resource base, lead to decline in social connections and reduce demand for urban food products.

The proportion of the population living in urban areas in SSEA has changed rapidly, and with it expectations and opportunities to meet food security and broader livelihood needs. Urban aquaculture must be competitive both as a source of employment and to supply products in demand if it is to be sustained. Changes to markets occur with urbanization and these are considered below.

The relative openness of many aquatic production systems, in which inputs and outputs can readily move in and out of the system, has particular importance in urban environments. Boundaries of aquaculture systems are often ill-defined or porous, leading to poor control of inputs and outputs. Key issues include the potential for contamination by toxic compounds, and losses of stock to predation and theft. The viability of urban aquaculture is greatly affected by these factors.

Physical changes to the urban landscape

The major factor that constrains the longer-term development of urban aquaculture is the physical process of urbanization which tends to lead to infilling of water bodies in and around cities as land values rise (Parenteau *et al.*, 1993). Furedy (1990) reported a decline in the use of wastewater aquaculture, usually in urban areas, in countries such as Japan, Malaysia and Taiwan. The construction of roads may have major impacts on the speed and nature of urbanization, particularly if they impinge on areas of established aquaculture. This has been the case in Kolkata, where pressures on the remaining areas of wetland have intensified greatly since the development of better road links into the area. Infrastructure development can affect urban aquaculture directly. This includes programmes to install new sewerage to transport wastes away from previous areas of local reuse, and the covering of drainage channels that previously supported cage culture or aquatic vegetable production.

As urbanization occurs, change and interruptions to the supply of nutrient inputs can also cause the decline of urban aquaculture. Difficulties in obtaining sufficient wastewater was a major problem of fish farmers in Kolkata, necessitating them to change to alternative and costly inputs such as feeds, and changing the ratio of species stocked (Bunting, 2002). Reduced access to nutrients may be linked to improved sanitation infrastructure being developed, or lack of maintenance of current systems. Inevitably, institutional issues are often critical, and these are discussed elsewhere at more length (Bunting and Little, 2003).

Bangkok's peri-urban catfish farmers are also moving out. Originally close to trash fish sources for feed, and the centre of demand, availability of alternative feeds and improved purchasing power in provincial markets make location close to the capital less attractive (Fig. 3.3). Lately, hybrid catfish have been increasingly fed with chicken by-products (Little *et al.*, 1994) and chicken processors have moved to provinces such as Lopburi, that produce maize and, increasingly, the chickens that consume it, with a concentration of catfish producers now located close by (Ingthamjitr, 1997).

Despite these changes, reductions in the size or extent of urban aquaculture, or difficulties in accessing inputs, do not necessarily signal the demise of the system. Indeed, despite a widespread perception that the Kolkata wetlands are in decline, it appears that improved productivity in smaller areas has permitted a growth in overall production with its many benefits over the last decades (Little *et al.*, 2002).

Perceptions of urban aquaculture and its products

The definition of an urban economy as one in which the workforce is primarily engaged in non-agricultural activity (Iaquinta and Drescher, 2000) can lead to the assumption that both recent migrants

Fig. 3.3. Proximity to broiler chicken slaughterhouse wastes supports intensive hybrid catfish production in the suburbs of Bangkok, but there is a tendency for aquaculture to follow relocation of the chicken processing industry away from the capital.

and long-established urban households are disinterested in food production as a major livelihood focus. This contrasts with the total employment potential of production, processing and trading of highly perishable aquatic products occurring in urban areas, often among the very poor. In many areas of the Indian subcontinent, employment within the fisheries sector is the preserve of low status, often low-caste people. However, the loss of skilled labour to more highly paid employment is believed to be a constraint to continued operation of aquaculture in the Kolkata wetlands (Kundu, 1994).

There is a close relationship with ethnicity, social status and acceptance of excreta reuse in food production, which is known to be highly variable in Asia (Edwards, 1992). Changes in perceptions are expected with improved education. Middle-class urbanites, though more distant from the source and background of the food they consume, are likely to become more selective and discerning in food choices, particularly those concerning health risk. In some areas there is widespread knowledge about, and ambivalence to, the use of wastes in aquaculture, such as in Kolkata, but even here there appears to be a trend towards the consumption of larger fish imported from rural areas of India (Little *et al.*, 2002). More typically, urban people will enjoy greater choice, and any urban food production has to deliver clear benefits such as freshness. In Kolkata, fish produced in urban systems are clearly identifiable by being small and marketed live, which many poorer and richer consumers appreciate. In Ho Chi Minh City, price clearly mediates choice; richer people eat wild fish from the Mekong Delta, the middle class choose marine fish and the poor eat cultured freshwater fish, often produced in urban areas, which are cheap, available and small (AIT/CAF, 1997). There are indications that the growing purchasing power of city populations in Asia may in fact undermine the continuance of urban aquaculture as larger, 'clean' fish are produced in rural areas more consistently and with a higher quality image than the fish produced close by. The import of large fish into Kolkata from Andra Pradesh, and the increasing quantities of fish from the Kolkata wetlands sold among the poor in provincial urban markets of West Bengal, indicates this trend, and a similar situation is apparently emerging in Hanoi (Fig. 3.4).

The openness of aquatic systems predisposes them to problems of both consumer perception and real risk of loss, which are particularly important in urban areas. These principally relate to the risks of contamination and theft.

A paradoxical effect of proximity to urban markets can be extensification of urban aquaculture that occurs for a range of reasons when producers seek to reduce

Fig. 3.4. Selling fish live requires sophisticated transport if fish are raised in rural areas. In Hanoi, better transportation infrastructure means that live fish from rural areas now competes with fish produced closer to the city. Conversely, more fish raised around the city are being sold in other provinces.

costs, particularly of nutrients. Many of these are linked to the opportunity for alternative, more lucrative livelihood activities and/or the risks associated with theft. In several urbanizing provincial centres in otherwise rural northeast Thailand, recent studies have confirmed earlier observations (Demaine *et al.*, 1999) that ponds closer to the urban centre are more likely to be used for trapping wild fish than the stocking and culturing of fish *per se* (AIT/SCAT, 2003). The conflicts in and around the Kolkata wetlands are a good example of how urban aquaculture can exist despite levels of theft that range from chronic loss to more acute law and order problems (Kundu, 1994).

The dynamic nature of settlement in urban areas in and around water bodies, insecurity of tenure and the nature of the community combine to make law and order a major issue affecting sustainable food production. Losses of stock to predators such as birds and mammals may be relatively less of a problem in urban than rural systems, but theft is likely to be exacerbated by settlement patterns of the poor and institutions/power relations that dominate poor communities.

Contamination

Contamination of water bodies, whether of a chronic or acute nature, can affect both the productivity of aquaculture systems and public health. The urban poor are disproportionately vulnerable to environmental and health hazards in urban areas (Wratten, 1995) and the reduction of such hazards is a major priority in most urban development programmes.

Difficulties controlling flows of both inputs and outputs can place increasing pressures on aquaculture in urbanizing situations. Thus, although productive aquaculture can be undermined by polluted water, aquaculture itself can be discouraged as a source of pollution. Discharges from intensive aquaculture may be increasingly subject to control in urbanizing contexts where water quality for multi-purpose use is a priority. Often this reduces viability of commercial systems or, in contrast, can lead to the development of zero discharge systems.

The relationship between aquaculture and wastewater use in urban areas is complex, partly as a high proportion of surface water is effectively contaminated in and around urban areas of developing countries because of inadequate sanitation. Additionally, wastewater may not be sewage alone, or at all, but rather constitute a mixture of run-off and local discharges. The best-developed and understood urban aquaculture in the SSEA region is based on semi-formal use of sewage that is introduced into ponds on the outskirts of cities. Areas outside of the city infrastructure that are urbanizing, but lack sewerage, frequently have *de facto* aquaculture from eutrophication of surface water from overflowing cess pits and septic tanks. Overhung latrines are also an important feature of urban life in some countries. The economics of zero nutrient costs are well understood by entrepreneurs who will site such ponds and their associated latrines on major routes out of cities and around areas of public congregation (e.g. bus stops).

Public health issues of such aquaculture are covered elsewhere (Edwards and Pullin, 1990; Edwards, 1992; Little and Edwards, 2003), but any holistic health assessment would rank these practices as more hygienic than alternatives, as all are better than direct contamination of surrounding land. Development towards pre-treatment of wastewater is considered a necessary basic safeguard for public health by most (WHO, 1999).

Using wastewater for controlled eutrophication of fish culture ponds may reduce overall public health risks from pathogenic organisms, disease vectors, even risks of heavy metals and agrochemical residues. In places where industrial pollution has undermined wastewater aquaculture, there is evidence that untreated discharge into local rivers has increased, for example in the To Lich River near Hanoi, Vietnam (Edwards, 1997) with consequent impacts on people's health. A number of reviews on the potential health hazards associated with urban aquaculture systems, especially those involving wastewater, have been produced (Mara and Cairncross, 1989; Edwards, 1992, 2001; Furedy *et al.*, 1999; Little and Edwards, 2003). Additionally, the health issues surrounding peri-urban natural resource development are reviewed by Birley and Lock (1999) and Lock and de Zeeuw (2003).

A major unresolved issue, where there are explicit risks from certain practices, is how the risk can be quantified and its relative importance assessed. In general, many of the risks associated with aquaculture are not specific to urban aquaculture – the contamination of culture water with human wastes or agrochemicals may be as likely in rural as urban contexts. Where urban environments tend to stimulate greater intensification and/or integration between waste reuse and aquaculture on one hand, or aquaculture and horticulture on the other, risks are probably relatively greater. Mara and Cairncross (1989) identified four groups of people directly at risk: field workers, crop handlers, local residents and consumers. Bunting and Little (2003) discuss the nature of the specific hazards to each group, describe the factors that influence the degree of risk and outline potential mitigation strategies. The importance of identified risk factors (Lock and de Zeeuw, 2003) is discussed specifically for aquatic farming systems.

The use of surface water in and around cities for fish culture may be particularly important for disease vector control, as productive semi-intensive fish ponds are essentially treatment systems of high biological activity. The clean, sun-lit shallow water required for an important malarial vector (*Anopheles gambiae*) is unlikely in the fertile fishpond environment, which is also contradictory to the needs of the filiariasis vector (*Culex quinquefasciatus*) that thrives in highly polluted water (Lock and de Zeeuw, 2003). These vectors are more likely to be a problem in aquatic vegetable production and unutilized polluted water bodies, respectively.

The risk associated with use of agrochemicals, particularly in horticulture, in and around aquaculture is serious and believed to be increasing. Pests on both terrestrial and aquatic vegetables in peri-urban areas around Bangkok are managed through intensive agrochemical control (AIT/KU, 2003). Recent and preliminary risk assessments have produced disturbing predictions on environmental and human health (Van den Brink *et al.*, 2003). The threats from contamination of heavy metals, which are common contaminants of wastewater in urban areas, are important, but their tendency to attach to organic matter may make the risks specific in terms of location, system and species of aquatic product (Bunting and Little, 2003). Knowledge of this type is more advanced for conventional vegetables than aquatic products (Lock and de Zeeuw, 2003).

Although insights into potential health impacts are growing, actual health impact data are scant (Lock and de Zeeuw, 2003) and are a major requirement for future research. Industrial contamination of water may also affect the viability of aquaculture by affecting product taste. In one region of China, production has shifted away from food fish, which became unmarketable because of their phenol taste, to nurseries (Edwards, 1992).

The risks from zoonosis are considered in Little and Edwards (2003). Speculation continues about the importance of aquaculture based on livestock waste as a factor in the emergence of new human viruses, but in reality these are more connected to the management of poultry and swine as separate or closely linked enterprises. Aquaculture linked with both on the same farm is rare, and the link has been overplayed and misinterpreted (Edwards *et al.*, 1988; Edwards, 1991).

The constraints to continued aquatic food production in urban areas often appear formidable, but clearly production is often being sustained, or even expanded, and new forms of aquaculture can develop. Conventional pond-based aquaculture appears resilient in urban areas and new approaches promise new opportunities.

Opportunities for Urban Aquaculture

Opportunities for urban aquaculture continue to develop as the nature of urban areas and production technology change. It could be argued that a high proportion of traditional aquaculture was urban. Application of knowledge, nutrients and markets is required for it to become established as a viable activity within any household. In contrast, rural aquaculture remains constrained in many contexts by these factors. Additionally, a preference among rural people to catch or purchase wild fish when seasonally abundant may also limit investment in aquaculture. Demand in urban markets tends to be more consistent, and historically aquaculture has developed to meet such demand. Recent work suggests the same is true of embryonic fish culture in parts of Africa.

Continuing opportunities

The opportunistic use of urban water bodies for food production will continue as an outcome of poorly managed cities where administration and/or enforcement of change is weak and investment in new infrastructure is poor. Enlightened governance of major conurbations in the developing tropics will necessitate careful analysis of priorities in urban renewal that take account of how city people, especially the poor, have survived. The assessment of the role of the city lakes in Hanoi, for example, appears to acknowledge that continued fish production is good for the environment and can be integrated with broader roles including recreation and flood control (see Chapter 7 of this volume). The concept of 'greener' cities situated in flood plains requires the retention, and indeed expansion of, water bodies and constructed wetlands within municipal boundaries of many of the fast-growing cities of SSEA.

The importance of the recreational aspects of urban aquaculture and its linkages with ordinary people can be shown by the growth in pay-fishing in many areas of SSEA, where fishing remains as much a pastime in rural areas as a necessity. The growth of recreational fishing can be related to either or all the opportunities identified as stimulating urban aquaculture. In many contexts it has developed as much as a desire to consume fresh fish by increasingly better off urban people, as a hobby interest similar to developed economies. In Brazil, fee fishing enterprises remain the major outlet for farmed tilapias, one which the industry has developed around (Lovshin, 2000). Opportunistic uses of urban water bodies, conversion of food fish enterprises and strategic siting close to high population densities have all occurred as stimuli in different parts of Asia. Opportunities for urban aquaculture will continue to occur as long as the current trend for rural people to seek a living in the city continues.

Urbanization and industrialization

Industrialization and urbanization occurring at the same time and place often stimulate the development of aquatic food production. Urban areas tend to have the most reliable infrastructure, or indeed the only infrastructure, in the SSEA that is conducive to industrial growth. A wide range of nutrient-rich wastes develop in association with rapid increases in population and the higher average purchasing power of urban compared to rural people. High levels of rural–urban migration by fish consumers stimulate demand. Industry concentrates nutrients of low opportunity value as the by-products of feeding a workforce (i.e. cafeteria wastes) and processing animal and human foods. This concentration makes waste reuse more economical and attractive to the entrepreneurs who typically innovate systems that are profitable locally. Increased consumption of livestock products tends to lead to intensification of poultry and pig production around urban areas, and these have often become major supports to urban aquaculture in SSEA.

Intensification

Aquaculture may have begun in cities as a process of holding wild fish at high densities in cages and tanks prior to sale, and thus the skills necessary to intensify culture may be well-developed. In contrast to the phenomenon of extensification that can be observed around some urban centres, intensification is a common adaptation to encroaching urbanization. Intensification makes better control and surveillance of stocks both possible and viable, and in doing so reduces the risk of theft, which can be a major constraint (Bunting and Little, 2003).

Intensification may be inevitable given changing costs and opportunities. Increasing costs of land make higher returns essential, and proximity to alternative resources and markets stimulates innovation. Thus, the extensive monocultures of the snakeskin gourami (*Trichogaster pectoralis*) in provinces close to Bangkok have become more intensive and robust in response to local industrial development. Yoonpundh (1997) found that farmers located closer to clusters of industry tended to diversify towards higher value systems and polycultures, despite less reliable water quality, than those in more rural areas. Intensification of the sewage-fed systems in Kolkata is also allowing production to increase despite loss of area and poor delivery of sewage (Little *et al.*, 2002; Chapter 5 of this volume). The intensification appears related to improved management and the introduction of species that are more productive in eutrophic systems, particularly tilapias and silver carp. A continuing demand for small, live fish has also been a major factor in expanding production. Raising small fish that poor consumers can afford, and that are easily and simply transported live, reduces risk and increases opportunities in the production chain (Little *et al.*, 2002). Continuous harvest and stocking improves cash flow and observability of stocks and reduces the impact of theft and occasional pollution events. Although improved road infrastructure has allowed imports from distant Indian states to compete for the markets in the city, it has also meant provincial urban markets in West Bengal could be exploited. The availability of ice and post-harvest expertise in and around such a large production centre has also increased the relative advantage of the producers in the Kolkata wetlands.

Evolution from food fish to fish seed and then to ornamental fish production is a common development track for aquaculture enterprises within urbanizing areas. In addition to an increasing cost of key resources as urbanization occurs, it also suggests how competition, knowledge and opportunities can stimulate a move up the value chain. Moving to production of juvenile foodfish with faster production cycles and improved cash flow is often an intermediate step, and supports a host of other niches such as broodfish and live feed collection or production, and hatchery suppliers. The major seed fish production centre of Jessore, Bangladesh is a good example of how a concentration of knowledge and resources has developed in a small urban centre linked in this case with knowledge and material inputs (e.g. dried pituitary glands from the foodfish markets) from Kolkata (Little, 2002). Mature examples of this evolution up the value ladder include Bangkok and Singapore, where high value ornamental fish enterprises are concentrated. Ho Chi Minh City is an example of a more domestic-orientated industry (see Chapter 6 of this volume). In addition to being well linked with juvenile producers on the outskirts of the city, poor slum dwellers living on the Saigon River are key actors in the network that produces and markets both juvenile food and ornamental fish through their role in harvesting tubifex worms from the highly polluted waterway (AIT/CAF, 1997; Fig. 3.5).

Processing

A potential strategy to maintain competitive aquaculture in urbanizing areas is to improve the efficiency of resource utilization. Thus, while feedlot pig production

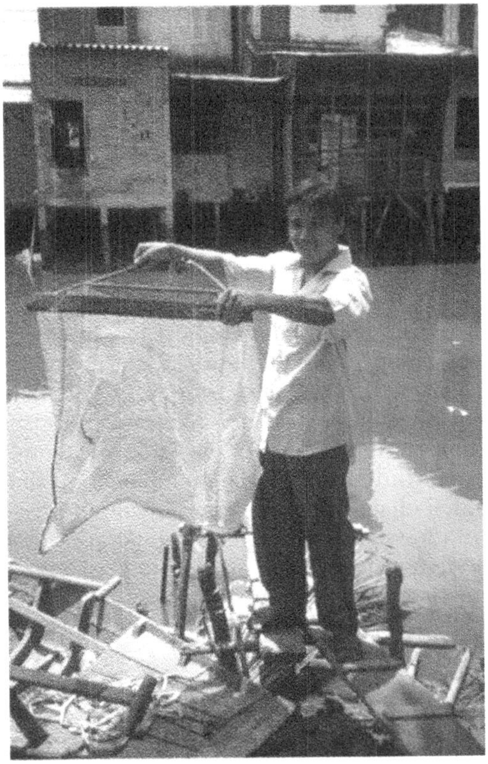

Fig. 3.5. Harvest and sale of live *Tubifex* worms from the heavily polluted Saigon River supports ornamental fish production in and around Ho Chi Minh City, Vietnam.

can remain competitive at a certain distance from the centre, semi-intensive fish culture that tends to have lower returns per unit area may be uncompetitive. One approach is the processing of pig manure to a suitable live feed and its use for feeding more valuable, air-breathing species of fish such as catfish. A system developed by one commercial operator used green blowfly larvae to process pig slurry, and reduced the pond area required by a factor of more than ten (Nuov *et al.*, 1995). Moreover, the process has been found to both reduce odour and fly problems, and increase the value of the residual manure for horticulture, in addition to greatly increasing the value of fish produced per unit area. Fish farmers in urban areas of Asia also innovate production systems to reduce risks from inconsistent water quality. The recycling of water and nutrients between higher input and semi-intensive systems has become commonplace on the periphery of Bangkok (Little and Griffiths, 1992).

Alternatively, wastes from urban production (livestock, human) can be traded on to fish producers further out of the city, where nutrients are more limiting and land prices lower. These trends are well established around Bangkok and Shanghai. Displacement of waste recycling and intensive food production from in and around such cities have led to waste disposal becoming externalized rather than, as formerly, invisible and internalized. Large quantities of wastes are now transported from Shanghai to neighbouring provinces for reuse (Whitney, 1991).

Research and Development Needs

A major shortfall in knowledge concerns detailed, joined-up analysis of how urban aquaculture benefits various stakeholders. The lack of clear responsibility within urban administrations is a major constraint in this regard, and urgently requires attention if holistic and sensible planning is to occur for Asia's emerging megacities and fast-growing secondary urban centres. The 'promotion of participatory, site specific and interdisciplinary field research on urban agriculture with a strong policy and action orientation' was raised as a major need recently (de Zeeuw *et al.*, 2000) and has guided development of a current research project in three countries of Southeast Asia. The Production in Aquatic Peri-Urban Systems in Southeast Asia (PAPUSSA) project is investigating the current status of urban aquatic food production systems at four sites in Cambodia, Thailand and Vietnam, and then piloting interventions with stakeholders in each. Preliminary assessment suggests that these systems are highly varied, and meet diverse needs in and around the major cities of each country.

Clearly a range of issues identified for ensuring the future of urban food production generally needs to be considered for aquatic systems. The sensitivity of environmental quality to the success or failure of aquaculture, both technically and in the perceptions of consumers, makes this a critical issue if aquatic production systems are to be sustained. Assessment of the current and future role of urban aquaculture within urban waste management is urgently required if many Asian cities are to develop, or improve, public health affordably in the near future.

A sufficient level of assured quality will rely heavily on improved education and enforcement of legislation. Potentially, the linking of improved quality of both water resources and food produced in cities could become the ultimate insurance for future farming of urban water bodies and waterways. Clearly, environmental and food quality awareness of urban populations will need to be radically increased in most contexts, but enforcing of rules that set levels of permissible intensification and acceptable eutrophication for multiple-use water will be critical. Furthermore, a better understanding of the nature of bioaccumulation of contaminants in aquatic farming systems, and their products, is urgently required if increasingly sophisticated and knowledgeable urban consumers are to continue to purchase them.

Conclusions

Urban aquaculture has been identified as both a means to increase quality food supplies and reduce organic wastes in South and Southeast Asia and these experiences inform potential elsewhere. Poor people appear to benefit from urban aquaculture, and a major challenge is to strengthen their position while supporting changes beneficial to other stakeholders. The benefits to the poor and challenges to sustainability are similar for urban aquaculture and urban food production as a whole, but the nature of aquatic systems in urban areas, particularly their inherent connectivity, makes these issues even more complex and challenging.

Acknowledgements

The authors acknowledge the support for their research from DFID-NRSP, specifically project R7872 and their colleagues Dr Nitai Kundu and Dr Madhumita Mukerjee and their staff. Interactions with colleagues as part of the inception of the EC-funded INCODEV PAPUSSA project have also been critical to the development of these ideas. Joseph Whitney, Professor Emeritus, Department of Geography and Institute for Environmental Studies, University of Toronto, and Professor Peter Edwards of Asian Institute of Technology, Bangkok, are thanked for critically reviewing the manuscript and providing reference material.

EC-funded INCODEV PAPUSSA project – Production in Aquatic Peri-Urban Systems in Southeast Asia – (ICA4–2001–10072) is being implemented by the Universities of Stirling and Durham in the UK; Royal Veterinary and Agricultural University, Department of Veterinary Microbiology (KVL), Denmark; Kasetsart University, Thailand; Royal University of Agriculture, Cambodia; University of Agriculture and Forestry, Research Institute Aquaculture 1 and National Institute for Health and Epidemiology, Vietnam.

References

AIT/CAF (1997) Hatchery development in and around Ho Chi Minh City. *State of the System Report.* AIT Aqua Outreach, AIT, Pathum Thani, Thailand.

AIT/KU (2003) Managing agrochemicals in multi-use aquatic systems. *State of the System Report.* Thailand report series 1/2003. Sisaket College of Agriculture and Technology, Sisaket and Asian Institute of Technology, Pathum Thani.

AIT/SCAT (2003) Improved resource use efficiency in Asian integrated pond-dike systems. *State of the System Report*. Thailand report series 1/2003. Sisaket College of Agriculture and Technology, Sisaket and Asian Institute of Technology, Pathum Thani.

Beveridge, M.C.M. and Little, D.C. (2002) History of aquaculture in traditional societies. In: Costa-Pierce, B.A. (ed.) *Ecological Aquaculture*. Blackwell Science, Oxford, pp. 3–29.

Birley, M.H. and Lock, K. (1999) *The Health Impacts of Peri-urban Natural Resource Development*. Liverpool School of Tropical Medicine, Liverpool.

Bunting, S.W. (2002) *Renewable Natural Resource Use in Livelihoods at the Calcutta Peri-urban Interface*. Final Technical Report. University of Stirling, Stirling.

Bunting, S.W. and Little, D.C. (2003) Urban aquaculture. In: *Annotated Bibliography on Urban Agriculture*. ETC Urban Agriculture Programme. SIDA/ETC, Leusden, Netherlands, pp. 447–477.

de Zeeuw, H., Guendel, S. and Waibel, H. (2000) The integration of agriculture in urban policies. In: Bakker, N., Dubelling, M., Gundel, S., Sabel-Koschella, U. and de Zeeuw, H. (eds) *Growing Cities, Growing Food: Urban Agriculture on the Policy Agenda*. Feldafing, Deutsche Stiftung fur Entwicklung (DSE), Germany, pp. 163–183.

Demaine, H., Innes-Taylor, N.L., Turongruang, D., Edwards, P., Little, D.C. and Pant, J. (1999) Small-scale aquaculture in Northeast Thailand. A case study from Udorn Thani. *Studies in Agricultural and Aquatic Systems 2*. Aquaculture and Aquatic Resources Management Programme, Asian Institute of Technology, Pathum Thani, Thailand.

Edwards, P. (1991) Fish–fowl–pig uncommon. *World Aquaculture* 22(3), 2–6.

Edwards, P. (1992) *Reuse of Human Waste in Aquaculture. A Technical Review*. UNDP-World Bank Water and Sanitation Programme, Washington, DC.

Edwards, P. (1997) Sustainable food production through aquaculture. *Aquaculture Asia* 2, 4–6.

Edwards, P. (2001) Public health issues of wastewater-fed aquaculture. *Urban Aquaculture Magazine* 3, 20–22.

Edwards, P. and Pullin, R.S.V. (eds) (1990) Wastewater-fed Aquaculture. *Proceedings of the International Seminar on Wastewater Reclamation and Reuse for Aquaculture*. Calcutta, December 1988. Asian Institute of Technology, Environmental Sanitation Information Center, Bangkok.

Edwards, P., Lin, C.K., Macintosh, D.J., Wee, K.L., Little, D. and Innes-Taylor, N.L. (1988) Fish farming and aquaculture. *Nature* 333, 505–506.

Edwards, P., Little, D.C. and Yakupitiyage, A. (1996) A comparison of traditional and modified inland artisanal aquaculture systems. *Aquaculture Research* 28, 777–788.

Edwards, P., Little, D.C. and Demaine, H. (2002) Issues in rural aquaculture. In: Edwards, P., Little, D.C. and Demaine, H. (eds) *Rural Aquaculture*. CAB International, Wallingford, UK, pp. 323–338.

Farrington, J., Ramasut, T. and Walker, J. (2002) Sustainable livelihood approaches in urban areas: general lessons, with illustrations from Indian Cases. Report to Swedish International Development Cooperation Agency, Working Paper No. 162.

Furedy, C. (1990) Social aspects of human excreta reuse: implications for aquacultural projects in Asia. In: *Proceedings of the International Seminar on Wastewater Reclamation and Reuse for Aquaculture*. Calcutta, December 1988. Asian Institute of Technology, Environmental Sanitation Information Center, Bangkok, pp. 251–266.

Furedy, C., Maclaren, V. and Whitney, J. (1999) Reuse of waste for food production in Asian cities: health and economic perspectives. In: Koc, M., MacRae, R., Welsh, J. and Mougeot, L.J.A. (eds) *For Hunger-proof Cities: Sustainable Urban Food Systems*. IDRC, Ottawa.

Haughton, G. and Hunter, C. (1996) *Sustainable Cities. Regional Policy and Development Series 7*. Jessica Kingsley Publishers, London.

Iaquinta, D.L. and Drescher, A.W. (2000) Defining periurban: understanding rural–urban linkages and their connection to institutional contexts. Paper presented at the *Tenth World Congress of the International Rural Sociology Association*, Rio de Janeiro, 1 August 2000.

Ingthamjitr, S. (1997) Hybrid *Clarias* catfish seed production and marketing in central Thailand and experimental testing of seed quality. D. Tech Sci. Thesis. Asian Institute of Technology, Bangkok.

Koc, M., MacRae, R., Welsh, J. and Mougeot, L.J.A. (1999) *For Hunger-proof Cities: Sustainable Urban Food Systems*. IDRC, Ottawa.

Kundu, N. (1994) *Planning the Metropolis, a Public Policy Perspective*. Minerva Associates, Calcutta.

Little, D.C. (2002) Freshwater fish seed-movements of technology and knowledge drive aquaculture development. *AARM Newsletter* 6(4), 5–6.

Little, D.C. and Edwards, P. (2003) *Integrated Livestock Fish Farming Systems: the Asian Experience and its Relevance for other Regions.* Food and Agriculture Organization, Rome.
Little, D.C. and Griffiths, D. (1992) Nile tilapia clean water from hybrid catfish farming. *Fish Farming International,* April 12–14.
Little, D.C. and Hulata, G. (2000) Strategies for tilapia seed production. In: Beveridge, M.C.M. and McAndrew, B.J. (eds) *Tilapias: Biology and Exploitation.* Fish and Fisheries Series 25. Kluwer Academic, Dordrecht, pp. 267–326.
Little, D.C., Kaewpaitoon, K. and Haitook, T. (1994) The commercial use of chicken processing wastes to raise hybrid catfish (*Clarias macrocephalus* × *Clarias gariepinus*) in Thailand. *Naga* 17(4), 25–27.
Little D.C., Kundu, N., Mukherjee, M. and Barman, B.K. (2002) Marketing of fish from peri-urban Kolkata. Working Paper 6. University of Stirling, Stirling.
Lock, K. and de Zeeuw, H. (2003) Health and environmental risks. Annotated Bibliography on Urban Agriculture. http://www.ruaf.org/bibliography-fr.html
Lösch, A. (1964) *The Economics of Location.* Translated from the 2nd rev. ed. by William H. Woglom, with the assistance of Wolfgang F. Stolper. Yale University Press, New Haven, Connecticut.
Lovshin, L.L. (2000) Tilapia culture in Brazil. In: Costa-Pierce, B.A. and Rakocy, J.E. (eds) *Tilapia Aquaculture in the Americas, Vol. 2.* The World Aquaculture Society, Baton Rouge, Louisiana, pp. 133–140.
Mara, D. and Cairncross, S. (1989) *Guidelines for the Safe Use of Wastewater and Excreta in Agriculture and Aquaculture: Measures for Public Health Protection.* World Health Organization, Geneva.
Marshall, A. (1919) *Industry and Trade.* Macmillan, London.
McGee, T.G. (1991) The emergence of desakota regions in Asia. In: Ginsburg, N., Koppel, B., and McGee, T.G. (eds) *The Extended Metropolis: Settlement Transition in Asia.* University of Hawaii Press, Honolulu, Hawaii, pp. 6–7.
McGee, T.G. (1995) Macrofitting the urban regions of ASEAN. In: McGee, T.G. and Robinson, I. (eds) *The Mega-urban Regions of Southeast Asia.* UBC Press, Vancouver, pp. 3–26.
McGee, T.G. and Robinson, I. (eds) (1995) *The Mega-urban Regions of Southeast Asia.* UBC Press, Vancouver.
Moser, C. (1998) The asset vulnerability framework: reassessing urban poverty reduction strategies. *World Development* 26(2), 1–19.
Mougeot, L.J.A. (1999) Urban agriculture: definition, presence, potentials and risks. In: Bakker, N., Dubelling, M., Gundel, S., Sabel-Koschella, U. and de Zeeuw, H. (eds) *Growing Cities, Growing Food: Urban Agriculture on the Policy Agenda.* Feldafing, Deutsche Stiftung fur Entwicklung (DSE), Germany, pp. 33–43.
Nuov, S., Little, D.C. and Yakupitiyage, A. (1995) Nutrient flows in an integrated pig, maggot and fish production system. *Aquaculture Research* 26, 601–606.
Parenteau, R., Gregoire, I. and Harnois, N. (1993) *Gestion integre des eaux de surface en Hanoi: problems enjeux et moyens.* Institut d'Urbanisme, Universite de Montreal.
Prax, J., Little, D.C., São, P.V. and Tuan, P.A. (2002) Seed production and distribution network in Mao Dien Community Red River Delta, Northern Vietnam. Working Paper, University of Stirling, Stirling.
Rana, K.J. (1998) Recent developments in aquaculture statistics. Fishery and aquaculture statistics in Asia. In: *Proceedings of the FAO/SEAFDEC Regional Workshop on Fishery Statistics. 19–21 August 1997.* Vol. 11. FAO, Rome, pp. 242–254.
Smit, J. (1996) *Urban Agriculture: Food, Jobs and Sustainable Cities.* Publication Series for Habitat II, Vol. I. United Nations Development Programme, New York.
Van den Berg, L.M., Van Wijk, M.S. and Hoi, P.V. (2003) The transformation of agriculture and rural life downstream of Hanoi. *Agricultural Transformations* 15(1), 3552.
Van den Brink, P.J., Sureshkumar, S.N., Daam, M.A., Domingues, I., Milwain, G.K., Beltman, W.H.J., Perera, M.W.P. and Satapornvanit, K. (2003) *Environmental and Human Risks of Pesticide Use in Thailand and Sri Lanka.* Alterra-rapport 789/MAMAS Report Series 2/2003. Alterra, Green World Research, Wageningen.
Von Thünen, J.H. (1875) *Der isolierte Staat in Beziehung auf Landwirthschaft und Nationalökonomie.* von H. Schumacher-Zarchlin Edition 3. Wiegandt, Berlin.
Whitney, J. (1991) Waste economy and the dispersed metropolis in China. In: Ginsburg, N., Koppel, B. and McGee, T.G. (eds) *The Extended Metropolis: Settlement Transition in Asia.* University of Hawaii, Honolulu, Hawaii, pp. 177–191.

Whitney, J.B.R. and White, R. (1992) Cities and the environment: an overview. In: Stren, R.E., White, R.R. and Whitney, J.B.R. (eds) *Sustainable Cities: Urbanization and the Environment in International Perspective*. Westview Press, Boulder, Colorado, pp. 229–232.
WHO (1999) *Food Safety Issues Associated with Products from Aquaculture*. WHO Technical Report Series 883. World Health Organization, Geneva.
Wratten, E. (1995) Conceptualising urban poverty. *Environment and Urbanization* 7(1), 11–36.
Yoonpundh, R. (1997) The reproductive biology of Snakeskin Gourami (*Trichogaster pectoralis* Regan) and intensification of seed production. PhD Thesis, Asian Institute of Technology, Bangkok.

4 Development Status of, and Prospects for, Wastewater-fed Aquaculture in Urban Environments

Peter Edwards
Asian Institute of Technology, Klong Luang, Thailand

Abstract

Informal reuse of wastewater in aquaculture is widespread in parts of Asia, although there has been limited introduction of formally designed and engineered systems. As there is interest in assessing the contribution of wastewater-fed aquaculture to the development of sustainable cities, a multidisciplinary typology is presented of factors to consider (i.e. associated sanitation technology, aquaculture systems, disposal of produce and effluent, and institutional aspects). An overview is also presented of the global occurrence of wastewater-fed aquaculture. Although most extant systems are threatened or in decline, wastewater-fed aquaculture has potential, as indicated by the recent introduction of schemes in Bangladesh and India. Design criteria are presented for the maximal production of fish safe for human consumption with minimal treatment of wastewater, and for the production of tilapia and duckweed as high-protein animal feed.

Introduction

Informal use of excreta and wastewater in aquaculture is widespread in parts of Asia, with relatively little practice in other parts of the world. However, there has been only limited introduction of planned reuse in aquaculture by national or local authorities using formally designed and engineered wastewater treatment and reuse systems (Edwards, 2000, 2002).

Little wastewater treatment takes place in urban areas of most developing countries, and two-thirds of urban wastewater receives no treatment at all – it is mostly discharged into the most convenient surface water source – as the cost of installing conventional mechanical wastewater treatment is prohibitive. Even when the latter is introduced, it rarely functions adequately due to high capital costs, high maintenance costs, the need for sophisticated equipment and qualified technicians. The development of sustainable technologies for wastewater treatment within the economic and technological capacity of developing countries that safeguard public health is required (WHO, in press). Interest is growing in inexpensive waste stabilization ponds that also achieve high rates of pathogen removal. There is need for a holistic approach based on natural treatment and reuse of both nutrients and water that is energy and cost effective, and provides employment and low-cost

 Urban Aquaculture
(eds B. Costa-Pierce, A. Desbonnet, P. Edwards and D. Baker)

food for the poor, while at the same time providing a financial incentive to stimulate effective operation and maintenance of wastewater treatment facilities.

Wastewater treatment and reuse schemes through aquaculture could play a major role in poverty alleviation programmes in urban areas. There is also growing interest in use of treated effluents in many parts of the world because of the rapidly increasing shortage of renewable water resources, especially in semi-arid and arid regions. The *Hyderabad Declaration on Wastewater Use in Agriculture*, an outcome of a recent (November, 2002) workshop entitled 'Wastewater use in Irrigated Agriculture: Confronting the Livelihood and Environmental Realities', sponsored by the International Water Management Institute based in Colombo, Sri Lanka and the International Development Research Centre based in Ottawa, Canada, recommended such an holistic approach to management of wastewater in aquaculture and agriculture (Anonymous, 2002).

This chapter presents a multidisciplinary typology of factors to consider in assessing the contribution of current and future potential of wastewater-fed aquaculture to sustainable urban development. The history and current status of wastewater-fed aquaculture are outlined, with specific examples from major world regions. Future prospects are discussed, with a design for the minimal treatment of wastewater with the maximal production of microbiologically safe fish for both improvement of existing schemes, and implementation of new schemes. Produce restriction through the production of high-protein animal feeds such as duckweed and tilapia is also described. These may provide alternatives to fish meal in formulated diets of both livestock and fish.

Typology

For wastewater treatment and reuse schemes involving aquaculture to play a greater role in sustainable urban development, there is a need to explore their potentials and constraints using a multidisciplinary system approach. A typology is presented to facilitate such a process (Table 4.1). The typology comprises: wastewater type, its conveyance to the aquaculture system and the degree of prior treatment; the aquaculture system with its status and trends, the resource system in which it is placed, and its technical and social aspects; disposal of aquaculture produce and the aquaculture system effluent; and institutional aspects covering its development and regulatory and policy frameworks.

Wastewater

Wastewater is used generically to cover the wide range of forms in which human excreta are used in aquaculture. Nightsoil is fresh human excreta. Contaminated surface water refers to natural water bodies, such as rivers and lakes, containing human faecal material because of absent or inadequate sanitation technology. Domestic sewage is water-borne excreta from flushed toilets, to which the term wastewater is often restricted, and it may be contaminated with varying amounts of industrial wastewater or sewage. Sewage may also contain urban run-off if the sewerage system is connected to storm drains. Septage is the sludge or slurry removed periodically from a septage tank which is associated with a flushed toilet. Sanitation technology options with reference to aquaculture are discussed in detail by Edwards (1992).

Nightsoil may be used directly in growout systems through an overhung fishpond latrine, in which case there is no conveyance as there is on-site use. Indirect use is use of contaminated surface water, in contrast to direct use, in which wastewater is transported either by cartage or conveyed by gravity flow or pumping through a reticulated network of sewage pipes or channels. Cartage of faecal solids in the form of nightsoil or septage may be done manually in containers transported on foot, bicycle or motor bike, or by boat or vacuum truck.

Table 4.1. Towards a typology for wastewater-fed aquaculture.

Wastewater type
- Nightsoil
- Contaminated surface water
- Domestic sewage
- Industrial sewage
- Urban run-off
- Septage

Conveyance to aquaculture system
- None/on-site use
- Indirect use
  - Natural drainage
- Direct use
  - Cartage
  - Reticulated sewage network

Prior treatment
- None
- Natural treatment
- Low-cost treatment
- Conventional sewage

Aquaculture system
- Status
  - Subsistence
  - Commercial
  - Experimental
  - Trends
  - Expanding
  - Stable
  - Declining
  - Conservation status
- Resource system
  - Agricultural land
  - Waste land
  - Wetland
  - Water body
- Technical aspects
  - Cultured organism
    - + Fish
    - + Aquatic plants
  - Culture facility
    - + Open water
    - + Pond
    - + Cage
  - Integration
    - + Polyculture of fish and aquatic plants
    - + Field crops
    - + Horticulture
    - + Rice
  - Wastewater application
    - + Gravity flow
    - + Pumping
    - + Sprinklers
  - Input supply

Aquaculture system *(continued)*
    - + Seed
    - + Feed
    - + Other
- Social aspects
  - Producer socio-economic characteristics
  - Tenure
  - Human exposure control
    - + Producers

Disposal of produce
- Intended use
  - Food
    - + Home consumption
    - + Local market
    - + Export market
  - Produce restriction
    - + Fish seed
    - + Ornamental fish
    - + Animal feed
- Social acceptance
- Human exposure control
  - Traders
  - Processors
  - Consumers
    - + Raw
    - + Cooked
    - + Processed
- Marketing and processing
  - Chain
  - Trader and processor socio-economic characteristics

Disposal of effluent
- Surface water
- Crop irrigation

Institutional aspects
- Development
  - Informal
  - Formal
  - Potential
  - Constraints
- Regulatory framework
  - Wastewater pricing
    - + Households
    - + Farmers
  - Water quality monitoring
  - Effluent standards
  - Public health guidelines
- Policy framework
  - Low-cost wastewater treatment
  - Benefits for poor producers
  - Benefits for poor consumers
  - Incorporation into urban planning

Indirect use is the planned application of wastewater from a receiving water body, such as a watercourse draining an urban area. However, in many areas where the river flow is mainly sewage, especially in the dry season, disposal in surface waters is in fact disposal in a 'natural conveyance channel' (van der Hoek, 2002). An example is the Hanoi wastewater-fed aquaculture system described in Chapter 7 of this volume.

Wastewater may be untreated or treated prior to use in an aquaculture system. Natural treatment refers to the treatment that takes place in contaminated surface water prior to use in aquaculture, in contrast to intentional treatment through low-cost or conventional sewage treatment. Low-cost treatment here refers to on-site treatment (pit-latrines, pour–flush latrines, biogas and composting latrines) and off-site treatment of nightsoil and septage. Conventional sewage treatment may be by mechanical treatment such as activated sludge, aerated lagoons or trickling filters or biological treatment in stabilization ponds.

Aquaculture system

With the exception of the overhung pond latrine, in which fish directly consume human faeces, wastewater is not used directly in aquaculture. The nutrients contained in wastewater are used as fertilizer to produce natural food such as plankton or benthos for fish, although nutrients are taken up directly by aquatic plants.

High organic matter loading rates of 200–300 kg BOD_5/ha/day are used to treat wastewater in stabilization ponds, as the aim is to minimize the amount of land devoted to treatment processes. However, organic loadings about one order of magnitude lower, 10–30 kg BOD_5/ha/day, must be used in wastewater-fed fish culture to avoid producing too dense a concentration of phytoplankton, which would deplete dissolved oxygen through respiration during the night. Thus, there is a need to reconcile the different approaches of sanitary engineers and aquaculturists to wastewater or stabilization pond design.

As wastewater-fed aquaculture often occurs in areas without a supply of relatively unpolluted freshwater, ponds for fish culture are filled initially with the only source of water available, generally contaminated surface water or raw sewage with a high percentage of organic matter. In this first step of this process, known as primary fertilization, the anaerobic water filling the pond is left to stand for 10–20 days until natural biological purification processes have reduced the biological oxygen demand and increased dissolved oxygen to a sufficiently high concentration for fish to be stocked. Subsequent additions of anaerobic water, intermittently in small ponds to more or less continuously in large ponds, maintain pond fertility to provide adequate natural food for fish in a process known as secondary fertilization (Fig. 4.1).

Regarding the status of aquaculture, the system may be at the household level for subsistence and/or sale of produce. Large-scale systems invariably are primarily commercial. A considerable number of experimental or hypothetical systems have been proposed (Edwards, 1992, 2000). Resource system refers to the larger system in which the aquaculture system physically operates, generally constructed either on agricultural or waste land, or in a wetland or water body.

There are various technical aspects in the typology. Wastewater may be used to grow fish or aquatic plants (freshwater macrophytes). Aquatic plants may be farmed in open water, such as faecally contaminated rivers, lakes or reservoirs, or in ponds, with the latter the major type of culture facility used to farm fish. Fish may be enclosed in cages in open water bodies.

Culture of fish in ponds may be integrated with aquatic plants in a polyculture. Pond water from the aquaculture facility may be used to irrigate crops, usually vegetables, on the dyke (Fig. 4.2). Effluent also may be used to irrigate crops, field crops, fruit trees, rice or vegetables. Wastewater may enter the system by gravity flow, by pumping or by sprinklers (Fig. 4.3).

Fig. 4.1. Wastewater being pumped into a wastewater-fed fishpond in Hanoi.

Fig. 4.2. Collecting lotus stems for human consumption in a pond fed with untreated conventional wastewater in Ho Chi Minh City, Vietnam.

Seed, in the form of larvae or fingerlings, is usually the only input to the system, although tilapias may not need to be stocked if self recruitment is adequate. Natural food in the forms of plankton and benthos produced as a result of the fertilization of wastewater usually provide adequate feed for fish. However, supplementary feed may be added in the monsoon season when natural food is limited by dilution, increased water turbidity and lower insolation.

Socially defining characteristics of wastewater use are the social and eco-

Fig. 4.3. A sprinkler introducing secondary treated wastewater diluted with river water into a wastewater-fed fishpond in Munich, Germany.

nomic status of the farmers or producers, including the involvement of women and children as well as men. Security of tenure of land (or water in the case of open water culture) will influence the degree of investment in the system by private farmers, but large schemes invariably need government involvement.

Product market

Wastewater is most commonly used to produce food for direct human consumption, in both the form of fish and aquatic plants. The main source of disposal of household-level produce is likely to be home consumption, although produce from larger and off-site aquaculture systems is often destined for the local market. Food produced in wastewater-fed aquaculture systems is unlikely to be exported because of its relatively low value.

Wastewater may also be used indirectly to produce fish seed or fingerlings, ornamental fish or high-protein produce for animal feed. Examples of the latter are wastewater-fed tilapias to feed high-value carnivorous fish or crustaceans farmed in separate systems and duckweed to feed herbivorous fish or livestock. Social acceptance of wastewater use is an important cultural criterion which may vary over time.

There are also public health considerations for people at potential risk from use of wastewater in aquaculture, especially producers, traders, processors and consumers of produce. Wastewater-grown food may be consumed raw, cooked or processed. There are obvious public health considerations, especially with consumption of raw and inadequately cooked or processed produce. There is a need to identify the marketing and processing chains, and the social and economic status of the traders and processors for wastewater-grown produce.

Disposal of effluent

Aquaculture effluent is usually discharged into adjacent surface water such as a river, although the quality of the effluent will be better than that of the input wastewater because of natural purification during the aquaculture phase. Occasionally, the effluent is used to irrigate a downstream crop, or even the pond water itself (Fig. 4.4).

Fig. 4.4. Cultivating vegetables on a wastewater-fed fish pond dyke at Mudialy Fishermen's Co-operation Society, Calcutta, India. The vegetables are watered with green, fishpond water.

Institutional aspects

Aquaculture wastewater use systems may be informal, with use of wastewater lacking permission or control by state agencies, or formal, with a certain level of permission and control by state agencies (van der Hoek, 2002). The terms 'planned' and 'unplanned' have been used to characterize direct and indirect wastewater use, respectively, but are of limited practical value for a typology (van der Hoek, 2002). The rationale behind use of the former terms is related to the involvement of the state in planning for wastewater use, but in wastewater-fed aquaculture farmers most often plan for use of wastewater without any state involvement. In many cases the wastewater may be the only source of water, as well as nutrients, available to farmers. As wastewater-fed aquaculture faces an uncertain future, potential and constraints warrant further elucidation.

In assessing the relevance of current and potentially new schemes for wastewater-fed aquaculture in peri-urban areas of cities, there is a need to investigate whether a regulatory framework exists that considers wastewater pricing for disposal cost of households, and purchasing costs by farmers, relative to water quality monitoring and effluent standards for wastewater treatment and use, and national public health guidelines for wastewater use. Furthermore, whether the benefits of wastewater use for poor producers and consumers and its role in low-cost wastewater treatment are recognized, possibly through incorporation into urban planning, should be determined.

History and Current Status

Overview

Wastewater use in aquaculture is widespread in parts of Asia, mainly informally, with relatively limited practice in other parts of the world. The practice has a long history in several countries in East, South and Southeast Asia, and especially in China where it is several centuries old

(Edwards and Pullin, 1990; Edwards, 1992, 2000). There has been only limited introduction of formal use of wastewater in aquaculture by central government based on engineered systems, as it has been mainly developed informally by farmers and local communities.

The primary aim of informally developed wastewater use has been to utilize both water and nutrients for aquatic farming rather than just the treatment or disposal of wastewater. Wastewater is almost invariably used in aquaculture without any formal treatment, with inadequate attention given to public health. This may be contrasted with the formal design of wastewater treatment systems by engineers in which the primary aim is treatment and disposal, with concern for public health.

Formal design of systems for wastewater treatment and use is relatively rare, even though it has long been demonstrated that fish can be cultured in maturation ponds in a conventional series of stabilization ponds, as they are typically aerobic at all times, and have low levels of unionized ammonia. About 90 engineered systems incorporating sewage use for fish culture were constructed in Germany from the late 19th century until the 1950s (Fig. 4.3), but almost all have shut down, with the notable exception of the ponds in Munich (Prein, 1990, 1996). Sewage-fed fish ponds were developed later in Asia than in Europe, initially in India in the 1930s, in China from the 1950s, and in Vietnam from the 1960s.

Numerous experiments have been carried out on culture of fish in engineered wastewater stabilization ponds around the world, even though it has rarely led to commercial practice: Africa (Egypt, Kenya, South Africa), the Americas (Dominican Republic, Peru, USA), Asia (China, India), Australasia (Australia, New Zealand), Europe (France, Hungary, Poland, Russia, UK) and the Middle East (Egypt, Jordan) (Edwards, 1992, 2000).

The great diversity of systems currently in use, characterized by wastewater type and conveyance system, culture facility and cultured organism, as well as geographical location, is outlined in Table 4.2.

Most wastewater is used directly to produce herbivorous and omnivorous fish, mainly carps, catfish and tilapia, or aquatic plants such as lotus (Fig. 4.2), water mimosa and water spinach for human food. There is also production of fish seed or fingerlings and aquatic plants, especially duckweed, to feed livestock and herbivorous fish.

Table 4.2. Types of wastewater-fed aquaculture systems (adapted from Edwards, 2002).

Wastewater type and delivery system	Aquaculture system	Cultured organism	Location
Nightsoil (overhung latrine)	Pond	Fish	China, Indonesia, Vietnam
Nightsoil (overhung latrine)	Pond	Duckweed	Bangladesh
Nightsoil, septage (cartage)	Pond	Fish	China, Vietnam
Contaminated surface water (waterborne)	Pond	Fish	Bangladesh, Indonesia, Vietnam
Contaminated surface water (waterborne)	Pond	Duckweed	China, Taiwan
Contaminated surface water (waterborne)	Cage in river	Fish	Indonesia
Contaminated surface water (waterborne)	Stakes in river, shallow pond	Aquatic vegetables	Widespread in Asia
Sewage (waterborne)	Pond	Fish	China, Germany, India, Vietnam
Sewage (waterborne)	Pond	Duckweed	Bangladesh

Specific examples

A brief overview follows of global wastewater use in aquaculture (Edwards, 2000). Most occurs in certain countries of Asia, with an almost insignificant occurrence elsewhere.

Asia

Wastewater use in aquaculture is not a traditional practice in Bangladesh, although widespread inadequate sanitation, as in many developing countries, leads to increased production of both capture fisheries and aquaculture through eutrophication of surface water. Fish were reported to be harvested from over 50 ha of ponds with contaminated surface water in peri-urban Dhaka (Rahman, 1992), but more recently this has been constrained by rapid urbanization leading to pond filling and to excessively high organic loadings of many of the remaining ponds, which preclude culture of carps and tilapias (Edwards, 2000). However, fish, especially tilapia, are currently cultured in several water bodies in the centre of the capital city of Dhaka, which contain water of an intense green colour because of inflow of contaminated water (Fig. 4.5). The largest are Golshan Lake, which is surrounded by high-class residential buildings and Banami Lake (Fig. 4.6). Duckweed wastewater treatment ponds also have been developed in Bangladesh over the last decade.

China has recently experienced a major decline of wastewater-fed aquaculture in the major production areas of the Pearl and Yangtze river basins (Li, 1997) due to economic growth, although it is still practised in less developed areas. Domestic wastewater has become increasingly mixed with industrial wastewater, causing undesirable odour and taste of cultured fish, which means that they are difficult to market. Furthermore, there is increasing consumption of higher value fish fed pelleted feed rather than low-value species from fertilized ponds, such as bighead and silver carp.

The world's largest wastewater-fed fishpond complex is in Kolkata, India, occupying almost 4000 ha. It has declined over the last few decades by about 50% because of urbanization (Creative Research Group, 1998), but its continued existence should now be assured as it has recently been assigned Ramsar status, an international convention for the conservation and sustainable use of wetlands. West Bengal is one of the few places in the world where new wastewater-fed aquaculture systems have been implemented. Inspired by lessons learned from farmers who developed the Kolkata system over the past few decades, three municipalities have installed what has been called an integrated wetland system for sewage treatment and resource recovery through aquaculture and agriculture, with pre-treatment of wastewater and fish pond effluent used to irrigate rice fields (Ghosh, 1997). Cultivation of vegetables on dykes recently has been introduced into the Mudialy Fishermen's Cooperative Society, a wastewater-fed fishpond system in the

Fig. 4.5. Wastewater flowing into a lake in Dhaka which functions also as a fishpond.

Fig. 4.6. Fish harvested in Banami Lake.

south of Kolkata. Irrigation of vegetables with relatively clean, green pond water, should lead to a considerable reduction in risk to public health for both farmers and consumers compared to crop irrigation with raw sewage (Fig. 4.4).

Fertilization of fishponds with excreta from overhung latrines is widespread in peri-urban as well as rural areas of Indonesia, particularly in West Java. In many villages, towns and cities, natural surface drainage from streams that are heavily contaminated with wastewater is directed into fishponds. Water spinach is also grown as a vegetable in some ponds (Strauss and Blumenthal, 1990).

Wastewater reuse in aquaculture is widespread in Vietnam, with overhung latrines on fishponds in peri-urban as well as rural areas, nightsoil cartage to fertilize both nursery and growout fishponds, and culture of fish and aquatic vegetables in surface water and sewage (see Chapters 6 and 7 in this volume for specific examples).

Africa

Wastewater reuse in aquaculture is not a traditional practice in Africa (Larsson, 1994), although fish are harvested from lakes contaminated with wastewater (Demanou and Brummett, 2003).

Europe

Although experimentation of fish culture in engineered wastewater treatment systems was carried out in several European countries, a viable operation remains only in Munich, Germany, where ponds stocked with fish are used now for tertiary and not secondary wastewater treatment as previously, and the area is a bird reserve rather than a commercial fish farm (Prein, 1996). A large-scale pilot project introduced in Hungary did not become commercial under the changing economic conditions in the country (Edwards, 2000).

Latin America

A United Nations Development Programme/World Bank sponsored project developed a pilot project that successfully demonstrated tilapia culture in tertiary treated wastewater stabilization pond effluents in Peru (Cavallini, 1996; United Nations Environment Programme, 2002).

Analyses showed that fish were acceptable for human consumption in terms of standards for viruses, bacteria and parasites, and for heavy metals, pesticides and PCBs. Furthermore, the fish were accepted by Lima city consumers, even when aware of the origin of the fish. There do not appear to be other cases of wastewater-fed aquaculture practice in this region.

South Pacific

There does not appear to be wastewater-fed aquaculture practice in the region.

Prospects for the Future

Improved designs

Human food

Conventional designs for fish culture in engineered stabilization pond systems such as in Lima (Cavallini, 1996) treat the waste more or less completely before the effluent flows into fish ponds. There is minimal reuse of nutrients in the wastewater because of the high degree of treatment before the effluent enters the fishponds.

A novel design takes into consideration the extremely rapid die-off of enteric viruses and bacteria in 'green water' fish ponds (Mara *et al.*, 1993; Mara, 1997; Mara and Pearson, 1998). Wastewater-fed ponds have large communities of protozoa and zooplankton (e.g. graze on microorganisms) and phytoplankton (e.g. photosynthesis raises the pH to lethal levels for enteric microorganisms). There is one day retention in an anaerobic pond, followed by only four days retention in a facultative pond, compared to conventional detention times of up to 30 days before the partially treated effluent flows into the fish pond. It was developed following a recommendation that environmental engineers and aquaculturists should unify their approaches to provide minimal (but adequate) treatment of wastewater, and maximal production of microbiologically safe fish for direct human consumption.

The design resolves the dilemma of simultaneously optimizing wastewater treatment and fish production in a practical way. The design steps using equations given in the manuals (Mara, 1997; Mara and Pearson, 1998) are as follows:

- Design an anaerobic and a facultative pond.
- Determine the total nitrogen concentration in the facultative pond effluent.
- Design the wastewater-fed fish pond which receives the facultative pond effluent, on the basis of a surface loading of total nitrogen of 4 kg total N/ha/day. This N loading rate provides adequate algal biomass in the fish pond for a fish yield of 5–10 tonnes fish/ha/year without the risk of severe dissolved oxygen depletion at night, which would cause fish kills.
- Calculate the number of faecal coliform bacteria in fish pond water to ensure that it is $\leq 1 \times 10^5$ per 100 ml (the previous WHO guideline for fish pond water was $\leq 1 \times 10^4$ per 100 ml, but a revision has been proposed based on research in Lima). If it is not, then increase the retention time in the fish pond (or consider the installation of a maturation pond before the fish pond).
- Determine the concentrations of NH_3–N to ensure that the concentration of toxic free ammonia (NH_3) is less than 0.5 mg N/l.

Animal feed

A further prospect is to use wastewater-fed fish as an ingredient for the formulation of livestock and fish feed. A significant percentage of the protein in formulated pelleted diets, especially for carnivorous fish and crustaceans, is fish meal, but there are severe limitations on marine capture fisheries which are the main source. Research has demonstrated the feasibility of using tilapia from wastewater-fed ponds as a high protein animal feed (Edwards, 1988; Edwards *et al.*, 1990).

The World Health Organization introduced the concept of crop restriction, for which a lower quality of wastewater is acceptable if there is no exposure of wastewater to the public, and protection is required only for agricultural workers, e.g. cereal crops, industrial crops, fodder crops and pasture and trees (WHO, 1989; Edwards *et al.*, 1990). Examples for aquaculture, for which the term produce restriction has recently been proposed by WHO, in an ongoing revision of public health guidelines, are production of high protein animal feed (tilapia or duckweed) and use of wastewater to produce fingerlings in aquaculture nursery operations. Wastewater reuse in produce restriction is indirect as animals destined for human consumption are raised in separate systems without the use of wastewater. As wastewater reuse in these systems incorporates an extra step in the food chain, they should be more readily acceptable socially than direct wastewater reuse (Edwards, 1990).

TILAPIA Research was carried out in Thailand on culture of tilapia in septage-fed ponds as a potential source of high protein animal feed (Edwards, 1988, 1990; Edwards *et al.*, 1987). Ponds were fertilized with septage at an organic loading rate of 150 kg chemical oxygen demand/ha/day, equivalent to a total nitrogen loading rate (the specific design criterion to consider when using other forms of wastewater) of 5–8 kg/ha/day. Fish harvested at 2–4 week intervals from the freely breeding population of Nile tilapia (*Oreochromis niloticus*) were small, but size of fish harvested for animal feed is unimportant. Mean net yields of tilapia averaged almost 7 t/ha/year.

Carnivorous walking catfish (*Clarias macrocephalus*) were fed experimental diets comprising whole minced fish with high concentrations of faecal coliforms because digestive tract contents were included. Catfish grew as well on minced septage-raised tilapia as on control diets of marine trash fish and fish meal in a formulated diet. Concentrations of microorganisms in *Clarias* were very low, with the exception of their digestive tract contents. Faecal coliform and bacteriophage indicator organisms for pathogenic bacteria and viruses, respectively, were completely absent from muscle tissue, blood and bile, and aerobic bacteria (standard plate count) were present in catfish muscle tissue in very low concentrations. These findings indicate that carnivorous fish fed with minced septage-raised tilapia were safe for human consumption.

DUCKWEED Cultivation of duckweed is a traditional Chinese practice to produce green fodder for grass carp fingerlings until they have grown large enough to consume coarse grass. However, there has been a tremendous amount of research over the last three decades on wastewater-fed duckweed cultivation (Edwards, 1990; Iqbal, 1999; Gijzen and Veenstra, 2001).

Duckweeds have several positive attributes: protein production up to 10 times greater than that of soybean; a high crude protein content of 25–45% on a dry matter basis; a growth rate of 10–40 t dry matter/ha/year; ability to grow in shallow water and shade; and ease of harvest by pole and net. Unfortunately, they have several disadvantages, such as growth adversely affected by both low and high temperature and high light intensity; occasional insect infestation; rapid decomposition; and are economically difficult to dry.

There is an impressive duckweed wastewater treatment and reuse system at Mirzapur, Bangladesh (Alaerts *et al.*, 1996; Gijzen and Ikramullah, 1999). A 0.7 ha duckweed covered pond, constructed in a serpentine lay-out to increase the wastewater retention time, was fed with effluent from an anaerobic settling pond receiving conventional wastewater. The pond complex received wastewater from 3500 persons and had a hydraulic retention time of 21 days. Duckweed was harvested and fed to fish in a nearby fishpond. In a financial evaluation of the combined wastewater treatment, the wastewater treatment/aquaculture system showed a net profit in the fourth and fifth year of operation of about US$2000/ha of total land area for wastewater treatment including duckweed and fishponds (Gijzen and Ikramullah, 1999).

Conclusions

Most wastewater reuse through aquaculture is threatened or in decline, with relatively few new schemes being implemented. The objective of internationally funded research and development initiatives in Peru was to develop a wastewater-fed aquaculture system for application in other countries in Latin America. Attempts to introduce it into other cities in the region, however, have failed (Moscoso, personal communication).

Major constraining factors are rapid urbanization, with consequent high opportunity cost and therefore limited availability of peri-urban land for relatively land-intensive reuse schemes; increasing industrialization, with mixing of domestic and industrial wastewater with toxic chemicals; and rising living standards, with more affluent urban consumers preferring higher value carnivorous and pellet-fed freshwater and marine fish, rather than relatively low-value herbivorous and omnivorous fish from wastewater reuse (Edwards, 2000).

The lack of knowledge of most planners and engineers of aquaculture, as a relatively low cost technical option in wastewater treatment and reuse, is an additional major constraint. The engineering profession promotes mechanical wastewater treatment such as activated sludge and trickling filters over stabilization ponds, regardless of aquaculture potential. Although mechanical treatment systems require much less land than stabilization ponds, such plants often do not function properly in tropical developing countries due to high operation and maintenance costs. Mechanical systems may not have a comparative advantage over ponds if broader economic analysis includes social and environmental factors, providing that land is available at reasonable cost.

Clearly, wastewater-fed aquaculture has potential in peri-urban areas of developing countries to provide employment and low-cost fish for the poor, as indicated by the recent introduction of schemes in Bangladesh and India. It may have particular potential in arid and semi-arid countries, where the need to reuse wastewater for water as well as nutrients is becoming urgent.

References

Alaerts, G.J., Mahbubar, M.R. and Kelderman, P. (1996) Performance analysis of a full-scale duckweed-covered sewage lagoon. *Water Research* 30(4), 843–852.

Anonymous (2002) The Hyderabad Declaration. *Urban Agriculture Magazine* 8, 4.

Cavallini, J.M. (1996) *Aquaculture Using Treated Effluents from the San Juan Stabilization Ponds.* Executive Summary. OPS/CEPIS/PUB/96.21. Pan American Center for Sanitary Engineering and Environmental Sciences (CEPIS), Lima.

Creative Research Group (1998) *Base Line Document for Management Action Plan (As per Ramsar Convention Guidelines), East Calcutta Wetlands and Waste Recycling Region.* Creative Research Group, Calcutta Metropolitan and Sanitation Authority, Calcutta.

Demanou, J. and Brummet, R.E. (2003) Heavy metal and faecal bacterial contamination of urban lakes in Yaoundé, Cameroon. *African Journal of Aquatic Science* 28(1), 49–56.

Edwards, P. (1988) Tilapia raised on septage as high protein animal feed. In: Pullin, R.S.V., Bhukaswan, T., Tonguthai, K. and Maclean, J.L. (eds) *Second International Symposium on Tilapia in Aquaculture.* ICLARM Conference Proceedings 15. Department of Fisheries, Bangkok, and ICLARM, Manila, pp. 7–13.

Edwards, P. (1990) An alternative excreta-reuse strategy for aquaculture: the production of high-protein animal feed. In: Edwards, P. and Pullin, R.S.V. (eds) *Wastewater-fed Aquaculture, Proceedings of the International Seminar on Wastewater Reclamation and Reuse for Aquaculture, Calcutta.* Environmental Sanitation Information Center, Asian Institute of Technology, Bangkok, pp. 209–221.

Edwards, P. (1992) *Reuse of Human Wastes in Aquaculture: a Technical Review.* UNDP-World Bank Water and Sanitation Program, Washington, DC.

Edwards, P. (2000) Wastewater-fed aquaculture: state-of-the-art. In: Jana, B.B., Banerjee, R.D., Guterstam, B. and Heeb, J. (eds) *Waste Recycling and Resource Management in the Developing World.* University of Kalyani, India and International Ecological Engineering Society, Switzerland, pp. 37–49.

Edwards, P. (2002) Wastewater reuse through aquaculture. In: *Environmentally Sound Technologies for Wastewater and Stormwater Management.* Compiled by UNEP International Environmental Technology Centre in collaboration with Murdoch University Environmental Technology Centre. IWA Publishing, London, pp. 81–106.

Edwards, P. and Pullin, R.S.V. (1990) *Wastewater-fed Aquaculture, Proceedings of the International Seminar on Wastewater Reclamation and Reuse for Aquaculture, Calcutta.* Environmental Sanitation Information Center, Asian Institute of Technology, Bangkok.

Edwards, P., Pacharaprakiti, C. and Yomjinda, M. (1990) Direct and indirect reuse of septage for culture of Nile tilapia *Oreochromis niloticus.* In: Hirano, R. and Hanju, I. (eds) *The Second Asian Fisheries Forum.* Asian Fisheries Society, Manila, pp. 165–168.

Edwards, P., Polprasert, C. and Wee, K.L. (1987) *Resource Recovery and Health Aspects of Sanitation.* AIT Research Report No. 205. Environmental Sanitation Information Center, Asian Institute of Technology, Bangkok.

Ghosh, D. (1997) Ecosystems approach to low-cost sanitation in India: where the people know better. In: Etnier, C. and Guterstam, B. (eds) *Ecological Engineering for Wastewater Treatment, Proceedings of the International Conference at Stensund Folk College, Sweden.* CRC Press, Boca Raton, Florida, 2nd edn, pp. 51–65.

Gijzen, H.J. and Ikramullah, M. (1999) *Pre-feasibility of duckweed-based wastewater treatment and resource recovery in Bangladesh.* Commissioned by The World Bank. IHE, Delft and PRISM, Bangladesh.

Gijzen, H.J. and Veenstra, S. (2001) Duckweed based wastewater treatment for rational resource recovery and re-use. In: Olguin, E.J., Sanchez, G. and Hernandez, E. (eds) *Environmental Biotechnology and Cleaner Bioprocesses.* Taylor and Francis, London, pp. 83–100.

Iqbal, S. (1999) *Duckweed Aquaculture. Potentials, Possibilities and Limitations for Combined Wastewater Treatment and Animal Feed Production in Developing Countries.* SANDEC Report No. 6/99. SANDEC, Duebendorf.

Larsson, B. (1994) *Three Overviews on Environment and Aquaculture in the Tropics and Sub-tropics.* ALCOM Field Document No. 27. FAO, Harare.

Li, S.F. (1997) Aquaculture and its role in ecological wastewater treatment. In: Etnier, C. and Guterstam, B. (eds) *Ecological Engineering for Wastewater Treatment, Proceedings of the International Conference at Stensund Folk College, Sweden.* CRC Press, Boca Raton, Florida, 2nd edn, pp. 37–49.

Mara, D. (1997) *Design Manual for Waste Stabilization Ponds in India.* Lagoon Technology International Ltd, Leeds.

Mara, D. and Pearson, H. (1998) *Design Manual for Waste Stabilization Ponds in Mediterranean Countries.* Lagoon Technology International Ltd, Leeds.

Mara, D.D., Ewards, P., Clark, D. and Mills, S.W. (1993) The performance of a series of five deep ponds in northeast Brazil. *Water Science and Technology* 19(12), 61–64.

Prein, M. (1990) Wastewater-fed fish culture in Germany. In: Edwards, P. and Pullin, R.S.V. (eds) *Wastewater-fed Aquaculture, Proceedings of the International Seminar on Wastewater Reclamation and Reuse for Aquaculture,* Calcutta, 6–9 December 1988. Environmental Sanitation Information Center, Asian Institute of Technology, Bangkok, pp. 13–47.

Prein, M. (1996) Wastewater-fed aquaculture in Germany: a summary. In: Staudenmann, J., Schönborn, A. and Etnier, C. (eds) *Recycling the Resource, Proceedings of the Second International Conference on Ecological Engineering for Wastewater Treatment.* School of Engineering Wädenswil-Zürich, 18–22 September 1995. Transtec Publications, Zurich, pp. 155–160.

Rahman, A.K.A. (1992) Tilapia in Bangladesh. In: Baluyut, E.A. (ed.) *Papers Contributed to the Workshop on Tilapia in Capture and Enhanced Fisheries in the Indo-Pacific Fishery Commission Countries.* FAO Fisheries Report No. 458 Supplement. FAO, Rome, pp. 139–142.

Strauss M. and Blumenthal, U.J. (1990) *Use of Human Wastes in Agriculture and Aquaculture, Utilization Practices and Health Perspectives.* International Reference Centre for Waste Disposal Report No. 08–90. IRCWD, Dübendorf.

United Nations Environment Programme (2002) *Environmentally Sound Technologies for Wastewater and Stormwater Management. An International Source Book.* Technical Publication Series 15. UNEP International Environmental Technology Centre, Osaka/Shiga. IWA Publishing, London.
Van der Hoek, W. (2002) Framework for a global assessment of the extent of wastewater irrigation. Paper presented at the *Workshop on Wastewater Use in Irrigated Agriculture: Confronting the Livelihood and Environmental Realities*, 11–14 November, Hyderabad, India.
WHO (1989) *Health Guideline for Use of Wastewater in Agriculture and Aquaculture.* WHO Technical Report Series no. 778. World Health Organization, Geneva.
WHO (2005) *Guidelines for the Safe Use of Domestic Wastewater and Excreta in Aquaculture.* World Health Organization, Geneva.

This page intentionally left blank

5 Peri-urban Aquaculture and Poor Livelihoods in Kolkata, India

Stuart W. Bunting,[1] Nitai Kundu[2] and Madhumita Mukherjee[3]
[1]Institute of Aquaculture, University of Stirling, Stirling, UK; [2]Department of Environment, Government of West Bengal, Kolkata, India; [3]Department of Fisheries, Government of West Bengal, Kolkata, India

Abstract

The origins of aquaculture in peri-urban Kolkata are described, and technical aspects of prevailing management regimes reviewed, including: the composition of species cultured; sources of water, including wastewater, used for culture; and feed, fertilizer and chemical application rates. Indian major carps and tilapia dominate production. However, several factors influence stocking regimes on individual farms. The nature of fish seed supply chains in the region is discussed. Employment practices, the role of labour unions and the livelihoods of those who depend either directly or indirectly on peri-urban aquaculture are presented. Employment in various capacities constitutes an important benefit of peri-urban aquaculture for poor people. However, labour unions are influential in setting terms and conditions, which largely dictate the widespread strategy of frequent stocking and harvesting. This in turn results in regular employment for those engaged in servicing the sector – for example, hatchery workers, seed traders, fish carriers and retail traders – and a year round supply of small, affordable fish to markets serving poor consumers. According to farmers, constraints threatening the viability of peri-urban aquaculture include uncertain wastewater supplies, high input costs, limited access to credit, poaching, disease and pest problems, inflexible labour arrangements and siltation of fishponds and feeder canals. Action to address such problems may help improve yields, and in doing so contribute to sustaining a number of poor livelihoods. However, action is also required to enable poor people to further diversify their livelihood strategies, and where necessary, or prudent, remove their reliance on insecure returns from natural resource-based activities.

The Origin of Aquaculture in Peri-urban Kolkata

The history of Kolkata since its founding in 1690 by the British East India Company as a trading post on the eastern banks of the Hooghly River is well documented. The current status of production systems in peri-urban east Kolkata may be attributed largely to critical historical developments (Furedy, 1987; Ghosh and Sen, 1987; Kundu, 1994). After a period of rapid and sustained expansion, recent change has seen a marked decline in the extent and apparent viability of traditional production systems.

Agriculture in peri-urban Kolkata has a long history dating back to the late eighteenth century. Shortly after founding the city, sewage and most solid waste were dis-

(eds B. Costa-Pierce, A. Desbonnet, P. Edwards and D. Baker)

posed in the river from a specially constructed jetty (ghat); some solid waste was also used as infill. Overall, this system proved unhygienic and it was widely acknowledged that a new strategy was needed. A system of burning solid waste, which was tried later, was not successful. Mr William Clark, Chief Sanitary Engineer, then proposed to remove municipal solid waste to the Salt Lake area, and during 1864 a low-lying area to the east of the city was acquired for the purpose of waste disposal. Dumping started in 1868.

In 1876 this land was leased for 3 years to Nandalal Das, and rights to capture fisheries in the area were given to Durga Charan Kundu for 4 years. However, unsatisfactory management by Nandalal Das forced the corporation to transfer the lease to Bhabanath Sen, who acquired the right of land, including the watercourses, for 20 years from 30 April 1879; the following year he acquired the fisheries. Under the supervision of Bhabanath Sen, horticulture in the area became established and better organized. In 1904, the sub-deputy Collector reported that of 220 ha taken on lease by Bhabanath Sen, 60 ha were being cultivated. Gradually, however, the entire area leased out to Bhabanath Sen was converted to horticulture, and this productive vegetable-growing area became known as 'Dhapa'.

In 1872 a fish jetty was constructed on Raja's drainage channel (khal), and this was closely followed in 1887 by the establishment of a flourishing fish market (hat) at Pagladanga. A navigation channel was constructed to connect the market to the town reservoir. Later, land taken to construct a storm water flow (SWF) canal caused disruption to both the fish-producing drainage channel and market. However, discharge of sewage carried in the SWF canal resulted in brackish lagoons becoming less saline, and freshwater fish colonized these lagoons. It is also likely that some informal stocking of fish was undertaken. An account given by Mr P. Ghosh, former Secretary of the Fish Producers' Association, suggests that the earliest attempt at formal aquaculture deliberately exploiting wastewater was undertaken by Mr Bidhu Bhusan Sarkar in 1918. Subsequent construction of the Dr B.N. Dey Outfall Scheme increased access of farmers in the area to wastewater, which in turn encouraged others to adopt wastewater aquaculture.

From a peak in 1945, when approximately 350 fish-growing sites managed for wastewater aquaculture covered 7300 ha, recent estimates have put the remaining pond area at around 3500 ha. In 1956 the Salt Lake Reclamation Scheme was formulated, and acquisition notices served on nearly half the farms managed for wastewater aquaculture. Between 1962 and 1967, under the direction of the government, about 1200 ha of ponds were filled with silt dredged from the Hooghly River. This reclaimed land was used to develop Salt Lake City, a major residential area to the northeast of Kolkata. From 1967 to 1972, a further 320 ha were reclaimed to extend Salt Lake City. In 1972, there were 4646 ha of fisheries. During the period 1978–1979 the East Calcutta and Patuli Townships were developed, converting 670 and 240 ha, respectively.

Following these major projects, recent pond conversions in the area have proceeded in a less overt fashion, with initial conversion to paddy rice farming, which is more easily developed for light industry or residential purposes. Since the early 1980s, roughly 2200 ha of ponds have been converted to rice farming. Paddy rice farming is widespread in West Bengal, and the influence of Kolkata on production extends throughout the region. Demand from the huge urban markets helps sustain higher than average prices. Nutrient rich wastewater is used to irrigate brackish-water rice cultivation and fish and shrimp ponds over 50 km from the city (Naskar, 1985).

Construction of the Eastern Metropolitan Bypass in the 1980s involved the conversion of selected wetland areas and portions of the solid waste disposal ground. Improved communications associated with the bypass have increased the attractiveness of the area to developers, and consequently the potential financial returns for fishery owners that sell their land has become substantial, making the temptation to sell stronger. Furthermore, the role of government in regulating conversion remains somewhat ambiguous. Although conscious of the need to retain peri-urban wetlands, some recent plans suggest that limited conversion to support infrastructure and economic growth may be desirable. The acquisition notice still threatens about 1100 ha of wetland at the Kolkata peri-urban interface.

Attempts have been made in recent years by fishery owners and others to stop further encroachment on the wetlands and, following a lawsuit filed by NGOs, the Calcutta High Court (1993) ruled in September 1992 that:

> No government or non-government body can reclaim any more wetlands, on the eastern fringes, where wetlands are defined in terms of being wet for six months or more in a year.

This was widely considered a landmark judgement, but doubts have been expressed regarding its implementation. Moreover, as yet no measures have been taken to properly conserve the remaining wetlands, even though the area has recently been declared a wetland of international importance under the terms of the Ramsar Convention.

Contemporary Aquaculture in Peri-urban Kolkata

Fish cultivation in ponds managed for wastewater aquaculture in peri-urban Kolkata is a practice that has been developed by local farmers and entrepreneurs during the past century. A recent report noted that there were 264 individual farms (bheries) operating on a commercial basis with a cumulative surface area of 2480 ha, and that these were distributed within four administrative wards, namely Bhangor, Bidhannagar, Sonarpur and Tiljala (CRG, 1997). This report noted that there were a further 22 farms with an area of 970 ha in this region that were either lying idle or only informally used for aquaculture. Fish culture is practised also in the numerous small ponds (jheels) in the Dhapa area, and in several homestead ponds concentrated in Sonarpur and Bhangor, some of which are managed for wastewater aquaculture. Preliminary stakeholder consultations suggested that around 400 households may be producing fish in homestead ponds receiving wastewater. The following sections draw on findings from a survey with 56 farm managers, and describe the current status of aquaculture in peri-urban Kolkata, including prevailing management practices and the key constraints to production. Further details of the survey and other activities undertaken during 2 years of research relating to 'Renewable natural resource-use in livelihoods at the Kolkata peri-urban interface' are given on the project website (Bunting, 2002).

Scale and distribution

Assessing the distribution of commercial farms based on location suggests that smaller farms (<2 ha) occur mainly in Sonarpur and Tiljala, while large farms (>40 ha) predominate in Bidhannagar. Bhangor is characterized by a mixture of small to medium sized operations (CRG, 1997). Subdivision of lakes in Bhangor, as a result of infilling with garbage and land reforms, has resulted in a large number of small ponds in this region. Although roughly 90% of the 264 commercial farms are below 20 ha in size, occupying 1080 ha, the scale of the remaining farms results in them accounting for a disproportionately large area. Pond depths in the region range from 0.15 to 1.2 m. Integration of additional production activities with aquaculture occurs on some of the larger commercial farms in Bidhannagar and Bhangor.

Management pattern

A diverse range of management strategies has developed in association with aquaculture in peri-urban Kolkata, the evolution of which has largely been governed by socio-political forces and legislative restructuring. Perhaps the key issue in this respect, however, has been land reform in West Bengal, which was implemented to increase access by poor families to scarce land resources. Traditionally, feudal landlords (zamindars) governed access to much of the land in the State. Key enactments were the West Bengal Estates Acquisition Act (1953) and West Bengal Land Reforms Act (1955). However, due to exemptions covering tea gardens, orchards and fisheries, the individual farms in peri-urban Kolkata remained largely intact until only recently. Amendments to the Land Reforms Act in 1995 brought fisheries within its jurisdiction, and consequently led to significant reorganization, with many private holdings being vested from their owners by the State and transferred to fishermen's groups (non-registered cooperatives). Having demonstrated their competency, such groups subsequently may apply to become registered cooperatives. The practice of vesting land has resulted in a decline in larger fisheries under private ownership. However, there are a number of smaller, independently managed household and individual ponds in the region. The majority of small ponds are found on land that was vested from feudal landlords as part of the land reform programme in West Bengal. Two of the largest farms in the region (Nalban and Goltala) remain under the direct control of the State Fisheries Development Corporation, Government of West Bengal.

Fisheries that remain in private hands are usually run on a partnership, shareholding or leasehold basis as owners rarely operate fisheries directly. In the case of partnerships, the landowner is paid a rent but may still participate in decision-making and take a share of the profits. In the shareholding system, profits are divided among the shareholders, although the owner generally takes 40% of the profits. When land is taken on lease, usually by five to ten people, the owner is paid an annual rent and profits are shared in proportion to the financial contribution of each leaseholder.

Water source

Access to wastewater draining from Kolkata was one of the primary driving forces behind the emergence of peri-urban aquaculture. Initially, opportunistic farmers exploited the wastewater resource to cultivate fish and vegetables. However, as siltation in the Hooghly River became more problematic, the authorities diverted a greater proportion of the silt-laden wastewater to agriculture and fishponds. This strategy required the construction of an extensive canal network that extended throughout the eastern peri-urban interface. Key arteries in this network include the fisheries feeder canal and Ghosh's canal. The drainage network was engineered to distribute wastewater under gravity to the majority of fisheries, but as silt has accumulated in the fishponds, an increased proportion of farmers find it necessary to pump wastewater from the canals. This problem has been further aggravated by the need to maintain a low hydraulic head at the pumping stations transferring wastewater from the city sewers to the drainage canals (Edwards, 2001). Recent research has shown that nearly one-third of pond managers must pump water from the drainage canals to their ponds during the year (Bunting *et al.*, 2002).

As part of the comprehensive survey of fish farms undertaken by the Creative Research Group (CRG, 1997), operators' access to sewage water was assessed. Respondents were requested to score the supply of sewage water on a five-point scale, where 1 indicated a more than adequate supply and 5 a highly inadequate supply. In the Bidhannagar region, perceptions of the sewage supply appear variable – 43% of respondents scored the supply as adequate or more than adequate, and an equal proportion scored it as inadequate or highly inadequate. Scores in Tiljala were also distributed across the possible range of responses, although the largest proportion (73%) classified the supply as inadequate or highly inadequate. Ten respondents in this region did not consider the question applicable, and 18 farms were rain-fed. The number of respondents in Sonarpur who classified the supply as inadequate or highly inadequate was also significant at 63%. No respondents in this area considered the supply 'more than adequate'. The modal score in both Sonarpur and Tiljala was 4, indicating an inadequate sewage supply. Overall, the majority (60%) of farm managers questioned regarded the sewage supply as inadequate to highly inadequate, a further 22% considered the supply adequate for part of the year, suggesting perhaps that the sewage supply is governed by seasonal changes. Only a relatively small proportion of respondents (18%) scored the supply as adequate to more than adequate.

The historical perspective presented by Furedy (1987) provides an insight to one of the key constraints in reusing wastewater for aquaculture. Siltation problems in the Kulti River were a driving force behind the diversion of wastewater to fishponds. However, since remedial action, no sustained effort has been undertaken to reduce or manage the high loading of suspended solids present in the wastewater. Therefore, the problem of siltation was merely transferred from the river to the secondary canal network and peri-urban fishponds. As the ponds and feeder canals have become silted, decreasing access to, and consequently reuse of wastewater in fishponds, siltation in the primary canals and river is again a problem for the municipal authorities. The fact that many fishpond operators experience significant problems in accessing wastewater results in an increased discharge volume to the Kulti River, which belies the myth that reuse in peri-urban fishponds treats a significant proportion of wastewater discharged from the city. Even when functioning ideally, wastewater flowing through the fishponds is not discharged directly to the Kulti River, but passes into downstream paddy rice fields and drains away from the primary canal network. The volume of wastewater that, having passed through the reuse system, is ultimately discharged to the river network has not been evaluated.

During the 1940s an ambitious scheme was implemented to intercept wastewater draining to the fishponds, and to treat it in primary settlement tanks situated at Bantala. The scheme also extended to transporting sludge, thickened in earthen basins, by train to tea-growing plantations in Darjeeling. Although successfully commissioned, the settlement tanks only remained functional for 1–2 years. Several reasons, including technical problems, high operation and maintenance costs and vandalism, have been given for the failure of the treatment plant. However, it is likely that a combination of factors led to such a rapid systems failure. The now defunct settlement tanks remain as a reminder to future planners and policy makers concerning the need to develop appropriate interventions that consider the range of stakeholder demands.

Problems with accessing sufficient wastewater from the secondary canal network have resulted in a significant number of operators using pumps. In Sonarpur, the majority of farmers have to pump wastewater to their fishpond, while only a couple are able to depend on gravity to supply their needs. Most managers in Bhangor rely largely on gravity, although some also need to pump water, though only a few depend solely on pumped water. Producers in Tiljala can still rely on gravity to supply wastewater, as can most in Bidhannagar, although a few are also required to supplement this by using a pump.

Although in common parlance, references to gravity being employed to supply wastewater to fishponds may, in some cases, be considered an oversimplification. Due to the nature of the sewerage system serving Kolkata, all drainage water must be raised by pump several metres so that there is a sufficient hydraulic head to enable drainage under gravity through the primary canal network to the discharge lock on the Kulti River and the secondary canals supplying the fishponds. Edwards (2002) discussed further problems that have emerged recently due to the technical and operational constraints associated with pumping large volumes of water from the city, and the differing demands of stakeholders in peri-urban areas. Furthermore, due to the topography of the region, and barriers such as roads, various siphons have been constructed. Controlling distribution through the drainage network means that several lock gates on the primary and secondary canals must be regulated and maintained. Unfortunately, farmers complain frequently that due to the prevailing systems for controlling the sluice gates, significantly less wastewater enters the secondary canal network than is actually required to sustain the fisheries. Several of the sluices in question are under the control of the Department of Irrigation and Waterways. Some individual fisheries are dependent on the maintenance of siphons to enable them to extract wastewater from the canals.

Maintenance of lock gates, siphons and the secondary drainage network by fishery managers, and the considerable cost of pumping wastewater from the feeder canals to the fishponds, constitutes a significant indirect subsidy supporting the managed disposal of wastewater from the city. However, the formal system draining Kolkata serves only a proportion of the population, usually middle class and more affluent households, together with businesses, that can afford a connection to the sewage mains. Greater understanding concerning the volume of wastewater used in fishponds, where pond water is discharged to downstream users, and the degree of treatment achieved, could be useful in evaluating the economic benefit of fishponds in terms of offsetting wastewater treatment costs. Furthermore, wider appreciation of the value of environmental protection, wildlife habitat, employment and food production associated with peri-urban aquaculture, together with other benefits and costs, may contribute greatly to informing a rational debate concerning the future of the system.

Pumping of wastewater from the drainage canals is undertaken routinely to sustain both water and nutrient levels in the fishponds. However, pumps are also employed on many of the larger fisheries to drain ponds and move water between ponds, often in an attempt to facilitate oxygenation. The need to pump wastewater to maintain water levels in the majority of ponds highlights the fact that many of the peri-urban wetlands are man-made, and that their existence depends on the continued functioning of the fisheries. Despite this dependence, many environmentalists do not perhaps appreciate this synergy. A greater awareness of this issue, might contribute to building consensus amongst environmental campaigners, producers and policy makers.

The necessity of having to pump wastewater to many of the fisheries also has other consequences: employing diesel pumps represents a capital and operating cost to producers, and also brings into question the environmental credentials of the system. However, owing to an agreement between the managers and labour unions, those fisheries required to pump wastewater are only required to engage labourers at a rate of one per 5 bigha or 0.7 ha, as compared with one per 3 bigha or 0.4 ha where water is delivered under gravity. Presumably the rationale is that the added cost of pumping means that the imposition of higher labour demands could threaten the viability of the farms.

Species cultured

Traditionally, aquaculture in Kolkata was dominated by the production of Indian major carps (IMCs), specifically a polyculture of rohu (*Labeo rohita*), catla (*Catla catla*) and mrigal (*Cirrhinus mrigala*). However, there has been a trend towards culturing tilapia (*Oreochromis niloticus*), usually as part of a polyculture with IMCs, but sometimes in monoculture. Several other fish species, including common carp (*Cyprinus carpio*), silver carp (*Hypophthalmichthys molitrix*) and grass carp (*Ctenopharyngodon idella*) are also produced. Tilapia culture is associated with several advantages: self-recruitment in ponds, which is absent for the IMCs; demanding less investment in stocking; rapid growth; and reportedly good disease resistance. Their omnivorous feeding behaviour also means they exploit various feeding niches, and the ability to feed efficiently on phytoplankton makes them highly productive in systems receiving significant fertilizer inputs, but where supplementary feeding is limited.

It is notable that IMCs are stocked predominantly in peri-urban ponds situated in Bidhannagar, a region characterized by larger individual farms, while in Bhangor, where the majority of water bodies are small, tilapia monocultures are more widespread (Bunting *et al.*, 2002). Several factors may be governing the decision as to which species or combination of species to stock. Investment costs associated with establishing tilapia monocultures may be less, making them more attractive to small producers; returns associated with culturing IMCs may be higher, but risks are also higher, thus deterring more vulnerable small-scale producers. In addition, water quality in ponds in Bidhannagar may better suit the requirements of the IMCs. This may also be the case in Sonarpur, where according to Bunting *et al.* (2002) six of the seven producers interviewed stocked mostly IMCs. In Tiljala, however, the number of farms stocking tilapia and IMCs were roughly equal. The decision-making process is likely to be influenced further by labour union demands concerning harvesting days per year, fish seed costs and market demand.

Fish seed and supply chains

Research has shown that the majority of producers purchase both fry and fingerlings, yet despite being less costly, only a small proportion stocked just fry due to the higher risk of mortality (Milwain, 2001; Bunting *et al.*, 2002). Fry are much more susceptible to poor water quality, disease, predation and environmental perturbations such as fluctuating temperatures. Farms buying fry tended to be larger, with more space available for separate nursery ponds in which the fish may be on-grown. The remaining operators that only purchase fingerlings tended to be smaller in size, hence with no scope for on-growing.

Regarding fish seed trading, it is important to understand the characteristics of the supply chains, including: (i) the nature and extent of the network, and hence vulnerability to external factors; (ii) the role of seed trading in supporting poor livelihoods throughout West Bengal; and (iii) technical, financial, institutional and social issues governing the shape and effectiveness of the network.

From those farmers surveyed, it was apparent that the majority (64%) depend solely on agents (goldars) to supply fish seed (hatchlings and fry) (Bunting *et al.*, 2002). Producers prefer to buy seed from agents to avoid the risks associated with transporting fish seed, plus they are not required to travel long distances from their farm to collect the seed. The agents are also better networked, enabling them to easily find the species, size and quality of seed required by the producers. Agents purchase seed from either Naihati fish market or from those in the Bankura district. It is then either delivered by truck, or agents commission carriers to transport it on their bikes, or manually on their heads, to the desired fishery. Managers requiring larger fingerlings tend to purchase them directly from producers. In Bhangor, where the majority of managers operate small fisheries and small ponds, fish seed are more likely to be purchased directly from hatchery operators; the hatchery manager will usually arrange for the delivery of seed to Sealdah fish market. Some fishpond managers are also involved with on-growing hatchlings to produce fingerlings that they then sell to other farms in the local area.

Milwain (2001) discussed in detail the organization and structure of fish seed networks supplying fisheries in peri-urban Kolkata. This author also makes a series of recommendations concerning possible improvements to fish seed production and distribution in West Bengal. Important areas identified for further consideration included improved water quality monitoring and management in hatcheries, particularly arsenic and iron concentrations, which may affect seed quality; better surveillance of tube-well water characteristics;

better information on risks from flooding; and assistance with marketing to reduce competition among producers. Milwain (2001) also proposes a workshop with government staff, specialists and business managers to develop improved management strategies. This could constitute an important aspect of future research and development in supporting poor livelihoods and enhancing the viability and efficiency of fish production around Kolkata.

From participant responses recorded in the study by Bunting *et al.* (2002), it is apparent that Nihati is the most common source for fish seed, mentioned by over a third of respondents, although sources in the local area were cited almost as often. Local sources probably include the purchase of fingerlings from other farmers in the area who may have on-grown a surplus of fish, either intentionally to sell to other producers, or inadvertently, owing to better than expected survival. Producers citing sources in the local area may, however, be buying from local traders but unaware of the origins of the seed they purchase.

Other significant sources for seed include Bankura, Sealdah, Bandel, Tribeni and Pandua, with 10–20% of producers interviewed purchasing seed from these regions. Less than one in ten respondents used suppliers in other regions, 18 of which were mentioned. From this study it is apparent that producers in peri-urban Kolkata use fish seed suppliers from a broad geographical distribution to meet their requirements. Although Milwain (2001) provides a preliminary review, further information would be required from participants to identify which factors, for example proximity, reputation, quality, cost or past experience, govern the decision concerning the selection of seed suppliers. The fact that 45 respondents identified 115 suppliers also suggests that producers purchase seed from several sources. However, more information would be required to assess whether this was due to producers trying to spread risks associated with buying seed of unknown quality from one source, or if producers were forced to use multiple suppliers to achieve their preferred stocking regime in terms of species, size and timing. It is also uncertain whether producers consider possible negative factors, such as potential increases in disease risk associated with sourcing seed from multiple suppliers.

Feed and fertilizer application

Survey findings showed a combination of mustard oil cake (a by-product of oil extraction from mustard seed) and mohua cake (a by-product of oil extraction from mohua tree, *Madhuca indica*, seed) was applied by almost two-thirds of producers, 16% and 7% of producers applied leather milk (a waste product of the leather industry) and hotel dust (organic waste from hotel kitchens) respectively, while 12.5% employed no additional feed or fertilizer. Mustard oil cake and mohua cake are traditional fish feeds. However, due to the cost involved, alternative nutrient sources such as leather milk and hotel dust are now being exploited. Middlemen collect leather milk from slaughterhouses and tanneries in containers, and transport it to the fisheries on trucks. Hotel dust is a recent innovation facilitated by local entrepreneurs who collect and transport it to farmers. Although obviously meeting a demand, potential negative environmental health impacts associated with these management practices remain to be assessed. According to some respondents, on individual farms where the integrated production of ducks and hens is practised, litter from pens and enclosures is used as a supplementary feed input. No detailed information, however, on the extent and management demands of this practice is available.

Integrated production

Farms commonly have facilities to on-grow seed to fingerlings in smaller nursery ponds prior to stocking in the main ponds. Further vertical integration, however, through the development of hatcheries and maintenance of broodstock, may be constrained due to poor access to good quality water in the area, limited financial resources and knowledge and competition from established commercial seed producers. More general farm-level integration was reported on some larger holdings, with activities such as growing vegetables, producing coconuts and rearing ducks and hens for eggs and meat. Such integration is usually undertaken on a relatively small scale, largely for self-consumption, although some produce may be sold at local markets. However, more widespread and intensive integration is reportedly constrained by a number of factors. One example reported was that of law and order problems – farmers are unwilling to risk the theft of livestock or unguarded crops from embankments, while the threat of poaching deters managers from planting trees along their embankments through fear of reducing visibility and consequently security. Restrictions imposed by labour unions limiting the types of work undertaken by some employees, and the added risk and demands of managing integrated systems, may also further constrain integration.

Disease and pest problems

Of those farmers interviewed, the majority (47) reported that production had been moderately affected by disease outbreaks. Only one reported severe effects on production, while eight did not consider disease to have an impact. Disease problems mentioned included fin rot, gill rot, dropsy, *Argulus* sp. (fish louse) and *Lernia* sp. (anchor worm). However, it should be noted that some of these terms refer to disease symptoms as opposed to causative agents. Summarizing the occurrence of diseases affecting fish in peri-urban ponds, CRG (1997) noted seven potential problems. The incidence of *Gyrodactylus* sp. and *Dactylogyrus* sp. parasites was reportedly insignificant, and that of epizootic ulcerative disease, tail rot, dropsy and fish leeches was low. The incidence of fin rot was reportedly moderate, occurring mainly from mid-July to mid-August and from October to February. This problem was reported to affect all fish species cultured. The usual treatment was the application of formalin. The most serious problems encountered were infestations with *Argulus* sp. and *Lernia* sp., which reportedly had a high incidence. The occurrence of these parasites was most common amongst IMC species.

Producers depend largely on their own experience in detecting and diagnosing disease outbreaks. A lack of scientific support from government and agricultural service providers was noted, as was the absence of a formal monitoring programme. Despite these constraints, producers reported that if detected in the primary stages, and appropriate treatments applied, mortalities could generally be avoided. Furthermore, producers reported that where fish receive proper nutrition, resulting largely from adequate access to wastewater, disease effects are minimized.

Chemical applications

Lime is generally applied to the sediments during fishpond preparation to kill pathogens, predators and small fish that may harbour disease. Lime is also applied to the water prior to stocking to reduce turbidity. Where there is a history of disease outbreaks, lime is used in conjunction with potassium permanganate for the regular treatment of pond water. Survey work showed that roughly 80% of respondents reported employing this management strategy. However, of these farmers, just under half reported using other treatments. Chemicals employed when fish show visible disease signs include terramycine (oxytetracycline) and metacid (methyl parathion). Treatments are usually purchased from retailers in local markets, who also provide advice concerning the appropriate compound and application rate. Rarely do retailers have scientific training, the materials or specimens to make valid diagnoses, or a complete understanding of the factors contributing to the disease outbreak. Consequently, disease diagnosis and identification is largely based on guesswork and consultation with other producers having experience of similar problems. Of those farmers interviewed, 18% reported that no chemical treatments were employed on their farms.

Finance

Regarding finance, most managers seemingly depended wholly or partly on advances from traders (dadan) to finance their production activities (Bunting *et al.*, 2002). Such arrangements usually involved managers entering into agreements with auctioneers (aratdars). Loans are usually taken in the form of subsidized fish seed, with an understanding that when the fish are harvested, they will be marketed through the same auctioneer, who will then take some fish as repayment and a commission of 3% on the remaining fish for providing the initial loan. Finance in the form of private loans or credit was reportedly used by 45% of respondents, and 43% utilized their own savings for investment. Only a small proportion of producers reported that they made use of formal loans from banks, cooperatives or local government departments. A couple had received loans from the Department of Fisheries, although this usually depended upon prior involvement of the department in developing a management plan for the fishery. While almost half the participants interviewed reported using more than one source of finance for investment, the relative advantages or problems encountered with each require further investigation.

Labour arrangements and unions

Employment opportunities associated with peri-urban aquaculture have been cited as one of the major reasons why the system is worthy of protection – reports have suggested that 8700 people are directly employed as labourers on fish farms (CRG, 1997). Furthermore, it has been proposed that operation of these largely commercial aquaculture ventures sustains several thousand jobs in allied support activities, such as the supply of inputs, including seed, and the marketing of produce. As with other examples of peri-urban farming, it was envisaged that work in these systems would enable recent migrants from rural areas to find employment. Furthermore, it has been reported previously that several groups of migrant labour have specialized skills, such as bamboo screen making, that ensure they are able to secure employment, although timing and duration may be variable. Despite the apparent benefit of peri-urban aquaculture in providing employment opportunities, arrangements, in the case of Kolkata, are largely governed by the labour unions. This suggests that lessons and recommendations drawn from an assessment of the current system concerning employment should be viewed cautiously, and that where labour markets are more flexible, practical and financial considerations may dictate that modified employment arrangements should be adopted. Labour costs in the case of peri-urban Kolkata represent the greatest operating cost (CRG, 1997), and the majority of farms are reportedly not financially viable (Mukherjee, 1996). Therefore, in the absence of enhanced production or higher fish prices, major restructuring of the labour force may be required to sustain the system.

Labour unions are well established in peri-urban fish farming and, on behalf of the labourers, negotiate terms and conditions for both temporary and permanent employees. Daily wage rates, benefits, job specification, minimum employment days per year, leave and the number of employees to be engaged are all governed by agreement between the labour union and farm managers. However, terms and conditions vary with respect to fishery size and key management aspects. The Dakshin 24 Parganas Zilla Bheri Mazdoor Union represents temporary labourers that constitute the majority of the workforce on the farms, and largely dictates the terms and conditions for employment. Periodically the unions and farm managers meet and negotiate terms and conditions. However, a number of disputes over employment rights resulted in a conflict resolution meeting between the union and managers. In 1992, a draft Code of Conduct was agreed upon at a meeting of the 24 Parganas Fish Producers' Association and district level union officials. According to the Code of Conduct, the size of an individual farm has a bearing on wages paid to labourers and the rate at which labourers should be employed. Small farms are classified as those between 1.3 and 13.4 ha, medium range from 13.4 to 33.4 ha and large farms cover an area above 33.4 ha. Aquaculture-based production systems below 1.3 ha in size are excluded from the Code of Conduct, as labourers working on operations of this scale are not unionized. Employment rates are based on the premise that individual farms that are not required to pump wastewater require a greater number of labourers (one per 3 bigha or 0.4 ha), while those that pump wastewater to the ponds require less labour (one per 5 bigha or 0.7 ha). The number of harvest days and quantity of fish to be marketed is fixed based on the area occupied by the individual farm. For further information on the process that resulted in the draft Code of Conduct see Bunting *et al.* (2002).

The majority of labourers are closely allied to party leaders and union representatives who, following consultation with the workforce, negotiate terms and conditions with members of the fish producer associations. Based on this arrangement, conditions such as the number of harvest days per year, minimum harvest quantity per day and daily wage levels are fixed. In many cases, managers have also been forced to create full-time employment for labourers, thus ensuring they get a regular income. However, due to the strength of unions, several managers reported that the prevailing terms and conditions were biased towards the demands of the workers. Farms with an area below 1.3 ha do not have to adhere to the Code of Conduct, and therefore all labourers employed by these farms are casual employees and lack the security of a permanent position. Larger farms sometimes hire casual labourers to undertake specific tasks, but it is difficult to judge the frequency of such opportunities, as such employment is usually in response to emergency situations, such as burst embankments.

Although many farm managers consider the Code of Conduct restrictive, it can be argued that the terms and conditions it sets out have been responsible for some key benefits attributed to aquaculture in peri-urban Kolkata. Stipulating a minimum number of harvest days per year and minimum harvest quantity contributes to the predominant stocking and harvesting strategy in the region, which in turn results in a constant supply of small fish to urban markets. The small size of fish cultured means they are affordable and generally sold in markets serving poor communities, while the year-round supply of fresh fish has benefits for food security in poor households. The consistent nature of stocking and harvesting also benefits small-scale suppliers and traders who are routinely involved in servicing the needs of producers and therefore receive regular commissions or wages.

Despite these benefits, producers face the same fixed employment costs, regardless of production rates or price levels. Therefore a decline in either may threaten the financial viability of the farm. The fixed-term nature of employment arrangements also means there are few incentives for labourers to achieve higher production or enhance the quality of fish produced. Introduction of practices such as profit-sharing or bonuses could contribute to enhanced production, better returns for managers and employees, more fish for sale in urban markets and greater employment opportunities in associated activities. Although some managers did consider the specified employment rates per unit area high, it was noted during other project work that some cooperatives reportedly provided employment for larger numbers (Edwards, 2001). A caveat concerning pumping in the Code of Conduct and its influence on employment rates is also significant in light of proposed development initiatives. Renovation of the canal network feeding the fishponds, and establishment of secondary pumping stations to enhance wastewater delivery, may mean more fisheries are required to engage one employee per 3 bigha (0.4 ha). This could almost double the number of people employed by these fisheries. Furthermore, added labour costs may negate any saving achieved through not pumping.

Employee livelihoods

Based on survey outcomes, it is possible to identify several categories of labour associated with fish production in peri-urban ponds. Labourers employed directly to manage and maintain the individual farms and small ponds include fishermen, who undertake the main work of netting, and fish carriers, land cutters and several others, who work clearing grass or applying lime, fertilizer and medications. The permanent workforce, which mainly lives on site, consists of the manager and those engaged in net making, construction work, weed clearance, harvesting and cooking. Members of the permanent workforce earn a regular wage and receive other benefits such as board and lodging, medical expenses and a small subsistence grant; they are also entitled to 4 days of leave per month. Casual labourers employed for netting, desilting, guarding, transporting, net repairing and various other activities usually live in villages adjacent to the ponds. General information concerning the location of these villages in relation to the fisheries was noted and, in some instances, arrangements made with employees for further visits to the local communities to initiate the process of focus group and household interviews.

Focus group interviews with women, men, girls and boys engaged in agriculture and fish farming have provided a valuable insight to benefits and limitations of occupation (Punch *et al.*, 2002). For all groups, the main benefit was the cash income generated. However, for many, seasonal changes in productivity and labour demands resulted in vulnerability due to widespread temporary employment arrangements. This reflects that poverty not only relates to economic resources, but also to expectations of income in relation to security and consistency. Women in focus groups expressed concerns over issues of family welfare and work-related health problems, but men did not. For many women, formal employment in fisheries meant there was less time for domestic chores and childcare, in some cases leading to informal childcare arrangements between local community members. For agricultural workers, the main difficulty related to the high cost of inputs, in particular seed for high yielding varieties, which they are forced to purchase owing to problems experienced with farm-saved seed. Wastewater use for irrigation during the dry months was perceived as advantageous, as less fertilizer is required. However, the cost of pumping this wastewater, which has increased due to siltation of the canals, combined with increased susceptibility to disease and a loss of taste, constitute problems requiring further assessment. Due to seasonal lulls in agricultural employment, women and girls interviewed recounted how they engage in supplementary activities such as collecting edible plants and snails and rag-picking. These associated activities, however, are dependent on access to common property resources and continued use of solid municipal waste on agricultural land in peri-urban Kolkata. Young people engaged in unpaid agricultural labour for their families recognized the contribution they make to sustaining the household. However, some suggested that being obligated to do unpaid work put them at a disadvantage as they are not free to earn money for their personal use.

For those engaged in fish farming-related activities, decreasing yields due to siltation in the ponds, disease and fluctuations in the wastewater supply constitute a serious threat to their livelihoods. Both men and women regarded permanent employment on large fish farms as the greatest benefit, providing security and the added benefit of sick pay, soft loans and half pay during periods when there is no work on the farm. These arrangements have been negotiated between the labour unions and fish farm managers and, although providing a form of social security, the farm managers perceive them as an added burden, particularly as farm productivity is apparently in decline. Young people working on fish farms recognized their contribution to enhancing the family income, but it is widely acknowledged that poverty and illiteracy combine to make child workers vulnerable (Ramanathan, 2000). Amongst the groups engaged in fish farming-related activities there was also fear that, owing to pressures to develop peri-urban land for residential and industrial purposes, access to employment opportunities and land for rent may be lost. Enhanced planning, regulation and better enforcement might contribute to reduced vulnerability, as would ensuring security of land tenure for poor farmers.

People in the focus groups identified a number of major constraints, including low incomes, problems with electricity and drinking water supplies, a lack of health care provision and alcohol abuse. Enhanced farming practices might be able to contribute to improved incomes, but risks and vulnerability associated with such enhancements may also be increased. Improved service provision is a responsibility of the municipal authorities in much of peri-urban Kolkata, and this suggests better development planning and coordination is required. Food security was not considered a problem, and perhaps reflects the productive nature of farms in the region and their contribution to providing affordable produce to poor communities in both peri-urban and urban Kolkata. However, potential health risks associated with farming, in particular contamination owing to the exploitation of waste resources, demands further consideration. Punch *et al.* (2002) presents a fuller description of key findings from the focus group interviews.

Household interviews with families engaged primarily in fish farming-related activities demonstrated that a lack of work during the winter resulted in seasonal vulnerability. Most households also farmed some agricultural land to produce rice and vegetables for subsistence, with any surplus being sold and, in some cases, younger male household members have entered into off-farm employment in the construction, manufacturing and service sectors. Despite widespread diversification of household livelihood strategies, families are still forced to take loans from moneylenders sometimes. For households engaged primarily in agriculture input, transport costs apparently constitute a significant burden, though problems with the availability of solid municipal waste, commodity price fluctuations and the seasonal lack of irrigation water are also of concern. During periods when work on their own farms is not possible, due to insufficient water supplies, household members look for employment opportunities on other farms in the region. Women and girls working for other farmers generally receive lower wages than men, and some women engage in scavenging over fields in the area, selling things of value they glean to supplement their income. However, declining use of solid municipal waste on agricultural land in the area, due largely to logistical problems, appears to threaten this coping strategy. Some farming families stated that due to the risks associated with the weather, disease and uncertain wastewater supplies, they were no longer intending to lease land to farm, but instead would work for other farmers or try to engage in a new occupation. Training for family members in appropriate skills and crafts may assist in the transition from agriculture to non-farm employment, while improved credit facilities might assist some in establishing new businesses.

Households whose primary income is derived from night guard duties regarded their salary, annual bonuses and weekly fish allowance as the most important benefits of this type of employment. Due to the nature of the work it is also possible for guards to engage in other work during the day, and in most cases this involves tend-ing their own land, ponds or livestock, or working in the local fish markets. Although households recognized that work as a night guard provided secure and consistent employment for a reasonable salary, there is always the threat of physical violence, and the risk of injury or death from tackling poachers. In one case a night guard household had purchased a mechanical tiller to hire out to others in the area, and while this might prove to be a good investment, there are attendant risks requiring consideration. Households dependent on seed trading face a high degree of seasonal unemployment, and consequently often engage in other income-generating activities, such as working in local auction markets for fish. Seed traders are also affected by periods of political instability that makes travel in the area difficult, and delays due to unforeseen accidents or breakdowns also constitute a serious problem.

Conclusions

The review presented above focuses largely on technical aspects of existing management practices for aquaculture in peri-urban Kolkata, and constraints to sustained or enhanced production. In particular, problems in accessing reliable wastewater inputs, dependence on externally sourced high-cost inputs, recurring disease problems and limited access to formal credit arrangements are highlighted. Households and individuals in peri-urban Kolkata engaged in aquaculture-related activities experience seasonal vulnerability, which appears most closely related to insufficient access to water during the dry winter months, despite the continuous discharge of wastewater from municipal Kolkata. Siltation in the primary and secondary feeder canals compounds this problem, and coordinated action on behalf of the local and municipal authorities is urgently required to address this constraint. Problems regarding delivery to farmers and diversion of solid municipal waste to a new composting plant are forcing producers in the area to switch to inorganic fertilizers, which not only has consequences for yields, soil condition, disease and financial risks, but also denies some of the poorest local people a valuable coping strategy, namely gleaning items of value from rubbish spread on the fields.

Considering the vulnerability of poor people in peri-urban Kolkata, particularly during periods of water scarcity or where traditional coping strategies such as scavenging are denied, either action should be taken to address the underlying problem, or to support those at risk in accessing alternative employment or income-generating activities. Findings from the research demonstrated that poor people were often keen to enhance their cash income, and that with adequate support this might be obtained from urban-oriented activities. Indeed, while enhanced production from aquaculture might contribute to sustaining poor livelihoods, to move out of poverty people in peri-urban Kolkata require access to alternative livelihood options and for young people the chance of a formal education appears to constitute the best route to accessing such opportunities. Providing infrastructure and services plays an important role in enhancing the livelihoods of poor people living in peri-urban Kolkata. However, this provision is not complete or seemingly equitable. Better communication and coordination by local bodies and municipal authorities appears to have a role to play in ensuring the demands and expectations of all community members are addressed.

From both focus group and household interviews it was apparent that improved infrastructure and service provision by the municipal authorities contributed to improved livelihoods for many people in peri-urban Kolkata. However, problems in benefiting from even basic service provision persist for many due to poverty denying access, and from incomplete geographical coverage. Such problems might be addressed through improved planning and resource allocation, including pro-poor policy, but the formulation and targeting of such initiatives would again require better communication and coordination of activities by local government bodies. Studies of market networks show them to be highly organized and efficient (Little *et al.*, 2002), but appropriate initiatives to help address concerns expressed by many regarding high input costs, and possible human health and animal welfare issues associated with marketing, appear warranted. Having attempted to outline above the pertinent issues related to the possible enhancement of poor livelihoods, it is necessary to note here that the complex physical, social and institutional nature of peri-urban Kolkata dictates that any attempt to address the constraints identified should be planned and implemented in consultation with a broad range of stakeholders.

The main technical issues constraining production have been identified, and targeted development projects to address these problems may help sustain and enhance yields and, in doing so, contribute to the continued support of many poor livelihoods, both directly and indirectly. Further action is required to enable poor people to diversify their livelihood strategies and, where necessary or prudent, remove their reliance on insecure returns from natural resource-based activities. Critical issues such as ensuring the safety of food from terrestrial and aquatic systems around Kolkata also deserve attention, as possible health hazards may not only affect consumer confidence, but may also result in detrimental health effects for producers, consumers and local residents, the very people reportedly benefiting most from aquaculture and agriculture in peri-urban Kolkata.

Acknowledgement

This publication is an output from a project funded by the UK Department for International Development (DFID) for the benefit of developing countries. The views expressed are not necessarily those of DFID. The project team expresses its gratitude to those farmers and labourers who gave their time to participate in this survey and thanks them for sharing their concerns and opinions, giving the outcomes of this work greater depth and meaning.

References

Bunting, S.W. (2002) Land–water interface production systems in peri-urban Kolkata. Website for DFID project R7872. http://www.dfid.stir.ac.uk/dfid/nrsp/kolkata.htm

Bunting, S.W., Kundu, N. and Mukherjee, M. (2002) *Situation Analysis: Production Systems and Natural Resource Management in Peri-urban Kolkata.* Institute of Aquaculture, Stirling.

Calcutta High Court. (1993) Ruling for matter #2851 of 1992 D/-24–9-1992, People United for Better Living in Calcutta (PUBLIC) vs. State of West Bengal, Umesh Chandra Bannerjee, Presiding Judge. *All Indian Review* 215–232.

CRG (1997) *East Calcutta Wetlands and Waste Recycling Region.* Creative Research Group, Calcutta.

Edwards, P. (2001) *Senior Stakeholder Meetings, Calcutta – 30 July to 6 August 2001.* Asian Institute of Technology, Bangkok.

Edwards, P. (2002) *Institutional Assessment: Reviewing Policies, Processes and Stakeholder Positions at the Kolkata Peri-urban Interface.* Asian Institute of Technology, Bangkok.

Furedy, C. (1987) From waste land to waste – not land: the role of the Salt Lakes, East Calcutta, in waste treatment and recycling, 1845–1930 In: Sinha, P. (ed.) *The Urban Experience: Calcutta.* Riddhi, Calcutta, pp. 145–153.

Ghosh, D. and Sen, S. (1987) Ecological history of Calcutta's wetland conversion. *Environmental Conservation* 14, 219–226.

Kundu, N. (1994) *Planning the Metropolis, a Public Policy Perspective.* Minerva Associates, Calcutta.

Little, D.C., Kundu, N., Mukherjee, M. and Barman, B.K. (2002) Marketing of fish from peri-urban Kolkata. University of Stirling, Stirling.

Milwain, G. (2001) Investigation into sustainable development and livelihood issues of the fish seed distribution network in West Bengal, India. Masters Thesis. University of Stirling, Stirling.

Mukherjee, M.D. (1996) Pisciculture and the environment: an economic evaluation of sewage-fed fisheries in East Calcutta. *Science, Technology and Development* 14(2), 73–99.

Naskar, K.R. (1985) A short history and the present trends of brackishwater fish culture in paddy fields at the Kulti-Minakhan areas of Sunderbans in West Bengal. *Journal of the Indian Society of Coastal Agriculture Research* 3(2), 115–124.

Punch, S., Bunting S.W. and Kundu, N. (2002) *Poor Livelihoods in Peri-urban Kolkata: Focus Group and Household Interviews.* University of Stirling, Stirling.

Ramanathan, U. (2000) The public policy problem: child labour and the law in India. In: Schlemmer, B. (ed.) *The Exploited Child.* Zed Books, London.

6 Wastewater-based Urban Aquaculture Systems in Ho Chi Minh City, Vietnam

Minh Phan Van[1,2] and Niels De Pauw[2]
[1]*Department of Science, Technology and Environment, Ho Chi Minh City, Vietnam;*
[2]*Ghent University, Ghent, Belgium*

Abstract

Steady increases in population and economic activity, as well as a lack of environmental infrastructure, have resulted in severe surface water pollution within Ho Chi Minh City. The clear boundary between urban and rural areas has faded. In this context, farmers and their counterparts have tried to set up appropriate wastewater-based urban aquaculture systems, which is a challenge that faces constraints. Eight major types of wastewater-based urban aquaculture systems occur within the urban area, mostly conducted on a trial-and-error basis. These are: (i) water morning-glory (*Ipomoea aquatica*), (ii) lotus (*Nupher lotus*), (iii) lotus and duckweed (Lemnaceae), (iv) water mimosa (*Neptunia oleracea*) and duckweed, (v) water mimosa, (vi) duckweed and growout of fish (tilapia, *Oreochromis* spp.), (vii) fish nursing and rearing and (viii) ornamental fish culture. Farming practices of these systems and their establishment and evolution under urbanization pressure are described. Generally, farming practice depends on cultured species, topographical characteristics of the landscape and pollution level of the water being used. Over the years, urbanization has not caused these wastewater-based urban aquaculture systems to disappear, but has pushed them away to more distant sites. Aquaculture products from wastewater-based urban aquaculture systems have been accepted in local markets as both animal and human food. Neither claims nor public health incidents associated with these products have been reported so far. The necessity of research for improvement of production efficiency and food safety of wastewater grown products, and the possible role of these systems within the context of sustainable water management in Ho Chi Minh City, are discussed.

Introduction

Ho Chi Minh City (HCMC), formerly Saigon, is located in the south of Vietnam and is the largest city in the nation. Its total area is 2053 km^2 with 18 districts, of which the urban areas, comprising 12 districts, cover more than 600 km^2. HCMC is situated 80 km from the coast, but within the city there are two major rivers – the Saigon River and the Dong Nai River – as well as a complex system of 27 smaller rivers and canals. Topographically, the general ground slope of the city is from the north-north-east to the west-south-west. The city is affected by semi-diurnal tides, which have the strongest influence in the south-south-western part.

The original urban sewerage and drainage system were constructed as a combined system during the French period in the 1870s, and later improved during the

 Urban Aquaculture
(eds B. Costa-Pierce, A. Desbonnet, P. Edwards and D. Baker)

wartime period of 1950–1975. The capacity of the system was, however, only sufficient for a population of 1.5 million as planned. No wastewater treatment plants were built during these periods.

From 1990 onwards, thanks to market-oriented policies, the economic activities of the city developed rapidly. The average annual economic growth rate of the city has been about 12.5% since 1994, although the rate decreased slightly in the period 1998–2000 (Ho Chi Minh City Statistics Office, 2002). Economic development, in turn, has resulted in a population increase, mainly by domestic immigration, from 5 million in 1996 to about 7 million in 2002 (Ho Chi Minh City Statistics Office, 2002).

While economic activity and population both increased, the city's environmental infrastructure has not kept pace. Consequently, many waters within the city are severely polluted by municipal wastewater discharges (Tables 6.1 and 6.2). Within this context, farmers are threatened by water pollution, which caused the collapse of many farming systems because neither information nor studies on farming with wastewaters were available at that time. Farmers therefore set up wastewater-based farming systems on a trial-and-error basis. A conversion of conventional farming systems into wastewater-based systems occurred as a consequence of the shift of the rural agricultural areas to the peri-urban areas having no water pollution control. Since wastewater treatment technologies are still unaffordable, farmers in the peri-urban areas of HCMC have been forced to create new wastewater farming systems in order to maintain their livelihoods. Over the past 10 years farmers have established a diversity of wastewater-based aquaculture systems, including production of aquatic macrophytes, as well as different fish species. This chapter gives an overview of this diversity of urban wastewater aquaculture ecosystems in Ho Chi Minh City.

Major Areas of Wastewater-based Aquaculture in HCMC

Today, wastewater-based aquaculture systems can be observed in many different places within the peri-urban districts of HCMC. However, there are only two areas

Table 6.1. The water quality of some major rivers and canals in the northern part of the city.

		[a]Tham Luong – VamThuat		[b]One tributary
Parameter	Unit	High tide	Low tide	High tide
Temperature	°C	28.2	30.1	30.0
pH		6.5	6.6	6.9
DO	mgO_2/l	0.7	1.0	0.6
BOD_5	mgO_2/l	152.0	181.0	137.0
COD	mgO_2/l	310.0	240.0	234.0
Total solids	mg/l	32.0	98.0	26.0
Total nitrogen	mg/l	4.8	2.0	4.0
Total phosphorus	mg/l	0.6	0.2	48.0
Cadmium	μg/l	<1	<1	<10
Lead	μg/l	4.2	<2	<10
Total chromium	μg/l	NR	NR	20.0
Cr^{6+}	μg/l	<0.04	<0.04	NR
Mercury	μg/l	<2.5	<2.5	NR

[a]Japan International Cooperation Agency–People's Committee of HCMC (1999).
[b]One tributary supplying water to a lotus farming system, the average figures of 6 months (PMU 415, 2002).
NR: Not recorded.

Table 6.2. The water quality of some major rivers and canals in the southwestern part of the city.

Parameter	Unit	[a]Tan Hoa – Lo Gom High tide	[a]Tan Hoa – Lo Gom Low tide	[a]Tau Hu – Doi – Te High tide	[a]Tau Hu – Doi – Te Low tide	[b]One tributary High tide
Temperature	°C	28.4	30.0	28.4	29.9	29.2
pH		5.9	5.8	6.7	6.6	6.6
DO	mgO_2/l	0.0	0.0	2.8	0.0	0.0
BOD_5	mgO_2/l	326.0	536.0	151.0	251.0	118.25
COD	mgO_2/l	1456.0	988.0	249.0	400.0	168.29
Total solids	mg/l	1420.0	272.0	70.0	216.0	82.75
Total nitrogen	mg/l	38.2	46.2	2.0	11.2	19.01
Total phosphorus	mg/l	2.0	2.9	0.1	0.6	4.33
Cadmium	μg/l	<1	<1	<1	<1	0.155
Lead	μg/l	2.6	5	<2	<2	7.78
Total chromium	μg/l	NR	NR	NR	NR	17.96
Cr^{6+}	μg/l	<0.04	<0.04	<0.04	<0.04	NR
Mercury	μg/l	<2.5	<2.5	<2.5	<2.5	0.085

[a]Japan International Cooperation Agency–People's Committee of HCMC (1999).
[b]The tributary supplying water to a fish culture system, average figures of 2001–2002.
NR: not recorded.

in which the wastewater-based aquaculture systems are concentrated. These are situated in the northern and the southwestern parts of the city, and comprise several hundred hectares (Fig. 6.1).

The northern part of the city

The northern part of the city is the catchment of the Tham Luong-Ben Cat-Vam Thuat River system. This river system has many tributaries through which it is connected to urban areas. The system receives municipal wastewaters from the central towns and residential areas of the districts, such as Tan Binh, Go Vap, Hoc Mon and part of Binh Chanh. The population of the catchment area is estimated to be more than 1 million. Topographically, this region is uplands in comparison to the southwestern area. Therefore, the influence of tides in this area is not as strong as in other parts of the city. The soils are rich in clay and sand, which facilitates pond-building. Before being polluted by wastewater, the farming systems in the area were mainly vegetable and rice farms. The current wastewater-based aquaculture systems are mainly aquatic macrophyte systems, especially lotus (*Nelumbo nucifera*) and water morning-glory (*Ipomoea aquatica*). The water quality of a major river and its tributary which supply water to a lotus farming system in the northern part of the city is summarized in Table 6.1.

The southwestern part of the city

In comparison to the northern part of the city, the southwestern area is low-lying, comprising a large catchment area of several rivers such as the Tan Hoa–Lo Gom Canal, the Tau Hu-Ben Nghe River, the Kinh Doi-Kinh Te River and the Ruot Ngua River. Within this network of waterways, many intersections with complex hydrodynamic and pollution-dispersal regimes exist. Consequently, pollution levels within the areas have increased, and although the influence of the tide in this part of the city is stronger, it is not able to adequately flush pollutants out to sea.

The catchment has an estimated population of 3 million, and receives wastewaters not only from its population (Districts 5, 6, 8 and part of Binh Chanh District), but also from central districts of HCMC (Districts 1 and 10).

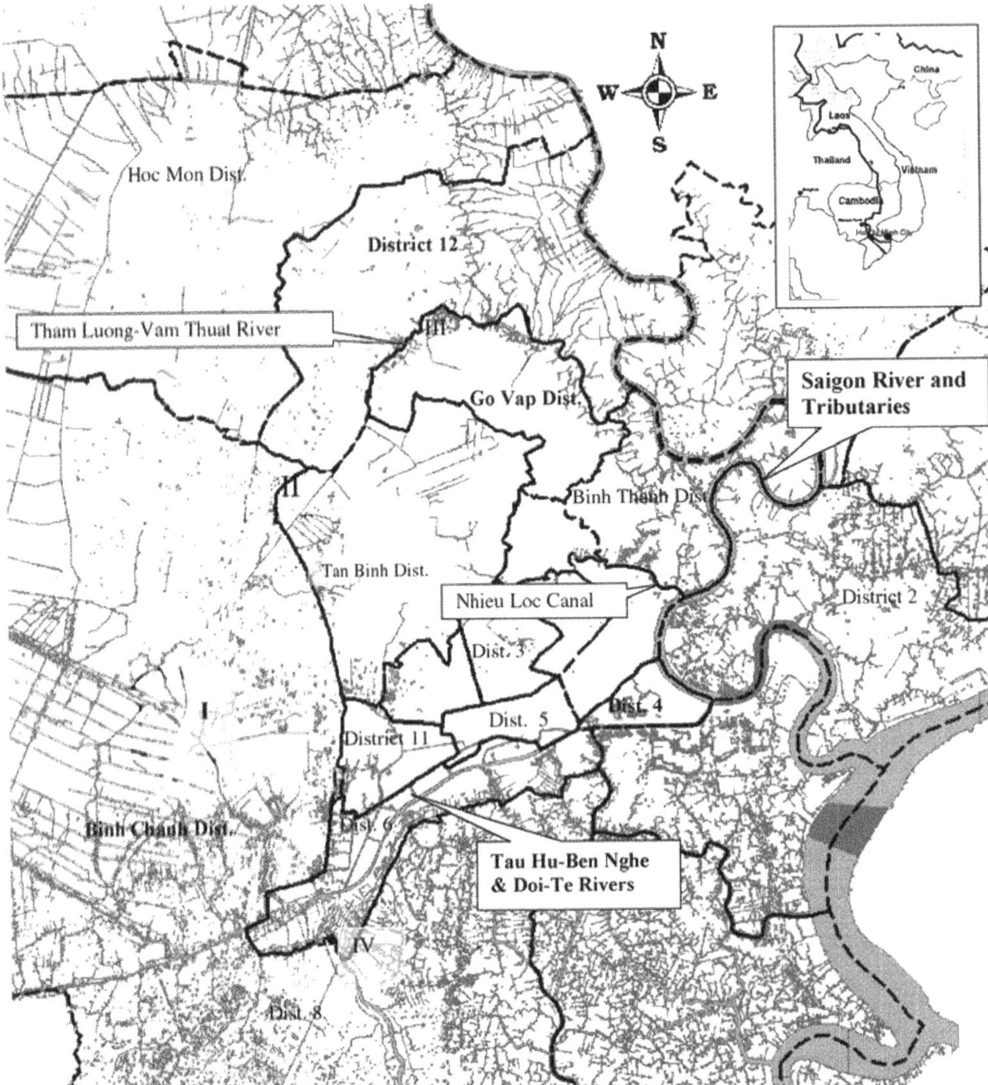

Fig. 6.1. Locations of wastewater-based farming systems in Ho Chi Minh City.

The soil in this part of the city is mainly muddy and its structure is very loose. Therefore pond construction in this area is costly. The wastewater-based fish-culture systems are common in this southwestern part of the city because farmers living in this area cultured fish for many years and have a strong background knowledge of fish culture so that they obtain higher profits, compensating the high capital investment.

The pollution levels of the major rivers and tributary supplying water to a fish-culture system in the southwestern part of the city are shown in Table 6.2. The data from Tables 6.1 and 6.2 illustrate that the water quality of the major rivers and canals in both parts of the city can be considered as domestic wastewater, if compared with the water pollution levels reported by Tchobanoglous (1991). The colour of these waters is black, and they have an intense odour that can be detected at a distance of several tens of metres.

Wastewater-based Farming Systems in HCMC

Table 6.3 gives an overview of the wastewater-based farming systems in HCMC. The historical context and present farming developments; the practices of farming and water management; as well as product preparation and prepared dishes are described.

Wastewater-based water morning-glory farming

History

Water morning-glory (*Ipomoea aquatica*) is an aquatic plant that can live both as an emergent plant as well as floating due to the cylindrical plant body, which is long, hollow and segmented with thin inner membranes between the segments. These membranes support the transformation of the segments into floaters. On the outer part, and at the dividing point of the segments, leaves and roots develop. The root system of the plant can grow in muddy soil to take up nutrients when the plant grows in shallow waters, but in deep waters the plant floats and its roots take up nutrients directly from the water.

The municipal wastewater-based farming system of water morning-glory in HCMC could have emerged initially in the Nhieu Loc–Thi Nghe Canal catchment. The whole canal passes through the densely populated, central districts of the city. Due to its central location, during wartime (1960–1970), refugees who were originally farmers from other provinces migrated into the city and took over land strips along the canal. The old government of South Vietnam did not issue any policy to prevent this situation, but instead supported it in an unofficial way. Those who came first took over the banks along the canal for settlement, and those who came later built over-hanging houses above the water. Consequently, slum villages were established along the banks, polluting and silting up the canal. 'Black waters' were formed under these slums, and floating clusters of water morning-glory around the stakes supporting the slums soon followed. These plants have been readily accepted as edible products by local markets. It was only in 1995 that these farming systems were removed from the canal as a result of a million-dollar project aiming at the sanitation and urban upgrading of the canal catchment.

During the period after the war, the rapidly increasing population and degraded and overloaded environmental infrastructure resulted in pollution of many other waters in the city. Based on the

Table 6.3. Different types and main characteristics of wastewater-based farming systems in HCMC.

Aquatic farming system	Culture type	Wastewater usage	Typical sites
Water morning-glory	Monoculture	Direct and continuous	Tidal flats/paddy fields
Lotus	Monoculture	Direct and continuous	Paddy fields/ponds
Lotus–duckweed	Polyculture	Direct and continuous	Paddy fields/ponds
Lotus–fish growout	Polyculture	Stabilizing and feeding	Paddy fields/ponds
Water-mimosa–duckweed	Polyculture	Stabilizing and feeding	Paddy field/ponds
Water-mimosa–duckweed–fish growout	Polyculture	Stabilizing and feeding	Paddy fields/ponds
Fish growout	Mono/polyculture	Stabilizing and feeding	Ponds
Fish nursing and growout tilapia fingerling production	Mono/polyculture	Stabilizing and feeding	Ponds
Ornamental fish production	Mono/polyculture	Stabilizing and feeding	Ponds

experience that could be learned from the Nhieu Loc–Thi Nghe canal, farmers introduced the municipal wastewater-based water morning-glory farming system throughout the city, initially in urban areas and later on in peri-urban ones. The development of the water morning-glory farming system in the urban waters was particularly notable in the 1980s and 1990s. Presently, however, these systems are found only in peri-urban areas which are receiving more and more municipal wastewater discharges. This is due to projects carried out in different phases for improving the water quality of canals in urban areas, and due to recent policies preventing the construction of slums and other facilities in the waterways of the city.

Water morning-glory is being sold to markets and consumed as human and animal food. Although Vietnam has issued standards for food safety, including salad crops, health-risk checks of water morning-glory have rarely been carried out.

Major areas of production

In the northern part of the city, wastewater-based farming systems of water morning-glory have been observed in the catchment of the Rach Cai River (Thu Duc District) and Tham Luong-Ben Cat–Vam Thuat River (Tan Binh, Go Vap and Hoc Mon Districts). These are upland districts, and for this reason water morning-glory has been cultivated mainly in paddy rice fields. These are irrigated by water from small adjacent tributaries and canals, which are connected with river sections receiving wastewater discharges from the main residential areas and markets. Originally, these water morning-glory paddy fields were not wastewater-based systems, but conventional ones. They were converted into wastewater-based systems because the main rivers were being polluted. The total area of these systems shows a decreasing trend under the strong urbanization pressure in the above districts.

In the southwestern part of the city, due to topographical characteristics, in contrast to the above-mentioned northern areas, water morning-glory is planted as floating clusters on the surface of small tributaries and canals without navigation routes, which flow slowly and receive wastewaters from the main residential areas and tidal flats of these waterways. Farmers in the region mentioned that they need a high investment to set up paddy fields of stabilized embankments in the area because of the muddy soil.

Farming practices

SEEDLINGS Depending on the type of farming system, there are two ways of propagating seedling water morning-glory plants. In the *paddy field cultivation type*, water depth of the field is maintained initially at 30 cm. Plants are cut into sections of 50 cm, and every three of these are inserted into one point of the muddy bottom of the field as a seedling, maintaining a density of three seedlings/m^2. After 2 weeks, many young buds shoot out from the old part of the plant to generate new plants. Wastewater is then routed into the paddy field to increase water depth to 50 cm, which is then maintained throughout cultivation. When new plants become older and roots on segments of the plants appear, plant tips are cut off. From these cuttings new plants of the next generation are produced. By doing so, farmers increase the density of the plants in the fields until they decide to harvest the first batch.

In the *floating-cluster cultivation type*, from the parental clusters many smaller ones of old plants are cut off as seedlings, and then carried to new planned sites. Owing to the netting of the plants, seedling clusters are kept on site with bamboo stakes in the middle of the clusters. The regeneration of the plant until the first harvest is manipulated as described previously. Floating bamboo frames are often used to distinguish the ownership of these 'floating fields' and to prevent the development of plants beyond limits of common boating routes for daily management activities.

MANAGEMENT The management of water morning-glory farming in paddy fields is easily carried out with proper embankments. Twice a week, after partially draining water from the field, wastewater is taken in to maintain a water depth of 40–50 cm. However, the frequency and amount of water taken into the fields can be changed, depending on the colour and size of the young plants. Farmers have used gradations of green colour and size of young plants as indicators. The bright green colour and big plant body are indicators. In the case that these indicators are not satisfied, farmers apply urea fertilizer once or twice, 10 kg/1000 m^2, as a supplementary nutrient source before harvesting.

The water management of the floating-cluster farming system does not require any special action.

HARVEST The generation of a new batch of plants by cutting off parts of plant bodies is a way that farmers promote increased density of plants in their fields, increasing plant yield. Only young parts of plants are harvested by cutting. Within a proper management practice, after harvesting three or four batches, the plant yield becomes stable. Normally, the harvest frequency is every three weeks, and an average yield of 1000 kg wet weight/1000 m^2 can be obtained with the paddy field type. The yield of the floating-cluster farming type is normally lower, around only 700 kg. When plants get too old to reproduce healthy young plants, normally after 2–3 years, the whole field is replanted with new seedlings in order to maintain high yield crops.

PRODUCTION RISKS Risks to both farming types are dramatic changes in the quality of the water on which the systems are based, and the development of unexpected insect infestations. It is difficult to assess pollution levels of the water when farmers only use gradations of the black colour of the water as an indicator of water quality. Nevertheless, when both types of farming are compared, paddy field farming is more advantageous. The paddy field farming type allows farmers more readily to control the quality of the water taken in, and to apply necessary pesticides, as needed, than in the case of the floating-cluster type. Occasionally, rapid mass mortality of plants in the floating-cluster farming type has been observed, which is thought to be due to the severely polluted water.

PRODUCT PREPARATION After harvesting, the plants are transported to farmers' houses to be sorted into two categories. The first is young plants which are bright green, soft and big, and have unbroken leaves without insect larvae, and which are sold for human consumption. The rest are plants which do not meet these criteria, and are used instead as animal feed. Because the first category is human food, the preparation of the product prior to selling is a time- and labour-consuming task. Normally, harvesting activities are undertaken in such a way that product preparation occurs in the late afternoon or evening, avoiding high temperatures which can wilt plants. In the preparation, attached debris is removed, and the product is washed and tied up into small bundles. Finally, they are rewashed before being submerged overnight in clean water to ensure a fresh appearance at markets the next morning.

VIETNAMESE DISHES Most Vietnamese dishes with water morning-glory are well cooked. There is only one dish in which water morning-glory is consumed as salad, but this dish is rarely prepared. After being washed carefully and re-selected to remove unwanted parts, the water morning-glory plants are cut into strips of 3 × 50 mm and submerged in clean water for several hours. With such a preparation, the water morning-glory strips become rolls, which are then mixed with vinegar before being mixed with other salad greens.

Wastewater-based lotus monoculture farming

History

Lotus (*Nelumbo nucifera*) is an emergent aquatic plant which roots in the bottom of shallow waters (Fig. 6.2). Its young leaves float on the water surface with long, weak stems, whereas old leaves do not float but project into the air with strong and spiny stems. The vegetative reproduction of lotus is by its roots, in tuber form, from which the next generation develops through rhizome growth. Young plants develop from the tips of these rhizomes.

In HCMC the lotus is, in conventional cultivation systems, normally planted in shallow ponds or swamps in lowlands. Vietnamese often use all parts of the lotus plant for domestic purposes. Its leaves are used for packing wet commodities at market, flowers for ornamentation and religious ceremonies, and seeds, tubers and rhizomes for human consumption. The lotus rhizome is processed for dishes quite popular in Vietnam. The market demand is different for different parts of the lotus plant and, accordingly, different farming practices are established to produce the different end products.

In the late 1980s, municipal wastewater discharges started polluting rice paddy fields in peri-urban areas. Because of a lack of clean water for irrigation, crops of rice often failed and many farmers abandoned their fields. However, some of them tried to convert them into other aquatic farming systems, and enabled the emergence of several wastewater-based aquatic plant farming systems, with wastewater-based lotus farming being one of them.

The wastewater-based lotus farming system possibly had originated in the upstream area of the Tan-Hoa Lo-Gom canal in the northern part of the city, as this area was recorded as the first severely polluted place in the city. In the area, before becoming polluted by municipal wastewater, several shallow swamps were used for lotus cultivation. It is told that farmers, by chance, observed that lotus could still grow well in some small, shallow swamps flooded by polluted water, and therefore in their efforts to transform rice fields into other crops, they tried farming lotus with the wastewater.

Today, lotus products, especially the lotus rhizomes, can be seen in many markets of the city. Although no public health incident has been reported so far for the rhizomes, a product from the bottom mud, some institutions of the city have become interested in investigating the likelihood of bioaccumulation of heavy metals in the rhizomes.

Fig. 6.2. Wastewater-based lotus monoculture farming system.

Major areas of production

Wastewater-based lotus farming systems occur in both the northern and southwestern part of the city, and appear to be occupying the largest area in comparison with other types of wastewater-based farming crops.

Presently, in the northern part of the city, there are more than 200 ha of wastewater-based lotus farming, mainly in Go Vap, Binh Chanh and Hoc Mon districts. In the Subdistrict Binh Hung Hoa alone, an area of 60 ha was recorded in 2000. However, the recent expansion of neighbouring residential areas has decreased the area of wastewater-based lotus farming in this part of the city. In the southwestern part, the Phong Phu Subdistrict, a lotus farming area of 245 ha was noted.

Farming practices

Wastewater-based lotus monoculture systems produce lotus rhizomes. These rhizomes develop very fast after growing from tuber roots, and can reach a length of 100 mm and a diameter of 10 mm after only a few days. When they attain their maximum length, buds of young plants will develop at the tips. This causes the weight of the tip to increase, and therefore to sink to the pond bottom, and young plants are then rooted in the mud.

The lotus rhizome has a larger market demand than other kinds of lotus products. Therefore, in areas concentrating on farming lotus with wastewater, the lotus monoculture systems normally occupy over 70% of the area.

SEEDLINGS Before starting to plant the seedlings, the water level of the ponds is decreased to a depth of 30–40 cm. If the water is 'new' wastewater, it will be stabilized for 1 week. Young plants with small tuber roots which are used as seedlings are then inserted into the muddy bottom of the pond at a density of one seedling/m^2. In case some of these young plants are too weak, bamboo sticks are used to support them. During the seedling period, water loss by evaporation is compensated by addition of wastewater. Two weeks after planting the seedlings, plants start to grow new leaves, which is an indicator of acclimation to their new environment.

MANAGEMENT When floating young leaves develop, they extend the length of their stems and move away from the original position of the plant. Observing this, farmers increase the water depth of the pond to a level of 70–80 cm with wastewater. Owing to the extended stems, leaves are still floating, and therefore photosynthesis is maintained.

Depending on the topography of the site, two types of water management practices are applied in wastewater-based lotus monoculture systems. In upland areas where tidal influence is small, feeding ponds with wastewater by gravity is impossible, and ponds are built in series and fed by a flow-through pattern. In lowland areas, where tidal height makes feeding the ponds with wastewater by gravity feasible, the system is fed automatically by an intermittent pattern owing to different levels between low and high tide. The flow-through pattern is normally applied in the northern part of the city, and the intermittent pattern in the southwestern one.

For the flow-through feeding system, at the upstream section of the wastewater-supplying source (canals/small tributaries), farmers construct small dams made of bamboo or wooden piles in order to increase the water level of the supplying source to create hydraulic head. The dam is built in such a way that most of the wastewater of the supplying source overflows the dam surface, and only a small part of it flows by gravity into a small ditch directed to the fields. Small amounts of wastewater flow continuously into the fields, and then run from field to field in series. The water then flows out of the fields, back into the downstream section of the supplying source. At a field which was fed by this practice, it was recorded that the quality of the supply water was above 100 mg/l BOD_5 and 20 mg/l total nitrogen. At the observed flow rate, the whole water volume of a 1500 m^2 field would be exchanged after 5 days.

In the intermittent feeding system, ponds are often not connected in series and are individually fed with wastewater by an adjacent ditch. They are fed intermittently, and automatically, by semi-diurnal tides. During high tide, the wastewater flows into the pond through a pipe of 20–30 cm diameter, then through the same pipe the water flows out of the pond and back to the source during low tide. Taking into account the tidal characteristics of the locality, farmers normally place the pipe 70–80 cm above the bottom to maintain this water level during low tide, and increase it to 90–100 cm during high tide. This means the water mass of the pond is fed with about 20–30% 'new' wastewater at each tidal cycle.

Thus, in both systems of pond feeding, the wastewater is taken directly into the ponds on a continuous or twice-a-day basis. This shows that lotus systems appear to have a high nutrient requirement, and are capable of adapting to highly polluted waters.

According to farmers' experience, to get a high productivity from lotus rhizomes, the wastewater should be rich not only in nutrients, but also in suspended solids. Sedimentation of these solids will build up an expected layer of mud at the bottom of the pond, which creates an environment favourable to the development of lotus rhizomes.

For obtaining higher yields of lotus rhizomes, besides good management of water supply as mentioned above, the removal of leaves shading the pond surface is also necessary. Normally, adult leaves floating far from the parental plant, or projecting up into the air, are removed in order to increase the penetration of sunlight into the water column. Therefore, a certain percentage of pond surface without leaves is always maintained. This is also believed to increase the water temperature. During the removal of old leaves, all unexpected aquatic plants, including duckweed, which grow naturally in ponds, are also removed. Because the lotus rhizomes are harvested on a continuous basis, the number of young plants is limited, resulting in a low density of plants in the pond.

HARVEST One month after planting the seedlings, farmers start harvesting the lotus rhizomes (Fig. 6.3). Lotus rhizomes shoot out so fast from tuber roots of plants that they become new plants in a few days if they are not harvested in time. At that moment, their colour changes from bright white to brown, which is not an acceptable market trait. As a result, every 3–4 days farmers walk into the ponds and harvest, by hand, the rhizomes submerged in the water. An average rhizome productivity of 2000 kg/ha/month has been recorded.

In the northern part of the city, lotus rhizomes can be produced throughout the year, but yield will fluctuate seasonally. The yields during the dry season are usually higher than during the rainy season, and it is believed that the higher temperatures of the dry season are favourable for the development of the plants. Conversely, in the southwestern region of the city, lotus rhizome production can be carried out only during the rainy season from May until December because lotus plants do not grow well when water salinity is 2–4 ppt.

In addition to lotus rhizome products, duckweeds can be considered a by-product of this type of system. Duckweed is harvested during the preparation of the pond surface to increase sunlight penetration. Although the duckweed yield is low in this type of system – only 100–150 kg wet weight/1000 m^2/month – this aquatic plant can offer farmers a secondary income because it is a commercial product to local markets.

PRODUCTION RISKS In comparison to other aquatic plant crops, it seems that there are very few risks for wastewater-based lotus farming. The only risk that farmers mention is the sudden breaking of pond dykes. This causes ponds to be flooded completely by the polluted water, resulting in the die-off of lotus plants if the water is heavily polluted. To prevent this problem, pond dykes are often re-stabilized once a year before the rainy season. Insects also damage the leaves of plants, but farmers do not consider the damage significant.

Fig. 6.3. Wastewater-based lotus monoculture system: harvesting products and preparing pond surface for sunlight penetration.

PRODUCT PREPARATION After harvesting, lotus rhizomes are transported to farmer or distributor houses for preparation. The objective of the preparation is to satisfy the product quality requirements for the market. The product has to be fresh – with its natural flavour – clean, brightly white, soft and brittle. The rhizomes are washed and attached debris removed. Then every rhizome is rewashed individually. Farmers brush the thin membrane that covers the outer part of the body of rhizomes by hand to eliminate even tiny attached particles to get a clearer, bright white colour. If a section of this membrane is brown, it is stripped off the body of the rhizome. Finally, they are tied up in bundles and submerged in clean water until being marketed the next day.

VIETNAMESE DISHES Before being prepared as food in the kitchen, the cook reselects the lotus rhizomes to get rid of unsatisfactory parts – for example, old and fibrous bases of the rhizomes – and washes them carefully. All dishes made from lotus rhizomes are cooked except one, which requires a careful hand-preparation process. Accordingly, the rhizomes are washed and cut into small strips of 3 × 50 mm, then submerged in water for hours. Finally, they are mixed with lemon juice or vinegar before being mixed with other food.

Wastewater-based integrated lotus and duckweed farming

History

The integrated wastewater-based lotus and duckweed farming system is a polyculture system. In situations of financial difficulties, or because of seasonal constraints, farmers use this type of system, or transfer the lotus monoculture system mentioned before, into the integrated one. This polyculture system produces mainly duckweed biomass, seed-pods and lotus flowers. As it is a temporary system, the area in use varies over time. In main areas of wastewater-based lotus farming, it often accounts for only 20–30% of the total area.

In recent years, this type of system is seen only when weather conditions are not favourable for lotus rhizome production. When farmers stop producing rhizomes, they do not clear off duckweed from the pond surface to get more sunlight; therefore, the lotus monoculture system is immediately converted to the lotus and duckweed system. Because of increasing land value and labour costs, this type of system shows a decreasing trend.

Major areas of production

The wastewater-based integrated lotus and duckweed farming system has been observed in both the northern and the southwestern parts of the city.

Farming practices

SEEDLINGS Seedling production for lotus in this system is carried out similarly to the monoculture system described previously. However, in contrast to the lotus monoculture system, farmers limit the initial density of seedlings to only 50 seedlings/1000 m^2 because the rhizomes are not harvested, allowing young plants to develop continuously, increasing plant density to a maximum in the system after a few months.

To develop duckweed mats in the pond, farmers collect duckweed in neighbouring ponds and stock them in the pond. They only try to cover the pond surface as much as they can, and do not take into consideration the initial density of duckweed because they think duckweed can spontaneously grow well in ponds.

MANAGEMENT The pond design and water management of the integrated wastewater-based lotus and duckweed system are the same as those applied to the monoculture system described previously. After planting seedlings, also by observing the development of young leaves, farmers increase the water level of the ponds to 70–80 cm to start the farming process. From then on, the ponds are fed with wastewater, and no further plant management is undertaken. Farmers harvest only flowers, seed-pods and duckweed biomass.

HARVEST Farmers harvest duckweed biomass weekly to sell to their clients according to their contract, or to local markets, or for their own use. Duckweed is mainly used for feeding fish, and any remainder for ducks and chickens. The duckweed productivity of this system is about 80–120 kg wet weight/1000 m^2/week.

Lotus flowers are sold mainly to serve monthly Buddhist religious days or other special occasions. Therefore, farmers normally collect them on a bi-weekly basis. The flowers which are not harvested on these days will be left for seed-pod production. A production of four flowers/m^2/month is needed to satisfy demand.

Seed-pods are the second major product of the system because not all flowers develop in time for religious occasions. Seed-pods are harvested when they become old and attain a yellowish colour, and are the most valuable commercial product of this type of system. An average seed-pod yield of 10 kg/1000 m^2/week was recorded. This equals 3 kg wet weight of seeds. In some places, farmers do not collect the lotus flowers for sale, but reserve them for seed-pod production. An average seed yield of 20 kg wet weight/1000 m^2/month was reported.

PRODUCTION RISKS Except for the initial planting of seedlings, farmers do not add any special inputs into these systems. As a result, possible production risks of the system are not considered.

Wastewater-based integrated lotus and fish growout farming

History

Wastewater-based lotus and fish growout is a polyculture system which produces lotus rhizomes and table fish. It is said that after becoming familiar with wastewater-based lotus farming, farmers thought that with a water depth of 70–80 cm, the system would also be appropriate for the integration of fish culture. This type of system is observed only in the southwestern part of the city, supposedly because the fish culture tradition of farmers in these areas had been established many years before local waters were polluted by municipal wastewaters.

Major areas of production

The system is concentrated mainly in Subdistrict 6 of District 8, and the Binh Chanh District, with hundreds of hectares in use in 2000. However, the current urbanization process within District 8 is pushing this type of system further into the Subdistricts Phong Phu and Binh Hung of the Binh Chanh District.

Farming practices

The farming practice of lotus rhizome production in the integrated system is similar to that of the wastewater-based lotus monoculture system described previously. However, there are some modifications in pond preparation and initial planting density of lotus. To facilitate the integration of fish, a surface of 100 m^2 without lotus in a 1000 m^2 pond is required. It appeared that this farming practice matches the feeding habit of the fishes – the planktonic feeds for the fish fingerlings develop because of more sunlight during the early stage of the cultivation and periphyton-based feed for adult fish develop because more (stem) substrates are available during the later stage. In addition, old leaf pruning is also more frequently required in this type of system.

After the first harvest of rhizomes, fish fingerlings of tilapia (*Oreochromis* spp.), kissing gouramy (*Helastoma temminckii*), silver carp (*Hypophthalmichthys molitrix*) and common carp (*Cyprinus carpio*) are stocked into the system. For a pond of 1 ha, an initial density of 80 kg of tilapia fingerlings and 20 kg of other species mentioned are normally applied. In comparison to the conventional fish culture systems, the size of fingerlings stocked in this type of system is larger (300–350 fingerlings/kg).

MANAGEMENT Unlike other wastewater-based lotus farming systems, water management in the wastewater-based lotus and fish culture system is a task that receives a lot of attention by farmers. Ponds used in this polyculture system are designed similarly to the ones fed intermittently by tides, as described previously. However, the pipe at the water inlet is placed at a higher elevation in order to reduce the volume of wastewater flowing in during high tide. The 'new' wastewater volume is only about 5–10% of the total volume of the pond for each tidal cycle. Reduction of the size of the pipe to only 15 mm in diameter is another alternative observed. The volume of wastewater intake is also gauged by observing the gradations of the black colour, and the odour of the water. In cases where the colour of the polluted water is very black and its smell quite strong, the pipe will be closed.

HARVEST Despite the fact that the initial density of plants is lower, and the amount of wastewater taken into the ponds smaller than in the wastewater-based lotus monoculture system, farmers reported that lotus rhizome yield is still as high as 1500 kg rhizomes/ha/month because the lotus plant quickly reaches its maximal density.

The system is run throughout the rainy season, from May until December. Fish are not fed with any feed except natural ones present in the system, though some farmers add duckweed collected from other ponds as a supplementary feed for fish at later stages of culture. When rhizome production ends, and the lotus plants die off in the early dry season because of too high salinities of the pond water, farmers still keep fish in the ponds for one or two more months, and continue to feed them with decomposed lotus plants. Such integration with lotus can yield 1200–1500 kg fish/ha. The majority of this yield, 80%, is tilapia.

PRODUCTION RISKS Careless water management, leakages of pond dykes and sudden changes of the wastewater taken in are the major risks to the system. Mass mortality of fish, because of the heavily polluted water, has often occurred in this type of system.

Wastewater-based integrated water-mimosa and duckweed farming

History

Water-mimosa (*Neptunia oleracea*) is a favourite vegetable of many Vietnamese due to its specific taste. It is a floating plant, and except for points from which leaves emerge, the whole body of the plant is covered by aerophora (aerenchyma), a cylinder-shaped floater. This floater, with a diameter of about 2 cm, is composed of thin, spongy, fragile, white fibres. If the floater is broken, the plant will sink and die. Under field conditions, impacts of wind, waves or other forces which can shake the plants strongly will break the floaters, causing mass mortality of the plants. To be able to cultivate such a kind of plant, farmers have to keep the plants stable on the water surface. Farmers have tried many different alternatives, from inefficient and labour-consuming ways such as stabilizing the plants with bamboo sticks and breaking strong winds by planting trees along the pond dykes, to the current simple way of applying duckweed mats around plants. The application of duckweed for stabilizing the water-mimosa is being used throughout wastewater-based water-mimosa farming systems. Besides the role of protecting the floaters of the water-mimosa, it is still not known whether the duckweed has any other role in this type of system or not. Farmers only mentioned that the integrated system with duckweed usually gives a higher yield of water-mimosa than without it, but the cause of the increased yield is not known.

Before the local waters were polluted by municipal wastewaters, in the southwestern part of the city some farmers were growing water-mimosa for their own consumption. According to these farmers, because of the labour-intensive management of the lotus rhizome production, and the increasing market price of water-mimosa, they attempted to cultivate the water-mimosa with polluted water. After successful trials in the late 1990s, many of them transformed their wastewater-based lotus farming system into an integrated water-mimosa and duckweed one. Originally, the wastewater-based water-mimosa farming system occurred in District 8. However, urbanization during the last three years in this district has pushed the water-mimosa farming system into the Binh Chanh District, which is located about 3 km away from its previous locations.

Major areas of production

It is interesting to note that wastewater-based integrated water-mimosa and duckweed farming is only occurring, and strongly expanding, in the southwestern part of the city. Presently, hundreds of hectares of other wastewater-based farming systems are being transformed into this type of system, especially in the Subdistricts Binh Hung and Phong Phu of the Binh Chanh District.

Farming practices

SEEDLINGS Wastewater is taken into the ponds and stabilized for 4–7 days, depending on the levels of water pollution. Then, bundles of three water-mimosa plants (1 m long each) are anchored with sticks, or with dried coconut leaf branches (a preferred material), and inserted into the pond bottom. The initial density of the water-mimosa plants is 1 bundle for every 1–1.5 m^2. Duckweed is stocked as much as possible to cover the pond surface. Thanks to large foliages of dried coconut leaf branches projecting into the air in many parts of the pond surface, the duckweed plants are protected from the wind while still being thin mats. Within a few days, the duckweed mats become thick enough to start stabilizing the water-mimosa plants.

MANAGEMENT Ponds used for farming water-mimosa are rather small, their area often less than 1000 m^2 in order to avoid strong impact of the wind. The water depth of the water-mimosa ponds is maintained at 80 cm (Fig. 6.4). On a weekly basis,

Fig. 6.4. Wastewater-based integrated water-mimosa and duckweed farming system.

before harvesting the plants, farmers decrease this depth to about 60 cm. After harvesting, a layer of 20–30 cm of the 'new' wastewater is taken into the ponds during high tide. Between harvests, the water supply pipe to the pond is closed tightly.

HARVEST Water-mimosa is a rapidly developing aquatic macrophyte. Young plants are reproduced from buds at connecting points between leaves and the parental plant body (leaf bases). Three weeks after seeding, plants can yield a first harvest by cutting off the young sections of the plants. The length of these sections is about 50–70 cm from their tips. This cutting stimulates buds on the remaining part of the plant body to develop into the next plant generation. While duckweed protects the water-mimosa plants from the wind, the development of young water-mimosa plants, which gradually establish a network throughout the pond surface, in turn give back a similar support to the duckweed. The development of this symbiotic relationship helps the whole planted water surface of the pond to stabilize when facing strong winds. However, it is necessary to control the over-development of duckweed in order to obtain high yields of the water-mimosa. To do this, the duckweed is harvested during the harvest of water-mimosa. In addition, whenever farmers observe that fronds of the duckweed become yellowish and smaller, an indicator of nutrient decline in the system, farmers immediately harvest duckweed to reserve nutrients for the water-mimosa. After harvesting the water-mimosa, the duckweed is again evenly distributed over the pond surface.

Farmers harvest water-mimosa plants weekly. From the third harvest onwards, an average water-mimosa yield of 100 kg wet weight/1000 m^2 is recorded. A similar yield of duckweed is also obtained. Traditional water-mimosa farming can be carried out throughout the year in some areas of the city where there is no increase in water salinity during the dry season. However, in the southwestern part of the city, wastewater-based water-mimosa farming is often ended in December, which is the beginning of the dry season, although the average salinity is then not yet higher than 4 ppt.

PRODUCTION RISKS Differing from wastewater-based lotus farming is the frequency with which wastewater is fed to the water-mimosa – once a week. It appears that water-mimosa is quite sensitive to frequent exposure to polluted water. In addition, the frequent water-flow of the wastewater during feeding can disturb the stabilized surface of the duckweed and water-mimosa mat, reducing water-mimosa yield. Insect damage is recorded as a secondary production risk against which farmers are applying insecticides when needed.

PRODUCT PREPARATION Harvested water-mimosa plants are washed with clean water. All old or broken leaves and attached debris is removed. Then, all floaters and hard and fibrous bases of the harvested plants are taken out of the plant body. Finally, the plants are rewashed and submerged in clean water before being marketed the next morning.

VIETNAMESE DISHES All Vietnamese dishes with the water-mimosa plant are well cooked. The most popular dish is the 'hot pot' in which the plant is submerged in a pot of boiling seafood soup.

Wastewater-based integrated water-mimosa, duckweed and fish farming

History

Based on their experience with wastewater-based integrated lotus rhizome production and fish culture, farmers have also established a wastewater-based integrated water-mimosa, duckweed and fish farming system. In this type of system, fish can help farmers by controlling duckweed development, and create an additional economically valuable crop.

Major areas of production

As the wastewater-based integrated water-mimosa, duckweed and fish farming system is a modification of the integrated water-mimosa and duckweed system, it is practised only in the southwestern part of the city.

Farming practices

WATER-MIMOSA AND DUCKWEED The farming practices of water-mimosa and duckweed plants in this integrated system are similar to the above-mentioned system. However, due to the presence of fish, the intake of wastewaters has to be carried out more carefully.

FISH CULTURE An area of one-tenth of the pond surface without any plant growth is prepared for fish culture. To create this area, a nylon mosquito net is used to divide the pond surface into two parts with the support of bamboo stakes (Fig. 6.5). The height of the net is only 30 cm, by means of which the duckweed and water-mimosa mat are separated from the area of fish culture, but fish can move under the net to the duckweed to look for prey.

Ten kilograms of tilapia and 1–2 kg of grass carp (*Ctenopharyngodon idella*), common carp and kissing gouramy of large size (300–350 fish/kg) are stocked into the pond. However, many farmers presently have selected the kissing gouramies as a target species to replace the tilapia because it grows better in this type of system, and its sale-price is increasing.

HARVEST Duckweed and water-mimosa are harvested weekly as described in the wastewater-based water-mimosa and duckweed system. The productivity of the water-mimosa is similar to the integrated system of water-mimosa and duckweed. For the duckweed yield, it was observed that it may be equal to that of the integrated system without fish at an early stage of the farming, but lower in the later stages because of fish feeding. Actually, farmers do not pay much attention to duckweed yield in this type of system.

After cultivation of water-mimosa and duckweed is finished, normally in December or January, fish are kept in culture for 1 or 2 more months to consume decomposed plants and other natural feeds which are left in the pond. With good management practices, the system can give a yield of 200–300 kg fish/1000 m^2.

Fig. 6.5. Wastewater-based integrated water-mimosa, duckweed and fish farming system.

PRODUCTION RISKS Because for this type of system there is no need to take in wastewater to the ponds continuously or on a daily basis, risks of fish mortality are minimized.

Wastewater-based fish growout farming

History

With the available tradition of fish-culture in manured ponds, the farmers in the southwestern part of the city were acquainted with the use of wastewaters for fish culture. However, failures experienced during many years limited the establishment of this type of system to only 20% of the total area of farms. The reason for this is that the duration of the fish growout is long, at least 6 months. During this period of time, fish mass mortality may occur at any time because of the intrusion of the wastewater into the ponds from uncontrolled entries. As a result, it is observed that this type of system is located mainly in farms in which other wastewater-based farming systems are also established, especially the system of fish fingerling production. Such horizontal integration facilitates farmers exploiting the resources of other systems. Preferred fish species in this system are tilapia, silver carp, grass carp and common carp.

Major areas of production

The southwestern part of the city was one of the major areas of fish culture before the waters became polluted. Therefore, farmers have readily shifted their fish growing-out system to a wastewater-based one. Farmers living in other parts of the city, for example the northern part, although polluted water is available, have not set up wastewater-based farming systems because of a lack of experience. Moreover, fish farming always requires a higher investment than aquatic macrophyte farming.

Farming practices

Tilapia is the primary target species and silver carp secondary. Grass carp and common carp are additional species often used to exploit natural feeds in the pond. For a pond of 2000 m^2, the stocking densities are 100 kg for tilapia (size of 120–150 fish/kg) and also 100 kg (size of 800 fish/kg) for all silver carp, grass carp and common carp, but the silver carp accounts for about 90% of stocking biomass.

MANAGEMENT Ponds used in this type of system have to be of large size, 1500–2000 m^2, with a water depth of 1–1.5 m if high yields of fish are to be expected. The pond is filled with polluted water and is stabilized until the water quality is assessed to be satisfactory for fish stocking. During the stabilization of water, the water supply pipe is closed. After fish stocking, the pipe is opened at an interval of 2–3 days to feed the pond with 'black' wastewater during high tide. The amount of black water fed into the pond is based on a farmer's experience. It is described that the duration of the water intake should be only 10 min, and the black colour of the pond surface, which is caused by the wastewater, should be equal to one-fifth of the pond area. The black water feeding should be done only at noontime or on sunny days to prevent oxygen depletion in the pond. Farmers mentioned that if the black water is not taken in, fish yield will be low.

FISH FEEDING A supplementary feed is applied to the system to ensure high fish yield, often only duckweed and rice bran. The feeding rate varies according to the duration of fish culture. During the first two months of culture, supplementary feed is applied at a rate of 50–60 kg duckweed and 7–10 kg rice bran per day. During the last months, the feeding rate of duckweed is kept stable, but that of rice bran increases to 15 kg per day.

HARVEST Normally, the culture cycle is 6 months, and fish yield varies from 1500 to 2000 kg/2000 m^2 of pond area. Tilapia and silver carp account for 90% of the yield, with a survival rate of 80–90% and an average size of 300 and 800 g/fish, respectively.

Wastewater-based fish fingerling production

History

Fish farmers in the southwestern part of the city had set up a wastewater-based fish fingerling production system when the waters in the area became polluted. This type of system was developed during the late 1980s for providing grass carp and silver carp fingerlings to neighbouring provinces. During that period these herbivorous species were especially preferred for culture because intensive fingerling production was still poor and industrial feeds for growout of the fish were lacking.

The wastewater-based fish fingerling production system is applied to rear fish fry until they reach the size of fingerlings. Fingerlings of many fish species, such as tilapia, silver carp, grass carp and common carp, can be produced within the wastewater-based system by the same technique. At present, due to the increased market for common carp, most farmers are only producing this species. The common carp fingerlings produced can be used for growout, or as bait for carnivorous ornamental fish.

Major areas of production

This type of farming system is observed only in the southwestern part of the city.

Farming practices

POND PREPARATION Ponds are drained and lime applied, and the pond bottom dried under sunshine for one day. These actions kill all unexpected fish remaining in the ponds from previous culture batches which could scavenge the stocked fry. Ponds are then filled with wastewater and stabilized until the water becomes clear and light green (Fig. 6.6). A density of 50,000 fry/1000 m^2 is often applied at restocking.

MANAGEMENT Water management practices, from filling the pond to the interval of supplementation with wastewater into the system is the same as that of the fish growout system. Normally, a rearing cycle of 4 weeks is enough to produce fingerlings of commercial size. During the first 2 weeks fry live on natural food only. Thereafter, rice bran and sludge worms, *Tubifex tubifex*, are used as supplementary feeds. They are supplemented respectively at a daily rate of 1

Fig. 6.6. Wastewater-based fish fingerling production. The limnocorrals are used for stocking fingerlings newly harvested in the same pond.

and 1.5 kg/day by placing the feeds on trays which are hung under the water surface in different corners of the pond.

HARVEST With this rearing technique, a common carp fingerling yield of about 100 kg/1000 m^2 is achievable, consisting of three different class sizes, 700 fish/kg, which accounts for 80% of the fish, and 450–500 fish/kg and 280–300 fish/kg for the rest. The survival rate of fish in the system is as high as 80–85%.

PRODUCTION RISKS In comparison to the wastewater-based growout of fish, this type of system appears to be safer from threats of uncontrolled intrusion of polluted water owing to its shorter duration of culture, only 4–5 weeks.

It appeared that this production system would be of great potential for reusing part of the wastewater in HCMC, and farmers prefer this type of system to others for the following reasons:

- The duration of production is short and therefore risks are small.
- Fish fry lives on natural food and in case of lack of natural food, sludge worms (*Tubifex tubifex*) are locally available.
- Turnover of capital is quick.
- Fish fingerlings, especially of the preferred species, have a big market in Vietnam.

Wastewater-based tilapia fingerling production

History

Like other wastewater-fed fish culture systems, wastewater-based tilapia (*Oreochromis* spp.) fingerling production systems are concentrated mainly in the southwestern region of the city, occupying an estimated area of more than 100 ha.

In the mid-1960s, a private tilapia farm was established in Binh Phu, presently Subdistrict 10 of District 6, in the southwestern part of the city. This was the first tilapia farm in Vietnam. Tilapia were imported from South East Asian countries, possibly from Malaysia and Thailand, mainly *Oreochromis mossambiscus* in initial years, and called 'African gouramy'. After that, due to its better growth rate, the

Nile tilapia, *Oreochromis niloticus*, was imported and locally named 'Taiwanese African gouramy'. Widespread development of tilapia culture in the southern provinces of Vietnam has occurred since then, especially after 1965. After 1975, the mentioned farm was transformed into a state-run one, namely the Binh Phu Fish Farm. Under state ownership it focused mainly on silver carp and grass carp fingerlings, which were originally transferred from the north of Vietnam, because of the market preference at that moment, though it still produced tilapia fingerlings.

Due to being adjacent to the central districts of the city, the waterways of the catchment in which the Binh Phu Fish Farm is based became polluted at an early stage of the economic recovery of the city. Also, the Tan Hoa–Lo Gom canal, which was the water supply source of the fish farm, could not avoid this problem. It was so heavily polluted that fish culture in the area nearly collapsed. In 1990 the fish farm was sold to an urbanization project. Under pressure of the urbanization process and the severe pollution of water, the tilapia production activities were pushed completely off this original site to Districts 8 and Binh Chanh.

To solve the problem and to recover the farmers' income, the city government in 1990 launched a research project on wastewater reuse in fish culture (Tuyen *et al.*, 1992). By observing changes over time of the polluted water in the ponds, some farmers started learning how to stabilize the polluted water for reclamation. This helped scientists direct their experiments on waste stabilization pond technology. However, due to financial difficulties, the experiments achieved limited preliminary results. Recently, the tilapia market started to develop strongly again, and the city therefore continues to fund a study on the reclamation of polluted canal water with integration of tilapia fingerling production. The major focus is on the integration of environmental technology, aquaculture and fish quality, in terms of public health concerns. The concrete objective of the study is to assist farmers in improving their current wastewater-based tilapia culture systems.

Major areas of production

The major areas of the wastewater-based tilapia fingerling production presently are located in the southwestern part of the city, comprising District 6 (Subdistrict 10), District 8 (Subdistricts 6, 7 and 16) and the Binh Chanh District (Subdistricts Binh Hung and Phong Phu). Today, the most important farms are found in the Binh Chanh District because of the heavy urbanization of Districts 6 and 8.

Farming practices

MANAGEMENT Wastewater is taken into one pond, called a sedimentation pond, for stabilization. During this period, the water supply pipe is tightly closed in order to avoid unexpected inflow of wastewater during high tide. When the water becomes clear and light green and its strong smell is no longer recognized, it is transferred to other ponds, called receiving ponds, in which tilapia brood fish will then be stocked. However, recently this water transferring technique has been abandoned because it takes too much time and/or energy. To fill the receiving ponds by gravity, the water depth of the sedimentation pond has to be always higher than that of the receiving pond and, therefore, the filling process of the receiving ponds has to be undertaken gradually unless energy is expended for pumping. On the contrary, according to the recent technique, the reclamation of wastewater and fingerling production can be carried out in the same pond (single pond system). The polluted water is stabilized for days in one pond, and then the brood fish are stocked directly into it.

FISH STOCKING Tilapia brood fish are stocked with a density of 80–100 kg/1000 m^2 of pond area. Two or three weeks after stocking, the brood fish are seined out of the pond, but fish fingerlings are kept in for 1 or 2 more weeks. The spawning of brood fish takes place during their stay in the pond. Fish fry and fingerlings live on natural food only, and no supplementary feed is given during production.

HARVEST The production cycle, from the polluted water intake until the harvest of fish fingerlings, is normally 5–6 weeks. When harvesting, fingerlings are classified by means of big boxes which are made of steel screens with different mesh sizes, between 6 and 12 mm. An average fish fingerling yield of 80–100 kg/1000 m^2 was recorded.

PRODUCTION RISKS All wastewater-based aquaculture systems as described are very vulnerable because of the polluted water. Mass mortality of fish has often been caused by these risks, especially that of broodfish at the moment of stocking. The mortality of broodfish at the moment of stocking is a threat to investment capital. Therefore, although the system is very profitable, many farmers are still reluctant to invest in it.

Wastewater-based ornamental fish culture

History

Before 1975, besides the production of table fish, the southwestern part of the city, especially District 8, was an area of ornamental fish production. It provided ornamental fish to the whole of South Vietnam and for export. In recent years, as the economic situation improved and the lifestyle of residents became more urban, the market for ornamental fish expanded and is now being extended to the whole country.

Wastewater-based ornamental fish farming is a system that can be applied to culture many different ornamental species. However, at present it is only used for the production of Japanese ornamental carp (koi; *Cyprinus carpio*), golden carp (*Carassius carassius*) and species of the genera *Poecilia* and *Xiphophorus*, the preferred species in the market, especially in the south of Vietnam.

Major areas of production

Wastewater-based ornamental fish culture is being developed in the southwestern part of the city, especially in District 8, although it is not as popular as it used to be. Many farmers who produced ornamental fish in the past are now returning to this business.

Farming practices

The culture of Japanese ornamental carp and golden carp includes two stages, the rearing stage (from fry to fingerling size) and growout (from fingerling to adult/commercial size). All practices of the rearing stage, such as pond preparation, pond water management, stocking density of fish fry and feeding, are carried out similarly to the wastewater-based rearing of common carp. However, the growout of ornamental fish is different from that of common carp described above.

STOCKING After 4 weeks of the spawning stage, most of the fish fingerlings harvested are in the size class of 280–300 fish/kg. A survival rate of 80% of the initial stocking density has been recorded. It appears that the growth rate of these fish species is higher than that of the common carp under the same rearing conditions.

Fish fingerlings are selected on the basis of the market criteria of original tail and body patterns, and preferred colours. Only 40% of the harvested fingerlings can satisfy these criteria. The unsatisfactory fingerlings are sold as live feed for other carnivorous ornamental fish species. The satisfactory ones are restocked into the same pond for the growing-out stage. The fish density applied in this stage should be 1.5–2 fish/m^2.

FEEDING The growout stage lasts 3 months, during which fish are fed mainly with chicken feed, and with duckweed for the Japanese carp on an *ad libitum* basis, or without for the golden carp.

HARVEST The survival rate of fish is 80–90%. An average body weight of 300–400 g/fish, exceptionally 500 g, has been recorded for Japanese ornamental fish, and 100–150 g/fish for the golden carp. Fish in this stage grow rapidly due to the low densities and the *ad libitum* feeding.

After harvest, fish are not sold but transferred to cages for a natural coloration process. Both Japanese ornamental and golden carps are kept in cages for 2 weeks and fed chicken feeds of higher protein content. When cultured in cages, these ornamental fish show better colours. It is supposed to be due to the prevention of the fish from frequent exposures to the muddy bottom or dark waters during their scavenging on natural food, and due to feed with a higher protein content.

Culturing practice of *Poecilia* spp. and *Xiphophorus* spp.

Poecilia spp. and *Xiphophorus* spp. are viviparous species. They spawn naturally in ponds, and fry live on natural food. Therefore, the culture of these species is relatively simple. All practices for culturing them are the same as those of the fish-rearing system described previously. However, the supplementary feed is rice bran only. Species of the genera *Poecilia* and *Xiphophorus* are reared in small ponds of a few hundred square metres.

Fish can be harvested daily by hand net, and a 200 m^2 pond can yield 4000 fish of 8–12 cm length monthly. The *Poecilia* spp. and *Xiphophorus* spp. rearing system exploits as much as possible the water surface area. Although this type of rearing system only aims at serving the school children's fish culture hobbies during vacation time, it can provide a small extra income to farmers.

Economic Comparison of Wastewater-based Farming and Paddy Rice Farming

It is not easy to estimate exactly the economic efficiency of the currently applied wastewater-based farming systems in HCMC because the inputs to these systems are mainly family-based labour and on-farm materials. For this reason, the estimations given are based only on inputs from outside of the farms, and on a basic area of 1000 m^2 – the practical size appropriate for comparing economic efficiency among the wastewater-based farming systems in HCMC. The possible profit is estimated on a monthly basis per 1000 m^2 in order to compare their economic efficiency (the 2000 exchange rate of Vietnamese Dong (D$) used here is D$15,000 for US$1).

In addition, the economic efficiency of paddy rice farming, a traditional agricultural farming system, is also estimated to introduce a comparative figure. Table 6.4 provides a summary. Traditional paddy rice farming is different from wastewater-based farming systems, so the economic efficiency estimation of rice paddy farming is based on 1 ha and 3 months because these are the practical field size and cultivation duration, respectively, of this type of farming (Table 6.4). Wastewater-based integrated water-mimosa and duckweed farming differ from other aquatic macrophyte farming systems in that all inputs are based on family labour and use of on-farm materials. This is reflected in Table 6.4.

Economic Efficiencies among Wastewater-based Farming Systems

From Table 6.5 one can see that the economic efficiencies of currently applied wastewater-based farming systems in HCMC vary. The highest possible profit per month per 1000 m^2 is obtained with fish culture, especially tilapia fingerling production. Also, the integration of fish culture and aquatic macrophytes results in a higher monthly profit than the monoculture of the aquatic macrophyte itself. This is indicated clearly by the integrated water-mimosa and fish culture system in which fish are fed on an *ad libitum* basis with in-pond produced duckweed.

Among the farmed aquatic macrophytes, water-mimosa offers the highest possible profit per month per unit area. The lotus rhizome culture, on the other hand, results only in a medium profit. It is perhaps for this reason that several farmers in the southwestern part of HCMC try to convert their macrophyte systems into water-

Table 6.4. Economic efficiency of various wastewater farming systems. Based on year 2000 exchange rate of D$15,000=US$1.

Traditional paddy rice		
Cost of inputs per 3-month crop per ha:	Rice seeds	D$300,000
	Fertilizers & pesticides	D$600,000
Labour	For harvest	D$700,000
	For weed control	D$90,000
	For seeding	D$120,000
Total		D$1,540,000
Income per crop per ha:	4 t x D$2000/kg	D$8,000,000
Possible profit per crop	D$6,460,000	
Possible profit per month per 1000 m^2	D$215,300	
Wastewater-based water morning-glory		
Paddy field farming type		
Costs of input per 1000 m^2:	Fertilizers and pesticides	D$40,000
	Labour for harvest	D$120,000
Total		D$160,000
Income per monthly harvest:	1000 kg x D$600/kg	D$600,000
Possible profit per month per 1000 m^2	D$320,000	
Floating cluster farming type		
Cost of inputs per 1000 m^2:	Labour for harvest	D$120,000
Income per monthly harvest:	700 kg x D$600/kg	D$420,000
Possible profit per month per 1000 m^2	D$300,000	
Wastewater-based lotus monoculture		
Cost of inputs per 1000 m^2:	Labour	D$60,000
Income per monthly harvest:	200 kg x D$2500/kg	D$500,000
Possible profit per month		D$440,000
Wastewater-based integrated lotus and fish culture		
Cost of inputs per 1000 m^2:	Labour per month	D$60,000
Fish fingerlings per six months		D$120,000
Income per monthly harvest of lotus and per 6-month fish culture:		
From lotus rhizome:	150 kg x D$2500/kg	D$375,000
From fish for 6 months:	135 kg x D$7000/kg	D$945,000
Possible profit per month per 1000 m^2:	For lotus rhizomes	D$315,000
	For fish	D$137,500
Total		D$452,000
Wastewater-based integrated water-mimosa and duckweed		
Income/possible profit per monthly harvest per 1000 m^2		
For water-mimosa	400 kg x D$1500/kg	D$600,000
For duckweed	400 kg x D$200/kg	D$80,000
Total		D$680,000
Wastewater-based integrated water-mimosa and fish-culture		
Cost of inputs per 1000 m^2:	Fish fingerlings	D$140,000
Income per monthly harvest of water-mimosa and 6-month fish culture:		
From monthly harvest of water-mimosa:	400 kg x D$3000/kg	D$600,000
From fish for 6-month culture:	200 kg x D$7000/kg	D$1,400,000
Possible profit per month per 1000 m^2:	For water-mimosa	D$600,00
	For fish	D$230,000
Total		D$830,000
Wastewater-based fish fingerling production		
Cost of inputs per 1000 m^2:	Fish fingerlings	D$50,000
Income per 6-week batch production:	100 kg x D$12,000/kg	D$1,200,000
Possible profit per month per 1000 m^2		D$766,400
Wastewater-based tilapia fingerling production		
Income/production batch (6 wk)/1000 m^2	100 kg x D$14,000/kg	D$1,400,000
Income/possible profit per month per 1000 m^2		D$920,000

Table 6.5. Economic efficiency described as monthly profit per 1000 m^2 of current wastewater-based farming systems in HCMC compared to traditional paddy rice farming. Based on year 2000 exchange rate of D$15,000=US$1.

Type of (wastewater) farming system	Monthly profit per 1000 m^2
1. Traditional paddy rice farming	D$215,300
2. Wastewater-based water morning-glory farming	D$300,000–320,000
3. Wastewater-based lotus monoculture farming	D$440,000
4. Wastewater-based integrated lotus and fish culture farming	D$452,000
5. Wastewater-based integrated water-mimosa and duckweed farming	D$680,000
6. Wastewater-based integrated water-mimosa and fish culture farming	D$830,000
7. Wastewater-based fish fingerling production	D$766,400
8. Wastewater-based tilapia fingerling production	D$920,000

mimosa ones. However, the market demand for water-mimosa is not necessarily larger than that of the lotus rhizome. Many dishes that are prepared with the lotus rhizomes are much more preferred culturally.

It can also be deduced from Table 6.5 that all of the currently applied wastewater-based farming systems have higher monthly profits than traditional paddy rice cultivation. Table 6.5 also shows that farmers would get a higher income if they knew how to rotate their farming between water-mimosa during the rainy season and fish fingerling production during the dry season, introducing a new wastewater-based farming system on a year-round basis.

Conclusions

The present urbanization process without any water pollution control because of lack of financial means in HCMC forces farmers and their counterparts to explore and attempt different wastewater-based farming systems. Farmers have created a diversity of wastewater-based farming systems with monocultures as well as integrated farming systems of several aquatic macrophytes and fish species. The set up of each type of wastewater-based farming system originated from earlier experiences with traditional farming of the same type of crops without the use of wastewaters. Hence, this explains why wastewater-based aquatic macrophyte farming systems are concentrated mainly in the northern part of the city, whereas wastewater-based fish culture systems are located in the southwestern parts. Thanks to these farming practices, the livelihoods of farmers in the peri-urban areas which are receiving municipal wastewater discharges are not only safeguarded, but also improved. A preliminary assessment of economic efficiency shows that the income from the wastewater-based farming systems described before is higher than that of traditional paddy rice farming. The products of these systems presently are accepted as human food as well as animal feed. No public health incident has been reported so far.

Technically, the farming systems with aquatic macrophytes, except for the water-mimosa farming system, are based on direct and continuous wastewater feeding with limited care for water management. In contrast, fish culture systems are based on the use of stabilized wastewater which is very carefully fed into the ponds.

Although wastewater-based farming systems in HCMC are producing acceptable and profitable commodities, it is clear that more research is needed to improve their production efficiency and to apply them more scientifically into the wastewater reuse/treatment context. For instance, studies on the biology and ecology of cultivated aquatic plants in the described wastewater-based systems may be necessary for improving current farming practices. Studies on changes in the quality of the wastewater throughout the production process are needed to better understand the efficiency of pollutant and nutrient removal. Economic aspects of the systems should be assessed to get insight into the potential and prospects for these wastewater-based farming applications. Food safety of products is another important aspect of the described systems which must be looked at carefully, though these products today are readily accepted. It is not yet known whether or not toxicant bioaccumulation occurs in wastewater-grown vegetables or fish. These products are indeed directly exposed to wastewater, although for quite a short time, and added in small amounts (fish fingerlings, lotus rhizomes, young sections of water morning-glory) or produced in stabilized wastewater (water-mimosa, table fish).

Undoubtedly, current wastewater-based farming systems in HCMC are pioneering experiments that lay the foundation of sustainable water management in the city in the future. Since no space is left for wastewater treatment plants in urban areas, it is proposed to construct them in the peri-urban areas (Japan International Cooperation Agency–People's Committee of HCMC, 1999). In such a context, the integration of wastewater-based farming systems and wastewater treatment plants will be a socially useful and technically justified solution. Socially, it has been observed that in HCMC, urbanization projects, including projects for urban environmental infrastructure, have been pushing the farmers out of their land. Despite the fact that they receive valuable compensation for their land loss, it is not easy for them to cope with the urban lifestyle. Consequently, these farmers become unemployed shortly after their compensation is swallowed up by an unusual urban life. Bearing this in mind, many farmers in HCMC have therefore used their compensation money to buy or rent new land in the vicinity where they think the 'black water' will come along for continuing and protecting their livelihoods. Technically, the integration of wastewater-based farming systems and wastewater treatment plants will also be sound environmental solutions to further explore in the future. The development of such alternative agricultural/aquacultural practices are examples of potentially novel productive uses of treated, nutrient-rich wastewater in peri-urban areas (Costa-Pierce, 1998). However, the levels of integration and the efficiency of these levels are still unanswered questions (Edwards, 1990). The reuse of treated wastewater within different levels of integration of aquaculture systems has to be further studied in order to set up systems of high productivity and acceptable products in terms of food safety.

Acknowledgements

The authors are grateful to the farmers living in Subdistrict 10 of District 6; Subdistricts 6, 7 and 16 of District 8; Binh Hung, Phong Phu, and Binh Hung Hoa Subdistricts of Binh Chanh District; Subdistrict 10 of Go Vap District; Thanh Xuan Subdistrict of Hoc Mon District; and Tam Phu Subdistrict of Thu Duc District for providing the information and data described in this chapter. Special gratitude is also due to the Department of Science, Technology and Environment of HCMC for granting their survey and study from which the chapter is introduced. Sincere thanks are also expressed to Professor Peter Edwards, Asian Institute of Technology, for his encouragement in reviewing the current experience with wastewater-based farming systems in HCMC .

References

Costa-Pierce, B.A. (1998) Preliminary investigation of an integrated aquaculture–wetland ecosystem using tertiary-treated municipal wastewater in Los Angeles County, California. *Ecological Engineering* 10, 341–354.

Edwards, P. (1990) General discussion on wastewater-fed aquaculture. In: Edwards, P. and Pullin, R.S.V. (eds) *Wastewater-fed Aquaculture.* Proceedings of the International Seminar on Wastewater

Reclamation and Reuse for Aquaculture. Environmental Sanitation Information Center, Asian Institute of Technology, Bangkok, pp. 281–291.

Ho Chi Minh City Statistics Office (2002) *Statistical Yearbook Ho Chi Minh City*. People's Committee of Ho Chi Minh City, Ho Chi Minh City.

Japan International Co-operation Agency–People's Committee of HCMC (1999) *The Study on Urban Drainage and Sewerage System for Ho Chi Minh City in the Socialist Republic of Vietnam*. Main report. Pacific Consultants International, Ho Chi Minh City.

Project Management Unit 415 (PMU 415) (2002) *Tan Hoa-Lo Gom Canal Sanitation and Urban Upgrading Project: Choice of Lagoon Options for the Treatment of Den Canal in Subdistrict Binh Hung Hoa, Binh Chanh District*. Technical report. People's Committee of Ho Chi Minh City, Ho Chi Minh City.

Tchobanoglous, G. (1991) *Wastewater Engineering: Treatment, Disposal, Reuse*, 3rd Edition. Metcalf & Eddy Inc., Singapore.

Tuyen, N.V., Trinh, L. and Minh. P.V. (1992) *Preliminary Results of Waste Reuse in Aquaculture*. Technical report. HCMC Dept of Science, Technology and Environment, Ho Chi Minh City.

7 Wastewater Reuse through Urban Aquaculture in Hanoi, Vietnam: Status and Prospects

Vo Quy Hoan[1] and Peter Edwards[2]

[1]*Hanoi Agricultural University, Hanoi, Vietnam;* [2]*Asian Institute of Technology, Klong Luang, Thailand*

Abstract

Wastewater was collected in Hanoi by a combined wastewater and stormwater sewerage system and discharged without treatment to four rivers. Farmers, through experience accumulated over the past 40 years, have developed wastewater-fed aquaculture which has passed through three stages of development according to socio-economic changes in the country: (i) early development, (ii) formation of cooperatives and (iii) dissolution of cooperatives. Four farming systems were practised in the wastewater-fed production area: fish only, rice–fish, fish seed and vegetables. In the fish growout systems (fish only, rice–fish culture), a polyculture of Mozambique tilapia (*Oreochromis mossambicus*) and Nile tilapia (*O. niloticus*), two Indian major carps (rohu, *Labio rohita* and mrigal, *Cirrhinus mrigala*) and Chinese silver carp (*Hypophthalmichthys molitrix*), was practised in wastewater-fed earthen ponds. While tilapia had precocious and uncontrolled breeding, slow growth rates and subsequent small size at harvest, rohu and mrigal were the most popular species in farmed areas both with and without a wastewater supply. Total fish production in 2002 from wastewater and non wastewater-fed systems was 4350 t. Fish yields in wastewater-fed fishponds have risen to nearly 7 t/ha during a 10-month growing season. The total area of wastewater-fed aquaculture declined to 417 ha in 2002 from 751 ha in 1985. Privatization of land use rights has made it difficult to supply wastewater economically to rice–fish culture areas located far from the main wastewater-carrying river. However, the area of more profitable, wastewater-fed fish culture has doubled from 154 ha in 1985 to 332 ha in 2002. The Hanoi authorities appreciated wastewater-fed fish culture as they included it in the Master Plan of Hanoi City Development and in the Master Plan for Fisheries.

Introduction

The growing discharge of wastewater from cities has become a major concern for developing countries, and the concept of wastewater reuse rather than simple waste disposal has received widespread attention. Wastewater-fed aquaculture is an important method to promote technical and economic efficiency in the allocation and use of resources. It has been tested and implemented commercially both in developing countries, such as China, India, Indonesia and Vietnam, and in developed countries, such as Germany (Edwards and Pullin, 1990; Edwards, 1992).

(eds B. Costa-Pierce, A. Desbonnet, P. Edwards and D. Baker)

Hanoi is the capital city of Vietnam, with a total area of 927 km^2 and an estimated population of 2.7 million people – 1.5 million in the central urban area and 0.7 million involved in agriculture and aquaculture in the peri-urban area – in 2002. Faced with a dynamic and changing city because of a high economic growth rate and rapid urbanization, Hanoi government authorities have initiated a Master Plan of Hanoi City Development (1993) and a Master Plan for Water Drainage System in Hanoi to 2010 (Quy, 1995). Wastewater-fed aquaculture was recognized in both plans, as well as in the Master Plan for Fisheries to Year 2010, a project of the Ministry of Fisheries (1998).

As only preliminary descriptions existed on the development, status and potential of wastewater reuse through aquaculture, which has been practised in Hanoi since the early 1960s, a field study was carried out to describe its historical development, to evaluate its current status and to assess its prospects in Thanh Tri District, the main wastewater reuse area of the city.

Methodology

The study was carried out in 1996 (Vo, 1996), and updated subsequently by area and productivity of culture systems to 2002 in the Thanh Tri District, Hanoi. The initial survey was of an exploratory type, and followed the methodology recommended by Weber and Tiwari (1992). A combination of personal observations and a standardized questionnaire was used to obtain primary data.

A stratified random sampling technique (Miah, 1993) was used to ensure representative sampling and precision. Almost all farms with wastewater reuse were interviewed. Each farm was a sampling unit. The head of each farm was interviewed, with a questionnaire that contained both closed and open-ended questions. Farm household survey data were organized for entry in a DBASE computer database package using methods recommended by Gonzales and Tolentino (1993). The Statistical Package for Social Sciences (SPSS/PC+) was used for statistical and economic analysis.

Wastewater Discharge and Quality

Wastewater disposal system in Hanoi and Thanh Tri District

The first 75 km of sewers and open channels were constructed in Hanoi from 1873 to 1915. From 1954 to 1996, another 70 km of sewers were constructed, making a total sewer length of 145 km. However, only approximately 28% of the total area of Hanoi was sewered.

Both wastewater and stormwater were collected in combined sewers which discharged into open channels, and then into four rivers (Table 7.1). The sewage flowed without adequate treatment into Thanh Tri District, which is located in a wetlands area to the southeast of the city, and between two large rivers – the Red River to the east, which was contained within a flood protection dyke, and the Nhue River to the west (Fig. 7.1).

Wastewater flowed either directly through Thanh Tri District in the four draining rivers, or was pumped through farming areas, after which it flowed back into the rivers. Discharge water drained either into the Yen So regulating reservoir and was then pumped through the Yen So pumping station into the Red River (at a flow rate of 90 m^3/s), or it drained through the regulatory Thanh Liet dam to the To Lich and Nam Dong Rivers and was discharged finally into the Nhue River. In the dry season, the rivers functioned like sedimentation tanks. According to the Master Plan on Urban Drainage and Wastewater Disposal System in Hanoi (JICA, 1994), these rivers were to be reconstructed as concrete canals to increase the velocity of wastewater flow.

Wastewater Characteristics

There were three treatment plants in Hanoi, but they were not in operation

Table 7.1. Physical features of the four main rivers draining Hanoi.

Rivers	Length (km)	Width (m)	Depth (m)	Urban catchment area (km^2)	Basin area (ha)	Average wastewater volume (1000 m^3/day)	Flow capacity (m^3/s)
Kim Nguu	12.2	25–30	3–4	5.24	1448	85–100	8–45
Set	6.7	10–30	3–4	6.65	658	60–65	5–7
To Lich	17.8	30–45	3–4	28.61	270	100–120	10–170
Nam Dong	5.8	20–30	2–2.5	6.60	354	45–50	3–12
Total	52.5	–	–	47.1	2730	290–335	26–234

Sources: Vo *et al.* (1990); Pham and Tran (1994); JICA (1993); Dang (1995).

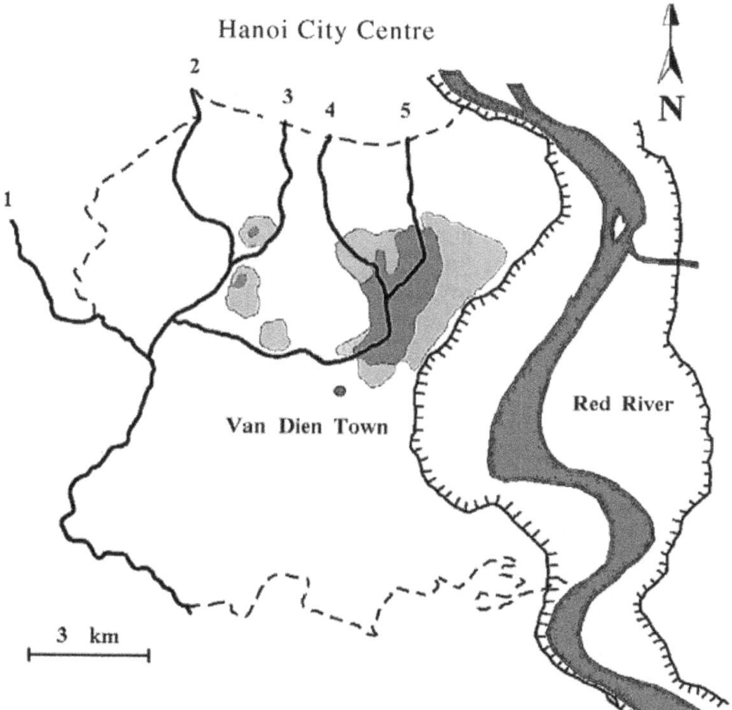

Fig. 7.1. Wastewater-fed areas in Thanh Tri District indicated by shaded areas in 1985, with the areas to which it was reduced in 2002 with darker shading. Source: Thanh Tri District Office. 1, Nhue River; 2, To Lich River; 3, Nam Dong River; 4, Set River; 5, Kim Nguu River.

because of the high operational cost and a lack of funds for adequate maintenance. Until 1998 almost all factories, hospitals and laboratories discharged wastewater to the rivers without any treatment. Following a new decree of the city authority in 2002, all factories, hospitals and laboratories should – theoretically – have had their own treatment plants to at least ensure treatment of the most dangerous wastes; however, there was no systematic control of their operation.

A study carried out by the Center for Water Resources Development and

Environment estimated that 75–80% of the wastewater was of domestic origin and 20–25% of industrial origin (Nguyen, 2001). Wastewater was heavily polluted, with a strong anaerobic smell, and was black or brown in colour. The averaged BOD_5 concentration was 100 mg/l (range from 30 to 150 mg/l) and the COD 173 mg/l. The COD/BOD_5 ratio of 1.7 indicated a relatively low industrial wastewater content of the overall wastewater.

Wastewater-fed Aquaculture Techniques

Historical development of wastewater-fed aquaculture in Thanh Tri District

Wastewater-fed aquaculture in Thanh Tri District has changed with time because of socio-economic factors, as well as improvements in technology during the long and unstable economic development of the country. Its development may be divided into three periods: (i) early development, (ii) formation of cooperatives and (iii) dissolution of cooperatives.

Early development period

In Vietnam the land market was well developed until the 1950s, when all land became state property and transactions in land were prohibited. During the 1960s, an extensive system of drainage and irrigation was constructed, and is considered one of the major successes of the Cooperativization Movement launched during that time period. A series of newly constructed pumping stations supplied wastewater to thousands of hectares of rice fields, and fish began to be cultured in combination with rice. During this period, farmers started to perceive the benefits of wastewater-fed aquaculture not only to satisfy the family's need for fish, but also to sell production surplus for domestic consumption. Fish culture became an obvious new venture to diversify the traditional rice-only farming system.

Formation of the cooperatives period

This period began in 1967 when all farm land was used communally, except for 5% of the total cultivated area, which was farmed by individual households. The cooperatives stabilized rice–fish culture in rice fields irrigated with wastewater, and in wastewater-fed, fish-only culture, in often flooded, low-lying land where rice could not be cultivated year round. Wastewater-fed aquaculture became one of the main livelihoods of communes, or cooperatives, located in the basin of the Kim Nguu River, and which had access to wastewater (e.g. Yen So, Thinh Liet and Thanh Mai). Fish culture later spread to another ten cooperatives in the district, but on a small scale in terms of their farmed area. The total area of rice–fish and fish-only production systems increased to nearly 900 ha by 1979. Fish production from Thanh Tri attained yields of more than 1000 t/year, with an average fish yield of 2.0–2.5 t/ha, and with a maximum of 3.0–3.5 t/ha.

Seed for fish culture in the district relied, in the beginning, on wild seed of several indigenous species (*Channa striata*, *Cirrhina molitorella*, *Ctenopharyngodon idella* and *Cyprinus carpio*) harvested from the Red River. Several fish species were introduced later to improve fish production: Mozambique tilapia (*O. mossambicus*) and Nile tilapia (*O. niloticus*) in 1963 and 1973, respectively; Chinese silver carp (*Hypophthalmichthys molitrix*) in 1964 and two Indian major carps (mrigal, *Cirrhinus mrigala;* and rohu, *Labio rohita*) in the late 1980s. The two Indian carps were the principal cultured species in wastewater-fed ponds at the time of the study as farmers had by then lost interest in tilapias, which had poor growth performance, uncontrolled recruitment and small size at harvest.

All farming activities were planned and directed by each cooperative, with households being remunerated according to registered working time. Lack of incentives to farmers led to a progressive decline in the efficiency of land use, culminating in food crises in the late 1970s.

Dissolution of the cooperatives period (or land reallocation period)

From the early 1980s land use was changed from cooperative to individual household management. The majority of cooperative land was shared among individual households for their use based on the number of household members. The rest of the land, about 10% of the total farmed area, was retained as cooperative land for private use through a bidding process, to defray public expenses, and to periodically readjust land allocation due to population growth. Cooperative land included large water bodies which were difficult to divide among households, and which required skill and/or large investments to be used efficiently. Cooperative land was allocated competitively to individuals or groups of farmers bidding the highest, for a period of five years. Farmers of medium to high financial status obtained access to open cooperative lands. However, lack of security over land tenure led to inadequate farm investments. This was rectified in 1988 with the introduction of the household contract system, in which land was allocated to farm households for 20 years. The privatization of land was accompanied by the end of the cooperative subsidies in wastewater distribution to fishponds. As each fish farm had to now arrange by itself to obtain wastewater for the fishpond, the common supply system did not work properly. Obtaining wastewater now became more complex and expensive for those fish farms located far from the river. As a consequence, many traditional wastewater distribution channels fell into disuse and deteriorated (Figs 7.2 and 7.3). However, those fishponds located along the rivers could still easily obtain a supply of wastewater, and became larger over time. In 2002 there were several large wastewater-fed fishponds, with the largest pond 18 ha in area.

With rapid changes in the economy as it moved to an open-market system, farmers took more initiative in land use because of economic incentives, were more efficient and benefited by applying better farm management techniques to increase fish yields. Total fish production has been increasing rapidly from about 1800 t with an average yield of about 2.5 t/ha in the late 1980s, to more than 3000 t and 3 t/ha in 1995. In 2002, total fish production was about 4350 t, and the fish yield in wastewater-fed fishponds had risen to a maximum of 7.5 t/ha, with an average yield of 4.2 t/ha.

Fig. 7.2. A disused wastewater pumping station.

Fig. 7.3. A disused wastewater distribution channel with fish-only production system on each side.

Production systems

According to the different types of land use, fish farming systems were differentiated into three types: fish only (production system I), rice–fish (production system II) and fish seed (production system III) (Table 7.1). This classification of farming systems supported those described by Pham and Vo (1990), Vu (1994), Dalsgaard (1996) and Edwards (1996). The areas of the different aquaculture systems in 2002 are presented in Table 7.1. The total area of wastewater-fed aquaculture declined to 417 ha in 2002 from 482 ha in 1995 and 751 ha in 1985.

Fish only: production system I

Only food fish were produced in this system, which was conducted mainly in large water bodies such as lakes or reservoirs, and low lying land which often flooded and was not suitable for rice cultivation. As it usually used wastewater only, this system occurred mainly in communes with a long tradition of fish farming and wastewater reuse, such as Thinh Liet, Tran Phu and Yen So.

The fish-only system was expanded into newly developed aquaculture areas in the district without a wastewater supply, such as in Dai Kim, Tam Hiep, Ngu Hiep and communes. Without wastewater, farmers had to fertilize the fish pond and feed the fish with other inputs, such as with supplementary feeds, either on-farm by-products or feeds bought from off-farm sources.

Nightsoil was bought from off-farm sources and used widely for all three production systems. Livestock manure was also used but in smaller quantities because of its lower nutrient content. The use of inorganic fertilizer was not observed in any production system.

Fish feeds were mainly farm by-products such as rice bran, broken rice of low quality, vegetable residues and assorted residues from cooking and food processing factories such as brewery waste, and wastes from the manufacture of noodles and fish sauce. Green fodder such as duckweed, water spinach and water hyacinth were used widely in ponds where grass carp were stocked. Pelleted feeds were not observed at all in 1995, but by 2002 had become very important feed inputs for newly intensified fish ponds.

Fertilization and feeding rates varied widely depending on the availability of inputs, the farmer's cash availability and his contacts – not everyone had the opportunity to buy feeds such as brewery waste at the low prices which made feeding profitable. Therefore, fertilization and feeding rates were based not only on technical principles, but more so on social and economic conditions as well as on the farmers' experience.

Farm areas ranged from 0.36 to 29.84 ha, with a mean of 7.1 ha. The major risk to operations was flooding. Farms required high inputs, but system management was less complex than for the rice–fish system. Most of the individual cooperative lands adopted this system.

Rice–fish: production system II

The rice–fish production system started with a combination of rice cultivation and fish culture at the beginning of the annual production season in early March, but was fish culture only after the rice harvest at the end of June to the end of the fish growing season in January of the following year.

Rice–fish systems are the oldest fish culture systems in the district (Pham and Vo, 1990; Vu 1994; Dalsgaard, 1996; Edwards, 1996). These systems were still common in 2002, but mainly on cooperative land owned by groups of farmers (Fig. 7.4). The farm area was generally larger than cooperative land owned by individuals. The mean farm size was 27.1 ha, with a range of 1.4–129.6 ha. The major risk to operations was flooding, which seriously affected production in 1994.

Rice–fish systems were popular for several reasons; they were suitable for lands which lacked a wastewater supply for the whole year, particularly areas far from rivers that supplied wastewater. Other advantages were that the systems created more job opportunities for farmers, since rice and fish were cultured in the same unit of farmed area. They also brought more income at lower risk than the fish-only system. The combination of the two different crops also benefited each other, with fish growing faster and rice having a higher yield and with lower inputs than for monocultures.

The main disadvantage of the system was that farm management was more complex, since there was separate ownership for both the fish culture and the rice cultivation operations. Farm owners needed to share interest and care in resource use for two different commodities raised in the same area. Therefore, these systems were suitable mainly for cooperative lands held by groups of farmers, which also enabled farmers of low economic status to adopt them.

Fig. 7.4. Rice–fish production system with wastewater canal in foreground.

Fish seed: production system III

This production system occurred in smaller areas, with a mean farm size of 0.48 ha and a range of 0.1–2.9 ha, and was operated by both individual households and by cooperatives (Fig. 7.5). Since fish seed production was almost entirely for the internal and planned demand of the district, any surplus production might have no market. Therefore, a major constraint of this system was market limitation, and thus it remained small in scale and with simple technology, even though it had the potential to generate high income and needed only low capital investment and relatively simple management. Most seed farms were located far from the main wastewater-carrying rivers and therefore used little to no wastewater. Farmers were concerned about using wastewater in nursery ponds because fry were sensitive to variable wastewater quality, and instead used alternative fertilizers and supplementary feed.

Changes in area

There have been significant changes in the relative areas of the different types of production systems with time (Table 7.2). The total aquaculture area in Thanh Tri District has increased only slightly from 936 to 1019 ha during period 1985–2002, but the total area of wastewater-fed aquaculture fell by nearly 50% from 751 to 417 ha, excluding a relatively small area of wastewater-fed seed production. The largest decline has been in wastewater-fed rice–fish culture, from 597 to 85 ha, because of abandoned wastewater pumping stations far from the main wastewater-carrying rivers (Fig. 7.2). Pumping costs were subsidized during the cooperatives period, but would be too high to justify rice–fish culture. However, there has been a recent doubling of the area of more profitable wastewater-fed fish-only culture, from 154 ha in 1985 to 332 ha in 2002, largely from the conversion of former rice–fish culture areas.

Pond characteristics and management

Ponds were earthen and rectangular or irregular in shape. Similar simple wastewater-fed pond construction occurred in China (Zhang, 1990) and India (Ghosh, 1990), in which adequate mixing of wastewater with pond water and degradation of

Fig. 7.5. Fish seed production system.

Table 7.2. Aquaculture area (ha) in Thanh Tri by year.

	Year			
Production systems	1985	1990	1995	2002
Wastewater-fed				
Fish only	154	142	166	332
Rice–fish	597	332	316	85
Subtotal	751	474	482	417
Not wastewater-fed				
Fish only	34	183	354	370
Rice–fish	71	247	205	142
Subtotal	105	430	559	512
Seed	80	80	80	90
Total	936	984	1121	1019

sewage also occurred. As ponds in production systems I and II were formed usually from the original low-lying flooded land or rice fields, most of them were irregular in shape. Rectangular ponds were found mainly in production system III, since generally they were dug ponds. The main characteristics of fish ponds in each production system are presented in Table 7.3.

Farm management was more complex in large rice–fish farms with cooperative land organized by groups of farmers, and in small area fingerling farms comprising several ponds. In contrast, large farms owned by individual farmers were the fish-only system, and comprised only a single pond, similar to some of the larger Calcutta wastewater-fed fish farms (Ghosh, 1990; see Chapter 5 of this volume).

Pond size was determined by the production system and the land available for its development. There was no significant difference between the rice–fish and the fish-only production systems in terms of pond size, but wastewater-fed fishponds were larger than non wastewater-fed ponds, similar to wastewater-fed ponds in China. However, while large fishponds of more than 10 ha might be difficult to harvest, they would have more buffering capacity against shock loads of wastewater influent (Wang, 1987; Zhang, 1990). It has been proposed that the way to increase productivity is by utilizing smaller (range from 0.5 to several ha) but better managed ponds despite the loss of water surface area from a larger area of pond dykes (Ghosh, 1990; Mara *et al.*, 1993).

Table 7.3. Pond characteristics in different aquaculture production systems in 1995.

		Wastewater-fed pond			Non-wastewater-fed pond		
	Ponds	Size (ha)		Depth (m)	Size (ha)		Depth (m)
Production system	per farm	Mean	Range	Range	Mean	Range	Range
Fish only	1	10.63[b III]	3.60–29.84	1.5–3.0	3.57[a II]	0.36–11.88	1.5–3.0
Rice–fish	2	8.94[b III]	4.86–32.4	1.2–3.5	4.82[a II]	1.44–60.88*	1.2–3.5
Fish seed	3	0.20[b II]	0.18–0.39	1.5–2.0	0.16[a I]	0.07–0.36	1.5–2.0

[a, b]Indicate significant difference between means of wastewater-fed and non-wastewater-fed systems (P<0.05); [I, II, III]indicate significant difference between means of production systems (P<0.05).
*Only one case observed.

The depth of the ponds varied widely from 1.5 to 3.5 m in the main area of growout ponds, which covered 90% of the total wastewater-fed area. But in the fry-rearing ponds, it ranged from only 1.5 to 2.0 m, similar to wastewater-fed fishponds in India and China. In growout ponds, 10–15% of the pond comprised an area with a depth of 3.5 m when the pond was completely full. Complete filling was used to hold fish in rice–fish ponds before the rice harvest and to facilitate fish harvest without completely draining the whole pond. Furthermore, deep areas were also for farmers to pump wastewater into the pond, and thus served almost like anaerobic and facultative stabilization ponds, having depths of 2–5 m (Edwards, 1992).

The schedule for pond preparation and maintenance depended on the scale, growing period and harvesting strategies of each production system. In the large areas of production systems I and II, pond preparation was carried out once a year after the last fish harvest when the ponds were dried completely. The main activities were repair of pond dykes, sun drying of the pond bottom and liming, together with the maintenance of the water supply and drainage channels, since the main risk of these systems was flooding.

Pond sizes of fingerling-rearing ponds were smaller, and ranged from several hundred to several thousand square metres. They required special pond preparation, such as complete drying and treatment, several times a year, after each culture period to control fish disease and predators.

Stocking and harvesting strategies

The common fish growout cycle was from March to December, about 10 months, but fish grew well, mainly during the hot season from late March to early November, about 8 months a year. The first stocking of fingerlings was generally in March at the beginning of the hot season (Fig. 7.6; Pham and Vo, 1990). Stocking and harvesting strategies are presented in Table 7.4.

Multiple stocking and harvesting were typical for the large growout fishpond systems, as reported from India (Ghosh, 1990) and Taiwan (Edwards, 1990). It ranged from 3 to 5 times per year, with a mean of 3 times per year in wastewater-fed fishponds, but was less (2–4 times) in non wastewater-fed fishponds. About 50% of the total fingerlings to be stocked during the growing season were stocked at the first stocking, and the other 50% were stocked later, after each partial harvest (Fig. 7.7).

Fig. 7.6. Releasing fingerlings in a growout fish pond.

Table 7.4. Stocking and harvesting strategies in different production systems in 1995.

Production system	Wastewater supply	Stocking density (number/m²)	Number of times fish stocked per growing period: Mean	Variation range	Harvesting times
Fish only	+	1.99[a]	3.0	3–5	3–5
	–	1.28[b]	3.0	2–4	2–4
Rice–fish	+	2.49[a]	2.5	2–3	2–3
	–	0.76[b]	2.2	2–3	2–3
Fish seed	+	188.3[a]	1.0	1	1
	–	229.9[b]	1.0	1	1

[a, b]Indicate significant differences between means ($P<0.05$).
+, – indicate with and without wastewater use.

Fig. 7.7. Woman transporting fingerlings to a growout pond.

In production system II, fish were stocked at first only in the deeper water area, because in the shallow part there was newly planted rice. Fish were released into the whole pond area after 1 month, when rice had become well established.

In production system III, there was a single stock and a single harvest only as small ponds enabled the farmers to drain them every 3 or 4 months to harvest fish, and to quickly start a new cycle.

Stocking sizes of the fish varied from small (150 tilapia; 35 Indian major carps and silver carp/kg of fingerlings) to large (4–20/kg of fingerlings for other Chinese carps) in production systems I and II.

Size of fish at harvest was relatively small (200–250 g for Indian major carps and silver carp; 300–500 g for common carp and grass carp) and very small for tilapia (only 60–80 g). Small size of fish at harvest was due mainly to harvesting 3 months after stocking; but more frequent harvesting may have increased fish yields (Edwards, 1990; Ghosh, 1990). A single stock and single harvest strategy in smaller growout ponds may achieve a higher fish yield of approximately two to three times that currently achieved in wastewater-fed fishponds (Mara *et al.*, 1993).

Stocking densities were always higher in systems fed with wastewater. In the growout ponds of production system I fed with wastewater, nearly 2–2.5 fish/m^2 of pond water surface were stocked, similar to those in Taiwan (Edwards, 1990) and India (Ghosh, 1990), compared with 1 fish/m^2 in ponds without a wastewater supply.

The main fish species cultured were the Indian major carps, rohu and mrigal, at nearly 35%, followed by tilapia (*O. mossambicus* and/or *O. niloticus*) and silver carp at about 30% each of the total fish stocked in the wastewater-fed fishponds. Similar observations were reported by Ghosh (1990) in India and by Wang (1987) and Zhang (1990) in China.

The two Indian major carps were the most popular species in wastewater-fed fishponds. Tilapia used to be the main cultured species at 50% of the total according to Pham and Vo (1990), and Vu (1994), but then was excluded from low or non wastewater-fed systems because of precocious and uncontrollable breeding, slow growth, small size at harvest, while taking up space and food and inhibiting the growth of carps. Silver carp was also a popular fish in the wastewater-fed fish ponds, but was cultured less than Indian major carps. Common carp (4%) and mud carp (1%) were also cultured; but grass carp was stocked only in non-wastewater-fed fish ponds.

Wastewater reuse methods

Raw wastewater was pumped directly from the rivers draining the wastewater from Hanoi, or through the canal system into which wastewater from the river was pumped into the deeper part (>3 m) of the fishponds. Meanwhile, pond water was drained by gravity through a single pond outlet. The first introduction of wastewater following pond preparation filled the pond with wastewater, which was left to stand for 10–14 days until the black-coloured wastewater became green and suitable for stocking fish. Natural purification occurred during this period with the development of plankton and benthos, which are natural foods for fish (Edwards, 1990; Ghosh, 1990).

Farmers had developed skills in recognizing the nutrient content of the wastewater, and regulated its volume and frequency of introduction into the pond. Volumes were larger during the hot summer months, but were reduced in winter. In large growout ponds, wastewater was pumped daily with rates varying from 5 to 15 h/day, but in small ponds (<1 ha in production system III), wastewater was pumped for only a few hours (3–4 h) every 2–3 days. Farmers observed the wastewater before and during the pumping process, but there were still cases of fish mortality – mainly recently stocked fingerlings – after pumping wastewater into the ponds.

The nutrient content of wastewater was based mainly on visual observation, such as water colour and transparency. Good wastewater for fishponds was black in colour. Strange or multi-coloured water indicated wastewater drained from textile factories or other industries, and was dangerous for fish.

Productivity

Fish production was higher in the three wastewater-fed production systems than in non-wastewater-fed systems (Table 7.5). The maximum fish yield in the wastewater-fed fish-only production system increased from 6.89 t in 1995 to 7.50 t/ha per growing season in 2002.

Similar yields were reported by Edwards (1996); but Pham and Vo (1990) reported lower yields, probably because of poor farm management in the past. Current fish yields were similar to those obtained in wastewater-fed fishponds in China (Wang, 1987; Zhang, 1990). Fish yields were lower in all three systems without a wastewater supply, probably because of the high nutrient content of water in wastewater-fed fishponds, and the low use of

Table 7.5. Mean fish yields (t/ha during a 10-month growing season) in different fish production systems from the farm survey in 1995.

	Wastewater-fed		Non-wastewater-fed	
Production system	Mean	Range	Mean	Range
Fish only	5.56[b II]	3.43–6.89	4.09[a II]	3.10–5.56
Rice–fish	4.74[b I]	4.16–5.86	2.64[a I]	1.59–3.51
Fish seed	7.65[b]	7.39–7.98	6.45[a]	4.24–8.61

[a, b] Indicate significant differences between means from wastewater-fed and non-wastewater-fed systems (P<0.05); [I, II] indicate significant differences between means from fish only and rice–fish systems (P<0.05).

fertilizers and supplementary feeds in fishponds without a wastewater supply.

Higher fish yields were observed in the field survey than in official district statistics, perhaps because farmers reported lower yields than those actually obtained to minimize the real value of the farm, and more recently the lease fees.

Social Analysis

Multiple social benefits of wastewater-fed aquaculture were observed. The system treated wastewater from the city and at the same time produced fish, provided income and employment, improved living standards of farming families and provided an opportunity for women to participate in the production process (Pham and Vo, 1990; Nguyen, 1995; Dalsgaard, 1996; Edwards, 1996).

Aquaculture production was one of the main production activities of Thanh Tri District, with about 19% of the total farmed area of the district occupied by aquaculture, from which 4350 t of fish, more than 10% of the total fish supply for Hanoi city, was produced in 2002. In the Master Plan for City Development (Hanoi People's Committee, 1993) aquaculture was to be retained in Thanh Tri District, and more recently it was planned to not only improve production in existing areas, but also to extend aquaculture to new farming areas. New aquaculture production areas were developed from rice-only production in 11 communes from 1996 to 1999, including using wastewater, although the area of wastewater-fed fishponds had only doubled from 1985 to 2002, compared to an increase in area of more than ten times for non-wastewater-fed fish culture.

Although only 1.1% (480 labourers) of the total workforce of the district were specialized workers in the operation of the district's total of 172 fish farms, more than 1000 labourers were involved either full or part time in aquaculture-related operations (with about 2.5 labourers for each specialized worker).

Aquaculture also improved the situation of women through new job opportunities and better working conditions. Traditionally, in agricultural production, women carried out heavier labour and more household activities than men (BINNIE *et al.*, 1993). According to the field survey, women were directly and actively involved in rice–fish, fish seed and vegetable production systems, which were more profitable and required less hard work than the rice-only production system. Furthermore, household income increased through aquaculture (Nguyen, 1995) and family nutrition improved, which would also benefit women and children. The wastewater-fed aquaculture systems also provided suitable employment for women in fish marketing, in which women dominated.

Conclusions

Wastewater reuse in aquaculture in Hanoi has been practised for several decades and has become an important production activity for a considerable number of people who live in the drainage areas of the city in Thanh Tri District. Wastewater-fed aquaculture, developed mainly by farmers through experiences accumulated over the past 40 years, was divided into three main periods of development according to socio-economic changes: (i) early development, (ii) formation of cooperatives and (iii) dissolution of cooperatives. The systems provide economic efficiency and job opportunities for farmers. Higher incomes from aquaculture when compared to rice monoculture undoubtedly improved family nutrition, and therefore also benefited women and children, who were most affected by the lower economic status of their households.

According to the Master Plan of Hanoi City Development (Hanoi People's Committee, 1993), Master Plan for Water Drainage System in Hanoi to 2010 (Quy, 1995) and Master Plan for Fisheries to Year 2010 (Institute for Fisheries Economics and Planning, 1998), Thanh Tri District retained the same land use pattern of agriculture and aquaculture. Planners gave a high priority for aquaculture development in the future development plans of both district and city authority levels, indicating their belief in the efficiency of the systems in resource recovery, as well as wastewater treatment.

Major constraints for wastewater-fed aquaculture were the greater complexity and costs of the wastewater supply under private management, so that almost all rice–fish culture located relatively far from the rivers could not obtain wastewaters, causing the total area of wastewater-fed aquaculture to decline from 751 ha in 1985 to 417 ha in 2002. However, there has been a doubling of the area of wastewater-fed fish-only culture from 154 to 332 ha. Techniques and skills developed by farmers were still far from optimal wastewater reuse (Mara *et al.*, 1993). Further studies and development of the systems should focus on improving the effectiveness of wastewater-fed fishpond systems in terms of wastewater treatment, wastewater reuse through aquaculture, rice and vegetable culture, and minimizing potential health risks.

References

BINNIE, SMEC, AACM, DELFT (1993) *Red River Delta, Draft Master Plan.* Ministry of Science, Technology and Development, UNDP/World Bank.

Dalsgaard, A. (1996) Wastewater-fed aquaculture in Vietnam. *Mekong Fish Catch and Culture* 1, 6–7.

Dang, T.S. (1995) The aquatic ecosystem and the role of wastewater fishponds in diminishing the potential pollution into water environment in Hanoi. In: Le, T.C. and Terry, R.A. (eds) *Some Problems of the Human Ecosystem in Vietnam.* Agriculture Publishing House, Hanoi, pp. 134–152 (in Vietnamese).

Edwards, P. (1990) An alternative excreta-reuse strategy for aquaculture: the production of high-protein animal feed. In: Edwards, P. and Pullin, R.S.V. (eds) *Wastewater-fed Aquaculture. Proceedings of the International Seminar on Wastewater Reclamation and Reuse for Aquaculture*, Calcutta, India, 6–9 December 1988. Asian Institute of Technology, Bangkok, pp. 209–221.

Edwards, P. (1992) Reuse of human waste in aquaculture: a technical review. *Water and Sanitation Report No. 2.* The World Bank, Washington, DC.

Edwards, P. (1996) Wastewater reuse in aquaculture: socially and environmentally appropriate wastewater treatment for Vietnam. *Naga* 19, 36–37.

Edwards, P. and Pullin, R.S.V. (1990) Wastewater-fed aquaculture. *Proceedings of the International Seminar on Wastewater Reclamation and Reuse for Aquaculture*, Calcutta, India, 6–9 December 1988. Asian Institute of Technology, Bangkok.

Ghosh, D. (1990) Wastewater-fed aquaculture in the wetlands of Calcutta – an overview. In: Edwards, P. and Pullin, R.S.V. (eds) *Wastewater-fed Aquaculture. Proceedings of the International Seminar on Wastewater Reclamation and Reuse for Aquaculture*, Calcutta, India, 6–9 December 1988. Asian Institute of Technology, Bangkok, pp. 49–56.

Gonzales, R.L. and Tolentino, J.M. (1993) *Data Analysis with SPSS/PC+4.0.* Programs in Computer Application Development, Asian Institute of Technology, Bangkok.

Hanoi People's Committee (1993) *Master Plan of Hanoi City Development to Year 2010.* Hanoi People's Committee, Hanoi.
Institute for Fisheries Economics and Planning (1998) *Master Plan for Fisheries to Year 2010.* Institute for Fisheries Economics and Planning, Hanoi.
JICA (1993) *The Study on Urban Drainage and Wastewater Disposal in Hanoi City, Inception Report.* JICA, Hanoi.
JICA (1994) *Master Plan on Urban Drainage and Wastewater Disposal System in Hanoi City, Main Report.* JICA, Hanoi.
Mara, D.D., Edwards, P., Clark, D. and Mill, S.W. (1993) A rational approach to the design of wastewater-fed fishponds. *Water Resource* 27, 1797–1799.
Miah, M.A.Q. (1993) *Applied Statistics.* A Course Handbook for Human Settlements Planning. Asian Institute of Technology, Bangkok.
Nguyen, T.T. (1995) Wastewater reuse through aquaculture and agriculture in Thanh Tri District: resources recovering and its influences. *Khuyen Nong* 3, 8.
Nguyen, Q.T. (2001) Impact of wastewater on water quality in irrigation system and treatment measures to reduce pollution: a case study in Nhue irrigation system. Working Paper 30. *Wastewater Reuse in Agriculture in Vietnam: Water Management, Environment and Human Health Aspects*, Proceedings of a Workshop held in Hanoi, Vietnam, 14 March 2001, pp. 18–19.
Pham, N.D. and Tran, H.N. (1994) Present environmental status in urban and industrial zones of Vietnam. *Hien Trang Moi Truong cac Do thi va Khu Cong Nghiep Vietnam.* Scientific Report. Hanoi University of Construction (in Vietnamese).
Pham, A.T. and Vo, V.T. (1990) Reuse of wastewater for fish culture in Hanoi Vietnam. In: Edwards, P. and Pullin, R.S.V. (eds) *Wastewater-fed Aquaculture. Proceedings of the International Seminar on Wastewater Reclamation and Reuse for Aquaculture*, Calcutta, India, 6–9 December 1988. Asian Institute of Technology, Bangkok, pp. 69–71.
Quy, H. (1995) US$ 1.2 billions for water drainage system project in Hanoi. *Vietnam Economic Times* 31 August, 1.
Vo, Q.H. (1996) Wastewater reuse through aquaculture in Hanoi: status and prospects. MSc. Thesis No. AE-96–26. AIT, Bangkok.
Vo, Q., Duong, D.T. and Tran, H.N. (1990) Wastewater treatment in Hanoi and its peri-urban area. Hanoi University of Construction (in Vietnamese).
Vu, V.T. (1994) *Investigated Results of Fish Culture System in Sewage Areas of Thanh Tri District.* Research Report. Research Institute for Aquaculture No.1. Habac, Vietnam.
Wang, B.Z. (1987) The development of ecological wastewater treatment and utilization systems (EWTUS) in China. *Proceedings of the Waste Reuse Symposium IV*, Denver, Colorado, 2–7 August, pp. 1263–1289.
Weber, K.E. and Tiwari, I.P. (1992) *Research and Survey Format Design.* Asian Institute of Technology, Bangkok.
Zhang, Z.S. (1990) Wastewater-fed fish culture in China. In: Edwards, P. and Pullin, R.S.V. (eds) *Wastewater-fed Aquaculture. Proceedings of the International Seminar on Wastewater Reclamation and Reuse for Aquaculture*, Calcutta, India, 6–9 December 1988. Asian Institute of Technology, Bangkok, pp. 3–12.

This page intentionally left blank

8 The Emergence of Urban Aquaculture in Europe

Stuart W. Bunting and David C. Little
University of Stirling, Stirling, UK

Abstract

A working definition for urban aquaculture is presented that invokes concepts of land use planning, notions of urbanism, where trade links and markets, nucleated settlements, administrative organizations and specialist labour guilds are considered indicative of urban communities, and encompasses aquaculture closely linked with industrial activity. Early examples of 'proto-urban aquaculture' developed in association with Roman villas, monasteries, castles, manors and millponds are described, and important factors in the emergence of such practices, including technological advancement, demonstration of social status, declining or unreliable capture fisheries, growing market demand for aquaculture products and use of unexploited resources by entrepreneurs assessed. Pressures and events that contributed to the abandonment of these systems, notably the contraction and eventual fall of the Roman Empire, and economic decline and depopulation in the medieval period, are reviewed. Following this historical enquiry, pertinent accounts relating to contemporary urban aquaculture operations throughout Europe are presented, including production in intensively managed recirculation units, horizontally integrated marine aquaculture facilities and systems exploiting industrial by-products, in particular thermal effluents. This review suggests factors underlying the emergence of contemporary urban aquaculture are similar to those that gave rise to proto-urban systems. However, drawing on lessons learned regarding earlier declines, when considering prospects for contemporary urban aquaculture in Europe, it is recommended that a systems-based perspective is adopted.

Origins of Urban Aquaculture in Europe

Aquaculture as we recognize it today has evolved, in general, from what have been termed 'rudimentary proto-aquaculture techniques' (Beveridge and Little, 2002). From the perspective of fishing communities, such practices might have involved the holding of captured fish in ponds until orders had been met, sufficient fish caught to warrant travel to market or demand and prices improved, impounding near shore areas to retain and shelter wild seed stock, or transplantation of fish eggs and relaying shellfish spat in attempts to improve survival (Balon, 1995; Beveridge and Little, 2002). Early examples of proto-aquaculture are more likely to have occurred in fishing and hunter-gatherer communities living in coastal or wetland areas. Considering Europe, shellfish farming was established

(eds B. Costa-Pierce, A. Desbonnet, P. Edwards and D. Baker)

in the Adriatic by the 2nd century BC, and probably developed several hundred years earlier (Beveridge and Little, 2002). Meanwhile, along the Adriatic and Tyrrhenian coasts in the 4th and 5th centuries BC, the Etruscans were actively managing coastal lagoons to enhance production, building embankments and incorporating sluices to trap fish inside. However, proto-aquaculture in Europe was not restricted to rural and coastal areas, but as will be shown, evidence of early aquaculture practices also comes from urban settings.

For this review a broad definition has been adopted to describe 'urban aquaculture'. However, considering the dichotomy between rural and urban land outlined by Best and Rogers (1973), where:

> In a land-use context, rural land encompasses areas which are under agriculture, forest and woodland, as well as wild, unutilized tracts in a natural or semi-natural state. Urban land comprises not only the sites of cities and towns with their associated features, like transport land, but also includes villages, hamlets and even individual or isolated dwellings which perform a similar function whether located in town or countryside...

it was considered prudent to offer some caveats. The working definition developed to guide this review therefore draws initially on notions of urbanism, where the development of trade links and markets, nucleated settlements, administrative organizations and specialist guilds of labourers are indicative of urban communities. Considering the historical inquiry, such conditions might exclude villages, hamlets and individual or isolated dwellings, but offer an opportunity to review aquaculture development in association with, for example, Roman villas, monasteries, castles and manors; establishments which were often the focus for regional industrial and economic activity. For the contemporary assessment, proximity to urban areas, namely larger towns and cities, is used primarily to define urban aquaculture. However, proximity to industry and associated exploitation of by-products is also considered here to constitute urban aquaculture.

Other authors have suggested urban farming activities are related to socio-cultural characteristics of producers, whereby they orient production toward meeting demand from urban markets, or where waste nutrients or water resources derived from urban communities are exploited to enhance production. However, in the majority of Europe, where market networks are far reaching, transport links well established and mass media all pervasive, it might be argued that even the most rural of producers are conscious of and responsive to demands made by urban consumers. Furthermore, restrictions on reusing wastewater resources, mainly due to health concerns and relatively widespread access to low cost inorganic fertilizer, processed feed and freshwater resources mean characterization of urban aquaculture based on exploitation of waste nutrient and water resources derived from urban communities is not perhaps appropriate in a contemporary European context. As a note of interest, however, aquaculture reusing domestic wastewater was relatively widespread in Europe during the first half of the 20th century. Prein (1990) documents 90 wastewater fishponds constructed in Germany, and reports similar systems from Austria, Czechoslovakia, Poland and the USSR. Wastewater aquaculture was also practised in Hungary, in the cities of Godollo, Balatonfoldvar, Szeged, Deszk and Fonyod (Olah, 1990).

What did the Romans ever do for us?

By the 1st century BC, Sergius Orata, gastronomy teacher to Cicero, had constructed saltwater ponds, or *salsae*, to hold fish destined for the kitchen, and a Roman nobleman, Lucinus Murena, had constructed several fishponds, or *piscinae*, close to the summer residence of Cicero at Grotta Ferraia, both to hold live fish, but also as a symbol of status and wealth (Balon, 1995). Following his example, several other nobles also reportedly constructed fishponds to enhance their status; Consul Lucullus (75 BC) dug through a hill near

Naples to enable water to be bought to his fishponds, an exercise that allegedly cost more than his villa (Balon, 1995). Common people also excavated freshwater ponds, or *dulces*, to culture fish for food and earn extra income (Zeepvat, 1988), the knowledge to construct fishponds and to maintain fish led to a permanent supply of various fish species, regardless of fishing success, weather conditions or indeed location.

As the Roman Empire extended into Northern and Western Europe, so the practice of aquaculture in marine and freshwater ponds stocked with indigenous species spread. The rate of knowledge transfer is illustrated by stratigraphic evidence from excavations that shows the earliest Roman fishponds in England to be from the 1st century AD, with one pond in Eccles, Kent dating from AD 65–120 (Dennison, 1989). One of the latest examples was recorded at Bancroft, Buckinghamshire, and dates from the end of the 4th century AD, toward the end of the Roman occupation. As on the continent, many of the Roman fishponds excavated to date in England are closely associated with villas or large country houses, either within or adjoining the main complex of buildings (Dennison, 1989). This author defined Roman fishponds as:

> artificially created pools of fresh or sea water constructed for the purpose of cultivating, breeding, and/or storing fish. Water enters and leaves the pond by the means of a series of channels and pipes, the flow of water being controlled by one or more sluices and overflow channels. Roman fishponds were dug into the ground and lined with stone; they are either square or rectangular, about 30 m by 10 m, and are usually associated with major villas.

Roman villas were centres for economic and industrial activity, and were often located near to towns. Therefore, fishponds closely associated with these settlements appear to constitute a rudimentary form of urban aquaculture. Roman fishponds were often associated with complex systems for water management. Ponds were constructed with sluices and overflow channels to control flow rates and prevent flooding, while supply channels conveying water from rivers, streams or springs often passed first through the villa, where the water was used for other purposes (Dennison, 1989). The reuse of such water in fishponds perhaps also constitutes one of the earliest examples of wastewater aquaculture.

Medieval urban aquaculture

For several hundred years following the decline of the Roman Empire, fishpond construction across Europe appears largely to have ceased, although knowledge on the subject persisted. Writings from the late 8th century (*c.*795) show Charlemagne, in *Capitulare de villis* talked of fishponds, or *wiwaria*, though on monastic estates 200 years later, fishponds were a rarity (Hoffmann, 1996). From the fall of the Roman Empire until the 11th century it appears that catches from wild fisheries satiated demand. However, with the rise in population during the 10th century, and its attendant economic growth, wild stocks of salmon, sturgeon, trout and whitefish began to decline due to over-fishing and environmental degradation (Hoffmann, 1996).

Early industrial and economic activity in many areas of medieval Europe were driven by power from watermills, and building dams and weirs to turn the overshot wheels stemmed the flow of rivers and streams, ponded water upstream and provided a barrier at which to set fish traps. Constructing watermills thus contributed to the decline of many anadromous and cold-water species by presenting a physical barrier to migration, concentrating fish and making them easier to catch, and producing standing pools of water where the water was warmer and oxygen levels lower.

Declines in certain stocks of wild fish, the competitive advantage of species tolerant of poorer water quality and low oxygen levels (bream, perch, pike, roach and tench), and the ability to grow such species in ponds, led to them dominating fish consumption and production. In 13th-century

Yorkshire, England, fish caught from royal and monastic ponds included bream, dace, pike, roach and tench. Evidence for the shift in species consumed has come from excavations across much of Western Europe and from contemporary documentary records (Hoffman, 1996). Archaeological remains from the Baltic demonstrate both a reduction in size and proportion of sturgeon consumed, falling from 70% of fish consumed in the 7th–9th centuries, to 10% in the 12th and 13th centuries. Fishery statistics and monastic records from the 13th to 16th centuries also recount declining catches and consumption of fish such as sturgeon, salmon, trout and whitefish (Hoffmann, 1996).

Declining wild stocks were accompanied by increased demand for fish from followers of the early medieval church and growing urban populations, and according to Hoffmann (1996):

> Medieval European awareness of fish as a nutritionally desirable substitute or supplement for meat is manifest in monastic customaries, in recipe collections and dietary treatises, and in literary allegories ...

The inability of capture fisheries to meet growing demand appears, in some regions of medieval Europe, to have bought about a resurgence in building fishponds. Hoffmann (1996) noted that fishpond construction across Europe began again in the 11th century, and increased rapidly during the 12th and 13th centuries. Furthermore, according to this author, despite the bias of the late medieval written record toward the church, both clerical and lay landowners constructed, owned and oversaw the operation of fishponds in 12th and 13th century Western Europe. From the late 11th to late 13th centuries, fishponds were constructed across England on estates belonging to bishops, monasteries and royalty (Hoffmann, 1996). Roberts (1966) noted for example, however, that during the 13th century many sub-manors were created in the Forest of Arden, south of Birmingham, through endowments by wealthy freemen, and that many fishponds were constructed in close association with the resulting manors. Fishpond construction by secular Norman aristocrats across newly subjugated England has been cited as a means to enhancing their status as landowners (Currie, 1991), and casts doubts over the ecclesiastical origins of medieval fish culture, at least in England.

The Domesday Book records the King's Fishpool, York, England, in 1086, constructed by the Normans in 1069 by damming the River Foss to form a moat (McDonnell, 1981). This waterbody covered an extensive area adjoining the city, and as its name suggests, further to its defensive role was used to culture fish. Moats, an important defensive structure throughout the medieval period, were constructed in association with castles across Europe. It is probable that these defensive structures, at the centre of many fortified urban areas, provided multiple-functions, including water for livestock, disposal sites for domestic waste and areas suitable for fish culture. The King's Fishpool, York, provided fish for consumption by the royal court, monastery and religious followers on days when eating meat was prohibited; at Cranborne Castle records show 50 live bream were stocked in the moat (Pounds, 1990). This author also notes that castles at Northampton, Marlborough and Windsor had stew ponds under their control, and fish from the king's *vivarium* at Kenilworth Castle were gifted to the Bishop of Worcester. At Knaresborough, fish were supplied from the millpond, highlighting the link between early industrial activity, aquaculture and food supply to proto-urban centres.

Strict adherence to avoiding meat resulted in many religious orders, and perhaps most notably the Cistercians, constructing fishponds in association with monasteries and their supporting granges. And although many monasteries were founded in relatively remote locations to ensure the religious community was isolated, the abbey and its granges often became the focus of regional industrial and economic activity, in what, at the time, was otherwise a largely rural or wilderness landscape. Although Cistercian communities were supposedly content to achieve self-sufficiency, by the 13th century notions of austerity held by those who, in 1132, founded Fountains Abbey, England, had been forgone, and economic activity was seemingly geared toward wealth creation (Dent, 1995). Considering production from fishponds associated with Fountains Abbey, it has been argued that these were exploited commercially to generate revenue, and that the abbey was able to profit from the example of continental houses, although strategies of fishpond management and exploitation were probably adapted to local conditions. Evidence from continental monasteries shows 13th-century fishponds were constructed downstream of abbey latrines at Vauclair, France and Maulbronn, Germany, where they received nutrient-rich water (Hoffmann, 1996). This arrangement highlights the close association between early centres of settlement in Europe and wastewater reuse to culture fish.

Another innovation contributing to the success of aquaculture in millponds, moats and monastic fishponds was the introduction across Europe of the common carp (*Cyprinus carpio*) originating from the Danubian basin (Balon, 1995). During the 11th and 12th centuries this species had been introduced to the upper Danube, Elbe and Rhine river systems, and during the late 12th and 13th centuries to the Maas, Seine and upper Rhone (Hoffmann, 1996). The first known record of common carp in the British Isles comes from financial accounts of the 1460s, which show the Duke of Norfolk introduced them to fishponds on his estate (Hoffmann, 1996). However, despite documentary evidence, questions remain over the motivations and mechanisms which gave rise to the pattern and timing of carp introductions.

Hoffmann (1996) surmised:

> How did long unwritten knowledge of pond design pass eastward and of carp pass westward? In medieval Europe, economic exploitation, ecological impact, and environmental mastery were, for good or ill, deeds of human agents.

For example, some suggest that knowledge and experience relating to engineering and animal husbandry, developed and transferred by one of the earliest 'multinational organizations', the Cistercian Order, were instrumental in spreading improved agricultural methods, including skills related to fish farming, throughout Europe (Dent, 1995). However, lay members in European society also built fishponds, suggesting knowledge either passed from the abbeys to the lay community, or that other knowledge sources existed. Circumstantial evidence and historical accounts suggest both mechanisms were operating. Discussing impacts on the lay community of the grange economy associated with Fountains Abbey, Dent (1995) noted that construction of two large dams must have involved the demonstration and transfer of engineering skills to the local workforce. Furthermore, under the supervision of the *magister piscium*, a lay brother skilled in fish-culture techniques, including cleaning ponds, re-stocking and harvesting, local labourers would also have gained knowledge and experience. Documentary evidence recounts how skilled laymen, whose expertise derived from knowledge of working with water and fish, accumulated over generations, travelled long distances to consult with landowners wishing to build sluices and weirs, manage fisheries and commission fishponds (Hoffmann, 1996).

Declines in early urban aquaculture

From the examples of early urban aquaculture presented above it is apparent that its rise was closely linked with technological innovation, entrepreneurial individuals, urbanization, the changing fortunes of capture fisheries and ultimately market demand. However, at this point it is also useful to consider what pressures and events resulted in the wane of urban aquaculture developed under the Romans and during the medieval period.

Roman decline

With the contraction and eventual fall of the Roman Empire, the market for produce from highly engineered and management-intensive urban aquaculture may largely have disappeared. Archaeological evidence, for example, from Britain, shows fishpond construction in the Roman style, and indeed the operation of such ponds, ceased toward the end of the 4th century AD. Although it is likely that by the end of the Roman occupation people from local communities had assimilated the knowledge to construct and manage fishponds, it seems the lack of demand for either the species or cost of fish produced in such ponds prohibited cultivation. Furthermore, dependence of fish culture in urban settings on the proper functioning of infrastructure, such as water supply canals and drainage channels, may have meant maintenance required to ensure continuity in production was beyond the means of individual producers or even communities, thus resulting in the forced cessation of production and abandonment of fishponds. However, until further evidence is available it would be wrong to assume that all fish culture in urban settings ceased immediately following the withdrawal of the Romans.

What is apparent from accounts and evidence of the wider declines in the post-Roman economy is that agricultural production in general became more of a subsistence-oriented activity. As Pearson (1997) notes:

> The marketplaces of Europe were still in their infancy or early adolescence during the sixth to ninth centuries, so that a general focus on self-sufficiency was less an option than a necessity.

Climatic change, most notably a general cooling in temperate regions between the 5th and 9th centuries, probably further constrained development of the agricultural sector, not least due to a shortened growing season. By the 8th and 9th centuries, it has been argued, sufficient and reliable food supplies for emergent emporia were most probably derived from large and well-managed royal or church estates. This redistribution was critical for the continued expansion of proto-urban communities, but may have resulted in the peasant population suffering food shortages (Pearson, 1997). Consequently it has been suggested that free settlements, even those on marginal land, might well have been better off than those being exploited on large estates. Further, according to this author, communities near waterways and seas, such as those at Wraysbury and Hamwic, England, benefited from greater nutritional options through access to varied fish and shellfish. The expansion of both freshwater and marine capture fisheries during the early medieval period also perhaps provides one explanation as to why renewed aquaculture development appears to have been restricted until the 11th century (Lepiksaar, 1985).

Medieval decline

Black Death during the mid-14th century devastated populations across much of Europe, and as a result further contributed to widespread economic decline and depopulation in many areas. The threat of disease transmission associated with communal living also resulted in the decline of many urban communities. Consequently, demand for fish farmed from moats, millponds and stews, which appears to have been synonymous with economically prosperous and growing settlements, probably decreased significantly. Indeed, limited evidence of urban aquaculture development in Europe from the 15th century suggests this was the case, as described later in this section.

With a reduced and more dispersed population it is reasonable to suggest that capture fisheries assumed a more important role, both in terms of overall fish supplies, but also in providing food security and income to people living in remote or isolated communities. The expansion of marine fisheries throughout much of Europe during the late medieval period, combined with better transport links and preservation methods, resulted in marine species meeting much of the demand from consumers. For example, Norse dried codfish was marketed throughout the whole of Catholic Europe (Lepiksaar, 1985). Heightened productivity, quality and dependability of supply from other agricultural sectors during the late medieval period would also have diminished the importance of farmed fish in providing food security and much needed protein in the diets of many.

Considering the case of Great Britain, dissolution of the monasteries by Henry VIII, starting in 1536, further contributed to the decline of aquaculture practised in communal settings. Physical disruption of fishponds and associated infrastructure at monasteries and associated granges might well have severely affected production. Meanwhile, the Reformation more generally may have affected the broader lay community, reducing demand for fish as adherence to religious practices of fasting was rejected and abandoned.

However, despite such severe upheavals in the late medieval period, in some areas of Europe growing urban populations and fluctuating capture fisheries appear to have resulted in fish becoming a luxury commodity – in some central European towns during the 15th century fish was 3–5 times as expensive as meat (Hoffmann, 1996). This stimulated fishpond construction on a huge scale. Widespread environmental degradation around many cities, however, combined with improved transportation arrangements and the ability to conduct complex and extensive hydraulic works, meant fishponds were constructed in rural areas to meet demand from urban markets. Between 1450 and 1550, Czech lords constructed an estimated 26,000 ponds covering several thousand hectares (Hoffmann, 1996). In 16th-century Poland, fishpond construction developed vigorously, apparently in parallel with general economic growth (Roberts, 1968).

Aquaculture in rural locations and the subsequent transport of harvested fish to urban markets while maintaining their freshness has provided a model for freshwater aquaculture development during the past 500 years. Although the environment around many towns and cities has improved markedly over recent decades, consumer attitudes throughout much of Europe now dictate that only fish perceived as coming from unpolluted and natural environments find a ready market. Consequently, demand for cultured trout, salmon and more recently marine fish, has shown a significant increase, while demand for fish such as carp has remained relatively static and geographically restricted. A tradition of eating freshwater fish, especially carp and catfish, persists in much of Eastern Europe. However, even here concern over the quality of production, in particular off-flavours, constitutes a significant constraint.

The historical accounts presented above of early aquaculture operations in what might be considered proto-urban settings gives a brief insight to basic factors stimulating development of such enterprises. Roman noblemen constructed fishponds as a sign of wealth and status, while their countrymen adopted the same techniques to enhance food security and earn money. Irrespective of motivations, varied accounts of fish culture demonstrate widespread technical competence and ability. Although knowledge of fish culture appears to have been disseminated throughout the lands under Roman occupation, the decline of the empire appears to have signalled a temporary cessation of building fishponds. It seems reasonable to assume, however, that abandoned fishponds would have been put to various uses, including the holding and growing of fish, and that knowledge of such practices would have been retained in local communities. With the loss of influence exerted by Roman villas and towns during the early medieval period, many communities appear to have reverted to a more rural existence, oriented toward self-sufficiency, not least because of the decline in markets for agricultural and other goods.

During the early medieval period when populations in Europe began to grow and urbanization resumed, it appears increased demand for fish from both lay and ecclesiastical communities was met largely by expanding capture fisheries. Environmental degradation and declining freshwater fisheries toward the middle of the medieval period, however, appear to have stimulated a renaissance in urban aquaculture. Population growth and urbanization, combined with greater wealth and the influence of various religious orders, appear to have further increased demand for cultured fish. Excavation of fishponds constructed at the time provides evidence for the rise of aquaculture, but documentary accounts also show fish culture undertaken in artificial environments resulting from early industrial processes, especially millponds, contributed to the supply of various fish species.

The rise and fall of proto-urban aquaculture has been associated here with various factors, including technological development, entrepreneurial activity, urbanization, capture fishery returns and market demand. These factors, however, do not act independently, and it is therefore difficult to say which if any is the most appropriate indicator concerning prospects for urban aquaculture. The changing supply of fish from capture fisheries has had a significant influence on the development and decline of aquaculture in both urban and rural settings. Sewage-fed fishponds that developed throughout much of Eastern Europe during the early 20th century, although reminiscent of earlier Roman and medieval developments reusing wastewater to culture fish, appear to have evolved out of harsh necessity at a time when food security was of concern and inorganic fertilizer in short supply. It is doubtful, however, whether such a production strategy would be acceptable in Europe today, regardless of the species or quality of fish produced. Thus, economic and social considerations have also played a major role in shaping the development of European aquaculture, and these factors will be considered further in the following sections concerning their influence on the evolution and potential of contemporary urban aquaculture.

Contemporary Urban Aquaculture

Over the past 30 years aquaculture development in urban settings in Europe has largely been at a pilot-scale, and supported by research and development funding. However, despite demonstrating widespread technical feasibility, there appears to have been relatively little commercial development. That which has taken place has been highly dependent on specific opportunities and local conditions. Furthermore, while markets for freshwater species traditionally eaten in Europe (e.g. carp, eel and catfish) have remained relatively small, and demand for marine species been met largely from capture fisheries, there has been little incentive for potential commercial operators to invest in urban aquaculture.

Commercial aquaculture development in Europe over the past century has largely occurred in rural and coastal areas. Most aquaculture in Europe occurs in nearshore marine environments, with Atlantic salmon (*Salmo salar*) and the blue mussel (*Mytilus edulis*) dominating production (Table 8.1). In Western Europe, farming rainbow trout (*Oncorhynchus mykiss*) has emerged as the largest freshwater aquaculture sector, although development has been largely restricted to rural areas with sufficient access to high volumes of good quality flowing water. In Eastern Europe, farming common carp remains important.

Based on import and export data from individual countries in Europe compiled by FAO (2003b), it is apparent that as a whole the continent is a major exporter of Atlantic salmon and trout and, to a lesser extent, of mussels and carp. Although nearly all production of European eels occurs in Europe, net imports to the continent, including other eel species and elvers, perhaps suggest an opportunity for increased domestic production. Culture of the European catfish (*Silurus glanis*), traditionally practised in fishponds throughout Eastern Europe, is largely semi-intensive in nature and results in a relatively small annual production. However, production

Table 8.1. European and global aquaculture production data and import volumes to Europe for selected species (statistics in metric tonnes for year 2001).

Species	European aquaculture (t/year)	Global aquaculture (t/year)	Net European imports (t/year)
Atlantic salmon	647,039	1,025,287	–146,800[a]
Blue mussels	403,063	426,301	–3,531*[,b]
Rainbow trout	317,901	510,055	–53,883*[,c]
Common carp	142,833	2,849,492	–1,535*[,d]
European eel	10,187	10,258	3,042*[,e]
North African catfish	2,695	6,941	1*[,f]
European catfish	576	648	1*[,f]
Tilapia	200*	1,385,223*	–

Source: FAO (2003a,b).
*Total for all mussel, trout (and char), carp, eel (and elvers), catfish and tilapia (and cichlid) species.
[a]Fresh or chilled, canned and frozen.
[b]Fresh or chilled, canned, frozen and dried salted or in brine.
[c]Live, fresh or chilled, frozen, smoked, dried or salted.
[d]Live, fresh or chilled and frozen.
[e]Live, fresh or chilled, frozen or smoked.
[f]Fresh, chilled or frozen.

of African catfish, cultured in more intensive systems, is higher. The apparent absence of net imports for all catfish species suggests demand from European consumers is being met by domestic production. Statistics for tilapia aquaculture in Europe indicate that domestic production is very low, yet the absence of imports suggests demand from European consumers is also low. This situation is in contrast to other regions, particularly Asia and Africa, where production of tilapia (and cichlid) species is an important part of the aquaculture sector.

Despite the dominance of marine and rural production in the European aquaculture sector there are examples of contemporary production strategies that may be considered urban in nature, or which have potential for development in urban settings. To demonstrate this, case studies dealing with some of the most interesting examples are presented. Recent experiences with recirculation systems, prospects for integrated marine aquaculture in urban areas and other opportunities for urban aquaculture, such as the reuse of thermal effluents to produce fish and marine worms, are discussed.

Recirculation units

Farms employing recirculation address some of the major problems facing aquaculture producers. Water reuse and limited waste discharges assist in reducing concerns over negative environmental impacts. Furthermore, the enclosed nature of the production unit limits the risk posed by disease and parasite transfers, or contamination with toxic substances. Also, the degree of control the operator has over water quality, temperature, photoperiod and aspects of husbandry such as feed management and cleaning, contribute to efficient production.

Currently in Europe a range of freshwater fish species are cultured in recirculation systems. Production of the European eel (*Anguilla anguilla*) and North African catfish (*Clarias gariepinus*) are established industries in the Netherlands (Eding and Kamstra, 2002). Recent reports have described European eel culture employing recirculation in Germany and Spain as well (Schmidt-Puckhaber, 2000; Torres, 2000). European catfish farming occurs mainly in France, Hungary and the Czech Republic, where 2001 production was 352,

116 and 51 t, respectively (FAO, 2003a). Culture has traditionally been in semi-intensive ponds, but now producers such as Svarc Fish Farm in the Czech Republic are culturing this species in recirculation units (Linhart *et al.*, 2002). Within Europe, pike-perch (*Stizostedion lucioperca*), sturgeon (*Huso huso*), Nile tilapia (*Oreochromis niloticus*), Atlantic salmon smolts and marine species such as turbot, cod (*Gadus morhua*), Dover sole (*Solea solea*) and seabass (*Dicentrarchus labrax*) are all grown in recirculation units, although little information is currently available regarding the exact volume and distribution of production.

Key parameters relating to current practices for producing European eel and African catfish in commercial recirculation units are presented in Table 8.2. Stocking densities possible for both the air-breathing catfish and eel are significantly higher than those considered appropriate for seabass or turbot. Feed conversion ratios of around 0.8 were reported for both catfish and seabass production units. However, the higher stocking densities and higher feeding rates possible in catfish culture mean production per unit volume (total culture system volume) is almost ten times greater. Furthermore, despite production per unit volume being 3–4 times that in the eel unit, only one exchange of culture water is required per hour, as compared with 2–4 in the eel unit, thus suggesting pumping costs are lower when culturing catfish. Water exchange in the turbot unit is also only once per hour; however, lower stocking densities and lower specific growth rates mean per unit volume production is less than one-quarter of that for catfish. Although the data outlined in Table 8.2 give an overview of operating parameters for recirculation units, a more thorough assessment of the management demands, finances and risks associated with each system would be necessary to develop comparisons to guide potential operators or investors.

Although commercial operators have developed recirculation units in most European countries, there is currently no means of establishing what proportion might be considered urban or rural in nature. Invoking the social-psychological based understanding of urban production outlined by Iaquinta and Drescher (2000), it might be argued that the highly intensive and market-oriented mode of production means all recirculation units are *de facto* urban in nature. However, this approach is perhaps not that useful as there are more than likely differences in, for example, the functioning, economics and perception of aquaculture in rural versus urban locations. Consequently, as outlined in the working definition for contemporary urban aquaculture, reference should be made to the geographical proximity to urban areas. One way of ascertaining this would be to conduct a survey with producers, but even then it may be difficult to delineate between urban and rural producers other than by using crude indicators such as distance from the urban centre or enclosure by administrative boundaries.

Considering prospects for aquaculture using recirculation in urban areas, there do not appear to be any technical problems to development. Limited water use means location is not dictated by natural hydrolog-

Table 8.2. Parameters characterizing European eel, African catfish, seabass and turbot production in recirculation aquaculture units in Europe.

Parameter	European eel	African catfish	Seabass	Turbot
Stocking density (kg/m^3)	80–250[a,b,c]	170[a]	50[d]	35[a]
Hydraulic retention time (h)	0.25–0.5[a]	1[a]	–	1[a]
Specific growth rate (%/day)	0.75[a]	1.7[a]	–	0.42[a]
Feed conversion ratio	1.3–1.4[a,b]	0.85[a]	0.8[d]	1.1[a]
Production (kg/m^3/year)	173–260[a]	740–1110[a]	118[d]	41[a]

[a]Eding and Kamstra (2002); [b]Schmidt-Puckhaber (2000); [c]Torres (2000); [d]Rigby (2000).

ical conditions and therefore there appears no physical reason why production may not be established near to processors and markets in urban areas. One major practical constraint to development in urban areas, however, will be the higher cost of land, as compared to most rural sites. Higher land prices may be compensated for by reduced transportation costs, improved access to labour and skilled technicians, and more reliable power supplies. However, only detailed financial assessment would demonstrate the relative trade-off.

Lack of value ascribed to reducing externalities such as nutrient discharges or water consumption also means it is impossible in most cases to consider such benefits in standard financial assessments used for investment decision-making. Introduction of a pollution tax or tradable permits might help translate environmental benefits to cost savings or even financial benefits. Other potential financial incentives to production in recirculation units, whether rural or urban, might include a price premium paid by consumers for environmentally friendly or locally produced fish. Assessment of consumer willingness-to-pay for such attributes, however, may often not be translated into price premiums that can be captured by the producer.

Aquaculture in recirculation systems is also at risk from system failures, and adequate safeguards and back-ups are necessary to try to minimize potential losses. There also remains a question, as with any investment, as to whether it is future proof, the assessment of which will require a broad-based systems perspective if outcomes are to be relied upon. Likely future events also become less predictable as the time horizon lengthens. Therefore, financial returns from recirculation systems must show worthwhile returns over relatively short operating periods, often less than 5 years. However, even over short time periods, sudden and unforeseen events such as changing consumer buying behaviour due to media coverage or health concerns, or competition from new producers, may severely affect sales and financial returns. And such issues are more likely to be of concern where aquaculture producers are supplying small niche markets, often with relatively novel species, as is the case with many recirculation units. Taking tilapia as an example, recent developments, such as that by Ocean Fresh Ltd near Cambridge, England (*Fish Farming International*, 2001), are state-of-the-art recirculation units to rear tilapia for the domestic market. However, as can be seen from Table 8.1, production of tilapia throughout the whole of Europe in 2001 was small, and net import absent, suggesting established markets are limited. Lower production costs in other regions, which contribute to the relatively huge volume of global production, also mean, even with the cost of processing and airfreight, tilapia produced outside Europe could reach consumers at a competitive price (*Fish Farming International*, 2003a). Furthermore, while European producers may aim to culture fish under biosecure conditions (Bebak-Williams, 2002) and to market the resulting quality fish at a premium, it remains unclear what level of price differential might be possible, and indeed whether consumers view intensive recirculation units in Europe more favourably than, for example, cages in Lake Kariba, Zimbabwe.

Horizontally integrated marine aquaculture

Opportunities for aquaculture in marine waters close to urban areas are currently being investigated and exploited in various locations throughout Europe. Cage-based salmon culture off the coast of Scotland, and seabass and seabream in the Mediterranean, are within view of certain coastal towns. While it may reassure some to see that near-shore waters are clean enough to produce food, others might be concerned regarding possible negative environmental impacts, including visual intrusion, that might, for example, deter tourists. Controversy surrounding the cage culture of gilthead seabream (*Sparus aurata*) in the Gulf of Aqaba, just offshore of Eilat, Israel, highlights potential problems of siting cage aquaculture operations in nearshore areas close to urban or tourist centres. Studies using fluorescent-tracers have demonstrated that particles originating from the cage site are transported as bedload and suspension-load to nearby coral reefs (Abelson *et al.*, 1999). Cage-based farming in the Gulf provides income and employment for some in the local community, both directly and indirectly. A hatchery situated on the outskirts of Eilat provides work for a kibbutz – a farm collective owned and managed by its members – and fish from the farm are sold in local restaurants. Onshore facilities such as those at the Aquicultura Balear hatchery and nursery only 5 km from the centre of Palma, Mallorca, Spain, provide employment for 20 full-time workers, and production of ten million large seabass and seabream juveniles annually is expected to benefit offshore producers (*Fish Farming International*, 2003b). However, benefits from aquaculture in either the Mediterranean or Gulf of Aqaba are likely to be small, as compared with revenue and employment generated by tourism.

Considering the importance of sensitive marine environments, both ecologically and economically, it is unsurprising that in countries such as Israel, zero discharge aquaculture systems, which restrict water use to an absolute minimum and avoid potentially damaging nutrient releases, are increasingly being promoted and developed (Rosenthal, 2003). Very limited water use in such systems also helps reduce pumping costs and limit the risk of introducing pathogens or contaminants. Development of recirculation units has contributed to culture practices that have helped reduce water use and nutrient discharges. Technical and financial limitations, however, have prevented their evolution to zero discharge. Filters required to maintain water quality produce nutrient-rich sludge, which requires disposal, while accumulation of nitrate, not removed by standard filters, means water exchange, although limited, is still necessary for dilution.

One approach that has been developed to address these problems, avoiding the production of sludge and accumulation of nitrate, is horizontally integrated aquaculture. In this example, wastewater from a primary culture unit is exploited in secondary production units to grow other species, either for sale or to facilitate vertical integration. This strategy is currently being employed by some producers in settings close to urban areas. For example, near Tel Aviv, on the Mediterranean coast, one producer is culturing abalone, followed by units containing fish, phytoplankton and oyster (Hussenot and Shpigel, 2003). Near Eilat, on the Gulf of Aqaba, a system integrating production of sea urchins, shrimp, seaweed and *Salicornia* has been developed (Hussenot and Shpigel, 2003). Approaches to horizontally integrated marine aquaculture are also being tested by producers close to La Rochelle, France, where water from ponds containing seabass is used to culture phytoplankton, which in turn is used to grow oysters and clams (Hussenot and Shpigel, 2003). Near Criccieth, Wales, research is underway to test the feasibility of growing marine worms, phytoplankton, oysters and clams in discharge water from a recirculation unit producing turbot and Dover sole (Hussenot and Shpigel, 2003).

While horizontally integrated production combining intensively managed systems with one or more secondary culture units may offer a promising strategy for optimizing resource use efficiency (Bunting, 2001a), questions remain concerning the management demands, financial performance and consumer perceptions associated with such systems. Bioeconomic modelling has been proposed as a useful tool for combining the assessment of biological and financial performance of horizontally integrated units (Bunting, 2001b; Bunting *et al.*, 2003). However, complementary assessments of likely price premiums, market conditions and policy developments are required to fully develop the economic aspects of such a modelling approach (Bunting *et al.*, 2003). Reliance on primary producers to assimilate nutrients into complex organic matter to feed other organisms in integrated systems is also light and temperature dependent. Therefore, such a strategy may not be appropriate for producers at higher latitudes in Europe. The very limited market for freshwater herbivorous fish or filter-feeding molluscs in Europe also means that horizontal integration may be restricted to marine systems. Aquaponics, or the integration of freshwater aquaculture with hydroponically grown plants, has been tested at the pilot scale. For example, integration of edible and ornamental plant production in wastewater from an eel unit in Denmark (Jungersen, 1997). The commercial viability of such systems in Europe, however, remains to be demonstrated.

Other Opportunities for Urban Aquaculture

Aquaculture, broadly defined, would include culture of ornamental species or stock for public aquaria, and although such production often occurs in urban settings, it was considered beyond the scope of this review. From the working definition presented earlier it was noted that aquaculture that is closely integrated with industrial processes generally might be regarded as urban in nature. One such strategy is the exploitation of thermal wastewater from power stations to facilitate aquaculture, which has been conducted in a number of locations throughout Europe.

Initial trials using cooling water from Hunterston power station, Scotland, showed farming species such as plaice (*Pleuronectes platessa*) and Dover sole was feasible (*Fish Farmer*, 1980). In Denmark, rainbow trout, turbot and oysters (*Crassostrea gigas*) were grown at the Asnaes and Skaerbaek power plants (Dahl-Madsen and Hoffmann, 1981). Turbot were grown in cooling water from the Wylfa nuclear power station in Wales (Jones *et al.*, 1981), while in Germany eels were grown in cooling water from the Emden power plant (Wienbeck, 1981), and in Belgium Nile tilapia were cultured in ponds and tanks receiving water from the Tihange nuclear power station (Melard and Philippart, 1981). However, when considering the reuse of thermal effluents, routine flushing of cooling systems with antifouling agents, contamination with heavy metals and increased risks from gas bubble disease and certain disease agents and pathogens constitute potential constraints (McVicar, 1975; Jones *et al.*, 1981).

Under the auspices of the Soviet Union, thermal effluents from power stations were exploited to culture fish in several Eastern European countries, and by the early 1990s over half the fish produced in Estonia, and 85% of carp, came from such systems. With the collapse of socialism, aquaculture in large production units using thermal wastewater and on collective farms in Estonia, as in many other Eastern European countries, declined significantly. In Bulgaria, however, cooling water from the Maritsa-Iztok 2 power station is still used by the Nomikom fish farm, established in 1990, to culture carp, catfish, sturgeon, trout, and more recently to produce ornamental species such as goldfish and koi.

In Western Europe, cooling water from power stations is still used for aquaculture. Perhaps due to concerns over possible health risks or negative consumer perceptions, however, recent developments have focused on producing fish seed for growout on other farms and marine worms for fishing bait and aquaculture feeds. At Gravelines, between Dunkirk and Calais in northern France, cooling water from a nuclear power station is used to produce juvenile seabass and seabream, mostly for the export market. In England, SeaBait Limited uses cooling water from the Alcan power station to culture polychaete worms which are sold domestically to fishermen as bait, and exported blast-frozen as feed for shrimp broodstock. In addition to pelleted plant material, spent brewery yeast is used to feed ragworms (*Nereis virens*), demonstrating a further link with industry.

Conclusions

The historical enquiry presented above provides a brief insight regarding factors that have promoted and constrained urban aquaculture in early European societies. From these accounts, it is possible to draw some parallels with contemporary developments and to identify some potential limitations. The earliest examples of aquaculture in proto-urban settings, conceived and implemented by Roman nobility, might be considered a demonstration of status, wealth and man's mastery over the natural world. For others, however, fish culture was associated with improved food security. The fortunes of urban aquaculture development in Europe also need to be considered with respect to the development and productivity of both freshwater and marine capture fisheries. Seasonal variations and limited returns from early capture fisheries in Roman times, and decimation of many fisheries as a result of over exploitation and environmental degradation in the early medieval period, stimulated interest in securing supplies of fish by means of urban aquaculture. Current over-exploitation of many fish stocks, both in Europe and in other regions, suggests to some that aquaculture will have an increasing role to play in filling any gap between supply and demand for fish. However, the relative role of urban as opposed to rural, and freshwater as opposed to marine, aquaculture in such a scenario remains unclear.

Demand for fish, both from capture fisheries and aquaculture during the medieval period, was also moderated to a large extent by prevailing religious beliefs. Early Christian doctrines meant demand for fish was high on days when consumption of meat was prohibited, while during the later medieval period, declining ecclesiastical influence resulted in fewer people adhering to such norms. Consumer attitudes, perhaps shaped by concerns about environmental degradation and over-fishing, animal welfare and healthy living, or perhaps due to ease of preparation, taste and price, also require consideration when assessing prospects for contemporary urban aquaculture. Broader social phenomena such as migration, and greater interest in exotic foods, perhaps stimulated by the media or overseas travel, and increases in disposable income, may also contribute to greater demand in Europe for novel aquatic products. This in turn might constitute an opportunity for urban aquaculture. Attributes that might be associated with aquaculture systems suited to urban settings, for example freshness and quality, or biosecure or environmentally friendly production, may also influence consumer buying behaviour, giving producers an advantage. However, it should be noted that attitudes expressed by consumers, in for example focus groups, do not necessarily translate into price differentials that can be captured by producers.

While the demise of urban aquaculture in Europe toward the late medieval period may in certain cases have been associated with particular events, such as dissolution of monasteries in England, the general decline might be attributed, for example, to deteriorating environmental conditions around newly burgeoning towns and cities, increased landings and distribution of fish from rapidly expanding capture fisheries, and greater competition from other agricultural sectors. Competition with other users of land and water around towns and cities must also have strongly influenced the siting of aquaculture operations, and played a key role in dictating the viability of established systems. The relative price of land close to towns and cities, as compared with that in rural areas, is also likely to strongly influence the nature and pattern of contemporary urban aquaculture development. While advances might be made in production efficiency, contributing to lower costs, there appears to be little reason to assume that operators of similar systems in rural areas would not share the same benefits. Therefore, those contemplating urban aquaculture need to consider the prospect of competition, from domestic producers, both urban and rural, and with increasing global trade, competition from other producers internationally.

The re-emergence of urban aquaculture during the 20th century appears largely to have come about out of harsh necessity. Reuse of domestic wastewater, for example, in Germany, Poland and Hungary, and cooling water from power stations in Estonia, Bulgaria and other Eastern European countries, were a consequence of having to make the most of available resources. Today such approaches could be considered as environmentally friendly, but concern over possible health and hygiene risks would probably make food produced in such systems unacceptable to many consumers. Despite this, opportunities have been identified regarding the reuse of cooling water to produce intermediate products, such as juvenile fish and marine worms, for use in aquaculture growout facilities. Research and development funding from national governments, and more recently the European Community, has in many cases supported the technical innovation that has enabled the exploitation of such opportunities. On a note of caution, however, having overcome technical engineering and husbandry problems does not necessarily guarantee the long-term viability of the culture practice. There are many factors, such as changing demand, increased competition, tighter regulation and unknown disease risks, that remain beyond the control of producers.

Any assessment regarding prospects for contemporary urban aquaculture in Europe should be conducted from a systems-based perspective. For individual producers and potential investors such an assessment would include not only technical aspects, but also consideration of likely financial returns over the short to medium-term and, to some extent, the level of risk associated with potential hazards facing the planned operation. In some instances, the financial assessment may be influenced by specific constraints or opportunities – the chance to make use of existing infrastructure such as vacant buildings, or to access research and development funding or regional development grants, may make a substantial contribution to establishing a financially viable business. Assessment of the market for products from urban aquaculture will also be necessary, and while knowledge of local markets and analysis of secondary data on prices and sales volumes might assist in identifying clear opportunities, more detailed analysis, for example, to consider possible consumer attitudes toward products from urban aquaculture, or evaluation of likely future competition from producers in other countries or other sectors, will most likely be beyond the means of most individuals. Consequently, producer associations, research organizations and government agencies appear to have a role to play in assessing prospects for urban aquaculture against a broader social, environmental, economic and political setting. It is also anticipated that the review presented here will be useful in focusing future research and development work on the most promising strategies and potentially limiting aspects relating to urban aquaculture in Europe.

References

Abelson, A., Shteinman, B., Fine, M. and Kaganovsky, S. (1999) Mass transport from pollution sources to remote coral reefs in Eilat (Gulf of Aqaba, Red Sea). *Marine Pollution Bulletin* 38, 25–29.

Balon, E.K. (1995) Origin and domestication of the wild carp, *Cyprinus carpio*: from Roman gourmets to the swimming flowers. *Aquaculture* 129, 3–48.

Bebak-Williams, J. (2002) Biosecurity in intensive aquaculture: don't build without it. *Global Aquaculture Advocate* 5(3), 80–81.

Best, R.H. and Rogers, A.W. (1973) *The Urban Countryside*. Faber, London.

Beveridge, M.C.M. and Little, D.C. (2002) The history of aquaculture in traditional societies. In: Costa-Pierce, B.A (ed.) *Ecological Aquaculture*. Blackwell Science, Oxford, pp. 3–29.

Bunting, S.W. (2001a) Appropriation of environmental goods and services by aquaculture: a reassessment employing the ecological footprint methodology and implications for horizontal integration. *Aquaculture Research* 32, 605–609.

Bunting, S.W. (2001b) A design and management approach for horizontally integrated aquaculture systems. PhD thesis, University of Stirling, Stirling.

Bunting, S.W., Bostock, J.C., Lefebvre, S. and Muir, J.F. (2003) Assessing prospects for horizontally integrated aquaculture using bioeconomic modeling. In: *Beyond Monoculture*. Abstracts and extended communications of contributions presented at the International Conference *Aquaculture Europe 2003*, Trondheim, Norway, 8–12 August 2003. Special Publication no. 33. European Aquaculture Society, Belgium, pp. 140–141.

Currie, C.K. (1991) The early history of the carp and its economic significance in England. *Agricultural History Reviews* 11, 97–107.

Dahl-Madsen, K.I. and Hoffmann, E. (1981) The potential for use of heated effluents for intensive mariculture in Denmark. In: Tiews, K. (ed.) *Proceedings of the World Symposium on Aquaculture in Heated Effluents and Recirculation Systems.* Volume 1. Stavanger, 28–30 May 1980. Heenemann, Berlin, pp. 139–148.

Dennison, E. (1989) *Roman Fishponds*. Monument Class Description, Monuments Protection Programme. English Heritage, London.

Dent, E. (1995) The impact of Fountains Abbey on Nidderdale. BSc thesis. University College of Ripon and York, St John, York.

Eding, E. and Kamstra, A. (2002) Netherlands farms tune recirculation systems to production of varied species. *Global Aquaculture Advocate* 5(3), 52–55.

FAO (2003a) *FAO Yearbook, Fishery Statistics, Aquaculture Production 2001.* Vol. 92(2). FAO, Rome.

FAO (2003b) *FAO Yearbook, Fishery Statistics, Commodities 2001.* Vol. 93. FAO, Rome.

Fish Farmer (1980) Dover sole is ready for the fish farm. *Fish Farmer* 3(2), 32–33.

Fish Farming International (2001) Bass and tilapia unit takes shape. *Fish Farming International* 28(3), 42.

Fish Farming International (2003a) Tilapia 'set to take off'. *Fish Farming International* 30(1), 30–31.

Fish Farming International (2003b) Mallorca produces first bass and bream juveniles. *Fish Farming International* 30(2), 24–25.

Hoffmann, R.C. (1996) Economic development and aquatic ecosystems in medieval Europe. *American Historical Review* 101(3), 631–669.

Hussenot, J.M.E. and Shpigel, M. (2003) Marine land-based integrated aquaculture systems for European countries and the EU INNOVATION project 'GENESIS'. In: *Beyond Monoculture.* Abstracts and extended communications of contributions presented at the International Conference *Aquaculture Europe 2003*, Trondheim, Norway, 8–12 August 2003. Special Publication no. 33. European Aquaculture Society, Belgium, pp. 31–36.

Iaquinta D.L. and Drescher, A.W. (2000) Defining periurban: understanding rural–urban linkages and their connection to institutional contexts. Paper presented at the *Tenth World Congress of the International Rural Sociology Association*, Rio de Janeiro, 1 August 2000.

Jones, A., Brown, J.A.G., Douglas, M.T., Thompson, S.J. and Whitfield, R.J. (1981) Progress towards developing methods for the intensive farming of turbot (*Scophthalmus maximus* L.) in cooling water from a nuclear power station. In: Tiews, K (ed.) *Proceedings of the World Symposium on Aquaculture in Heated Effluents and Recirculation Systems.* Volume 2. Stavanger, 28–30 May 1980. Heenemann, Berlin, pp. 481–496.

Jungersen, G. (1997) Resource-saving and ecological aspects of using wastewater from eel breeding as a source of nutrients, water, and carbon dioxide for plant production. In: Etnier, C. and Guterstam, B. (eds) *Ecological Engineering for Wastewater Treatment.* Lewis Publishers, London, pp. 127–134.

Lepiksaar, J. (1985) Fisheries, marine. In: Strayer, J.R (ed.) *Dictionary of the Middle Ages.* Volume 5. *Famine in the Islamic World – Groote, Geert.* Charles Scribner's Sons, New York, pp. 66–73.

Linhart, O., Stech, L., Svarc, J., Rodina, M., Audebert, J.P., Grecu, J. and Billard, R. (2002) The culture of the European catfish, *Silurus glanis*, in the Czech Republic and in France. *Aquatic Living Resources* 15, 139–144.

McDonnell, J. (1981) *Inland Fisheries in Medieval Yorkshire 1066–1300.* Borthwick Papers No. 60. University of York, York.

McVicar, A.H. (1975) Infection of plaice *Pleuronectes platessa* L. with *Glugea* (*Nosema*) *stephani* (Hagenmuller 1899) (Protozoa: Microsporidia) in a fish farm and under experimental conditions. *Journal of Fish Biology* 7, 611–619.

Melard, C. and Philippart, J.C. (1981) Pisciculture intensive du tilapia *Sarotherodon niloticus* dans les effluents thermiques d'une centrale nucleaire en Belgique. In: Tiews, K. (ed.) *Proceedings of the World Symposium on Aquaculture in Heated Effluents and Recirculation Systems.* Volume 1. Stavanger, 28–30 May 1980. Heenemann, Berlin, pp. 637–659.

Olah, J. (1990) Wastewater-fed fish culture in Hungary. In: Edwards, P. and Pullin, R.S.V. (eds) *Wastewater-fed Aquaculture. Proceedings of the International Seminar on Wastewater Reclamation and Reuse for Aquaculture,* Calcutta, 6–9 December 1988. Environment Sanitation Information Center, Asian Institute of Technology, Bangkok, pp. 79–89.

Pearson, K.L. (1997) Nutrition and the early-medieval diet. *Speculum* 72(1), 1–32.

Pounds, N.J.G. (1990) *The Medieval Castle in England and Wales: a Social and Political History.* Cambridge University Press, Cambridge.

Prein, M. (1990) Wastewater-fed fish culture in Germany. In: Edwards, P. and Pullin, R.S.V. (eds) *Wastewater-fed Aquaculture. Proceedings of the International Seminar on Wastewater Reclamation and Reuse for Aquaculture,* Calcutta, 6–9 December 1988. Environment Sanitation Information Center, Asian Institute of Technology, Bangkok, pp. 13–47.

Rigby, M. (2000) Sea bass farming in Europe: recirculation system technology, a viable alternative. *Abstracts from the Third Annual C-Mar PESCA Aquaculture Workshop*, Portaferry, Northern Ireland, 7–8 September 2000. Centre for Marine Resources and Mariculture, Portaferry.

Roberts, B.K. (1966) Medieval fishponds. *Amateur Historian* 7, 119–126.

Roberts, B.K. (1968) Fish culture in sixteenth-century Poland. *Agricultural History Review* 16, 161–163.

Rosenthal, H. (2003) Aquaculture in Europe: from carrying capacity considerations towards environmentally compatible zero discharge recirculation systems. In: *Beyond Monoculture.* Abstracts and extended communications of contributions presented at the International Conference *Aquaculture Europe 2003,* Trondheim, Norway, 8–12 August 2003. Special Publication no. 33. European Aquaculture Society, Belgium, pp. 67–68.

Schmidt-Puckhaber, B. (2000) Intensive fish culture in recirculation systems in Germany: fish farming at the highest level of productivity. *Global Aquaculture Advocate* 3(3), 52–53.

Torres, J. (2000) Fish culture in Spain: recirculating systems. *Global Aquaculture Advocate* 3(3), 77–78.

Wienbeck, H. (1981) On the oxygen balance of an experimental eel farm operated in thermal effluents of a conventional power station. In: Tiews, K. (ed.) *Proceedings of the World Symposium on Aquaculture in Heated Effluents and Recirculation Systems.* Volume 1. Stavanger, 28–30 May 1980. Heenemann, Berlin, pp. 91–98.

Zeepvat, R.J. (1988) Fishponds in Roman Britain. In: Aston, M. (ed.) *Medieval Fish, Fisheries and Fishponds in England.* British Archaeological Reports, British Series, 182(i). British Archaeological Records, Oxford, pp. 17–26.

This page intentionally left blank

9 Competitive Potential for USA Urban Aquaculture

Michael B. Timmons
Cornell University, Ithaca, USA

Abstract

Since 1970, total meat consumption in the US has risen from 80 kg to 90 kg per capita per year. As standards of living increase, meat consumption also increases. Chicken consumption has risen at a faster rate primarily due to affordability and because chicken is viewed as a generally more healthy meat product than beef and pork. Fish consumption has remained fairly constant for the last 20 years, primarily due to relatively high cost of fish in comparison to beef and poultry. Current costs of fish will, if anything, rise relative to other meat products due to the lack of availability of alternatives to traditional forms of fish capture or production.

Recirculating Aquaculture System (RAS) technology has the potential to dramatically alter the current distribution of meat production away from beef, poultry and pork. Producing fish using RAS technology can be done in a sustainable and environmentally friendly way. Fish production using RAS technology can be much more efficient in converting less desirable foodstuffs into higher quality proteins, particularly if a herbivorous-type fish is used as the species of choice.

The economics of fish production using RAS are reviewed and projections are made as to the near-term potential for producing fish protein using current state-of-the-art technology. The broiler industry is discussed as a case example of how a commodity was increased from less than a few kg per capita per year to being the dominant form of meat protein being consumed on a worldwide basis.

The role of university research and its effective coupling to an outreach programme that will accelerate technology adaptation by the private sector is discussed. Finally, the economic advantage of fish production using RAS technology in urban versus rural locations is quantified.

Introduction

The United States imported over US$10 billion of seafood in 2000, of which US$4.6 billion was imported shrimp, Atlantic salmon and tilapia. Of this total, US$3.8 billion was shrimp. To put this in perspective, the cost of these three aquacultural products in 2000 was equal in value to the combined exports of the US broiler and hog industries (US Department of Agriculture (USDA) Economic & Research Service data). The total trade deficit related to seafood trade is US$6.2 billion (US Department of Commerce (DOC) data). Currently, the US aquaculture industry

 Urban Aquaculture
(eds B. Costa-Pierce, A. Desbonnet, P. Edwards and D. Baker)

generates about US$1 billion each year, with 70% of this being from catfish production (National Marine Fisheries Service (NMFS) data). Marine aquaculture currently accounts for about one-third of commercial aquaculture production. The US DOC (approved 10 August 1999) promulgated an official Aquaculture Policy to increase the value of domestic aquaculture production to US$5 billion. Six goals were established within a series of policy implementation efforts. DOC was to develop a partnership effort with US Department of Agriculture (USDA), US Department of Interior (DOI), and other federal, state and local agencies to achieve these goals. Minimal progress to date has been made in achieving DOC goals. The basic thesis of this chapter is that recirculating aquaculture system (RAS) technology is the key technology that will allow the US to approach the goals set by the DOC in 1999.

Recirculating Aquaculture System (RAS) Technology

Fish is the last mass-marketed food being supplied to consumers by 'hunter–gatherers'. This method of bringing product to market is rapidly becoming obsolete and is no longer able to meet current US market needs. As a result, aquaculture is the fastest growing segment of agriculture, now supplying 30% of all seafood consumed. Conventional aquaculture methods, such as outdoor pond systems and net pen systems, are *not* sustainable in the long term, due to significant environmental issues and their inability to guarantee the safety of their products to the consumer. Conversely, indoor fish production using recirculating aquaculture systems is sustainable, infinitely expandable, environmentally compatible and has the ability to guarantee both the safety and the quality of fish produced throughout the year.

Outdoor pond (warm water systems, e.g. catfish) and net pen aquaculture systems (cool water, e.g. salmon) are disadvantaged by their large footprint requirements, limited appropriate natural sites, environmental issues with respect to the management of fish excrement, geographical limitations due to the need for a perfect growing climate, and because they takes place in an uncontrolled outdoor environment subject to disease, predators and other vectors. Outdoor pond and net pen-based systems are very much at a disadvantage with respect to a greater potential for diseases. Disease in fish systems is transferred by direct water contact with diseased organisms. Indoor systems use potable water and unless diseased fish or fish carrying diseases are introduced into the culture system, there is almost no potential for disease. And if there is a disease event, effective treatment is much more manageable than in traditional outdoor systems. The outdoor pond and net pen-based systems are also disadvantaged by inconsistent supply available to the market, due to more difficulties with control of the growing cycles which creates peaks and valleys of supply available. Finally, issues related to escape are of major concern, particularly with the potential for such events to happen with biotechnology-modified species. In such cases, RAS becomes the only acceptable culture technique because the animals cannot escape an indoor RAS and therefore will not have any impact on natural populations.

Indoor RAS offers the advantage of raising fish in a controlled environment, permitting controlled product development and harvesting. RAS conserves heat and water through water reuse after clean-up by biological filtration through a biofilter. RAS technology allows effective economies of scale, which results in the highest production per unit area and per unit worker of any aquaculture system. RAS is environmentally sustainable, it uses 90–99% less water than conventional aquaculture systems and provides for environmentally safe waste management treatment. RAS-designed aquaculture systems are infinitely scalable. There are no environmental limitations to the size of the intended fish farm to be built.

RAS offers a high degree of environmental control. This not only mitigates the risks of outdoor aquaculture (i.e. natural disaster, pollution and disease) but also allows for optimized species growth on a year-round basis. A similar optimization can be observed in the domestic poultry industry, where chickens were brought indoors and the cost of environmental control was more than recovered by higher growth rates, improved feed conversion and more efficient use of labour. For example, broiler growers produce 1,000,000 kg of chicken per man-year of effort. In addition to the growth advantages afforded by RAS technology, the low environmental impact of these systems means that they can be built closer to the consumer and replicated rapidly.

Indoor aquaculture also means a 100% guaranteed safe source of seafood, free from all chemicals and heavy metals. With increasing consumer concerns about food safety, producers using RAS have an unprecedented opportunity to meet the demands for safe seafood. Attributes of fresher, safer and locally raised product are clear advantages for RAS-produced seafood. Because RAS can be set up to produce the same volume of fish every week, week in and week out, they have a competitive marketing advantage over outdoor tank and pond systems, which are seasonal and subject to environmental disaster beyond the control of the operator.

The advantages of RAS-produced tilapia compared to traditional forms of aquaculture are summarized in Table 9.1 (Timmons *et al.*, 2002).

Aquaculture Production

Aquaculture production generally has been provided through pond-type systems or flow-through production systems. Both are extremely constrained and limited in terms of site location (you must be located near a suitable water source) and do not allow marine culture by their inland nature. Growth of aquaculture in the US has remained constrained for a variety of reasons, with environmental concerns and permitting processes being dominant. These constraints naturally increase the advantages of recirculating aquaculture system (RAS) technology, as RAS can eliminate potentially any negative environmental impact from the production operation. RAS also conserves water, eliminates escape of cultured animals and is basically site independent. The recycling nature of RAS also permits culture of marine or freshwater species, and allows the farms to be located primarily to the benefit of market proximity, as opposed to being sited based upon the availability of natural resources such as high volume water or open ocean sites. Timmons

Table 9.1. Water and land requirements to produce specific seafood species.

Species and system	Production intensity (kg/ha/year)	Water required (l/kg)	Land or water use needed to match RAS output	
			Land	Water
O. niloticus (Nile tilapia) ponds	17,400	21,000	77	210
I. punctatus (Channel catfish) ponds	3,000	3,000–5,000	448	400
S. gairdneri (rainbow trout) raceways	150,000	210,000	9	2,100
Panaeid shrimp pond (Taiwan)	4,200–11,000	11,000–21,340	177	160
O. niloticus (Nile tilapia) RAS	1,340,000	100	1	1

et al. (2002) demonstrated the large reductions in both water required per unit of production and land area required (Table 9.1).

RAS technology is now being successfully used to raise tilapia in northern climates at costs of less than US$2.20/kg (whole fish basis). Outdoor intensive production in Central America has reduced costs to the US$1.00–1.30/kg range (Table 9.2). The proportional costs that make up production are summarized in Table 9.3 for a current commercial tilapia operation in the Northeast producing over 230,000 kg/year (feed at US$0.55/kg, electric rates are US$0.03/kWh, gas heating at US$0.0085/MJ and oxygen at US$0.09/kg delivered). The perception that heating and pumping costs make RAS-produced fish high cost is not valid, as shown in Table 9.4 where these costs are 15% of the total direct costs of production.

There are no commercial recirculating aquaculture operations in the Northeast that have sufficient scales of production or processing to compete with large-scale aquaculture or offshore commercial fishing operations in the food service markets or wholesale. To be competitive with international tilapia producers, a farm would need to have approximately 2.3×10^6 kg/year annual production (to justify an automated processing plant) and direct costs of production of approximately US$1.10/kg. Similarly, to compete with the Idaho trout production systems, the scale of operation must be at the 0.5×10^6 kg/year level.

Table 9.2. Costs of tilapia production in selected countries (Johannsen, pers. comm.)

	Cost of production	
Country	US$/lb	US$/kg
Brazil, Ecuador, Cuba	0.50	1.10
Costa Rica, Jamaica	0.55	1.20
Colombia, Mexico	0.68	1.50
USA	0.91	2.00

Table 9.3. Component costs to produce tilapia (water and sewer supplied by a public utility).

Cost of goods sold	% of total
Purchases – fish stock	7
Purchases – feed	28
Purchases – oxygen	13
Direct labour	29
Supplies	3
Utilities	
Natural gas	9
Electricity	6
Water and sewer	5
Product delivery costs	1
Total	100

Our Sources of Seafood

Wild capture has been the long-standing method of providing the seafood we eat. As time has gone forward and world population has increased, our natural fisheries have become strained. This has led to concerted efforts to better manage our natural fisheries. Macinko and Bromley (2002) argue that US policy makers must recognize:

- The American public owns the nation's fisheries and all they contain.
- Poor management of fishery resources is at the heart of the fishery crisis.
- The US should manage its fisheries as it manages its other natural resources.

Macinko and Bromley acknowledge that managing through an individual fishing quota approach (IFQ) will only ameliorate the race for fish. It is well known that the percentage of seafood being supplied from the ocean continues to drop, with the difference between demand and supply being made up by aquaculture products (Table 9.4).

Even with improved management of ocean resources, the void in needed product to maintain similar per capita consumption levels as in 1999 will require an increase from aquaculture supply of 16.2 million tonnes. Some argue, and can present convincing data, that the amount of product that can be obtained from our oceans is even much less than what it was in 1999, meaning that aquaculture will need to supply an even greater percentage than the 35% predicted in Table 9.4.

Table 9.4. Contributions from wild catch and aquaculture (Food and Agriculture Organization, 2002).

	Million tonnes						
Production	1994	1995	1996	1997	1998	1999	2010 estimated
Capture	91.4	91.6	93.5	93.6	86.3	92.3	93.0
Aquaculture	20.8	24.6	26.8	28.8	30.9	32.9	49.1
Total	112.3	116.1	120.3	122.4	117.2	125.2	142.1
% Aquaculture	18.5	21.2	22.3	23.5	26.4	26.3	34.6
Reduction to fish meal/oil	32.5	29.6	29.6	28.5	23.9	30.4	
% Reduced of capture only	35.6	32.3	31.7	30.4	27.7	32.9	
World population billions	5.605	5.685	5.764	5.844	5.923	6.002	6.812
Per capita food fish supply (kg)	14.3	15.3	15.8	16.1	15.8	15.4	

Population in 2002 is 6.228 billion people.

Competing Meat Products

Aquaculture products – however they are produced and at whatever scale – must ultimately compete against other choices of meat proteins. Table 9.5 shows the quantities of meat consumed since 1960 (USDA Economic and Research Service data).

The broiler industry has shown a steady increase in per capita consumption over the last 40 years, and overall meat consumption has risen from 75 to 97 kg per capita per year, while the US population has increased from 181 to 280 million people. Conversely, beef consumption peaked at 42.9 kg per capita in 1976 and has dropped since to the current level of 30.0 kg per capita in 2001. Based upon current population levels, this is a loss of 12.8 kg per capita or 3.6×10^9 kg of product. If seafood is to exert a similar influence in diverting meat consumption, then the only way to do this is via market price. The ageing population and the accepted health benefits of eating seafood should drive the consumption curves. To make the dramatic changes, as were exerted by the poultry industry, the market price of seafood must be less than competing quality meats. The market share for expensive protein (the current condition for seafood) is essentially maximized at around 6.8 kg per capita (Fig. 9.1).

Can this happen? Table 9.6 provides a summary of wild catch in US waters. Note that the average price paid for seafood at the processing plant is US$1.11/kg. However, the price in New England is nearly double the national price. This gives strong credibility to the notion that locally produced product can command premium pricing. It is not surprising though that when these dock prices for wild catch are converted into fillet costs, the retail price makes the fish fillet an expensive piece of meat.

Hicks and Holder (2001) gave salmon production costs for net pen operations that produce on average 1.36×10^6 kg per cycle of 18 months. This farming approach continues to increase in scale. Productivity per person for a net pen operation is 136,000–204,000 kg/person/year. The present workload per net pen farm is about five individuals that care for placement of 500,000 smolt that will be reared to a market size of 4 kg, with culling and mortality paring that down to 400,000 animals. This translates into 213 t/person/year. But in addition, there are net changing crews and divers, so the support would be about 1–2 additional full-time people. Thus, it requires a maximum of seven people, which reduces the net productivity for a net pen operation to 152 t/person. This is why the scale is being driven to even larger placements of smolts, so that the labour cost per unit of meat produced is further reduced.

Using the Hicks and Holder numbers, a comparison is made to large-scale tilapia RAS production and to the current most efficient salmon producers (Table 9.7). Note that tilapia is actually more economical to

Table 9.5. Per capita consumption (kg) of various meat products[a] 1960–2001 (Delmarva Poultry Industry, 2002 and USDA Economic Research Service).

Year	Beef	Pork	Total red meat	Broilers	Other chicken	Total chicken	Turkey	Total poultry	Total red meat and poultry	Commercial fish and shellfish
1960	28.7	26.8	59.7	10.7	2.0	12.7	2.9	15.6	75.3	4.7
1961	29.7	25.7	59.6	11.8	1.8	13.6	3.4	16.9	76.5	4.9
1962	30.0	26.0	60.1	11.8	1.9	13.7	3.2	16.9	77.0	4.8
1963	31.9	26.5	62.1	12.3	1.7	14.0	3.1	17.2	79.3	4.9
1964	33.9	26.5	64.1	12.6	1.6	14.3	3.4	17.6	81.7	4.8
1965	33.9	23.5	60.8	13.6	1.7	15.3	3.4	18.7	79.5	4.9
1966	35.5	23.0	61.8	14.5	1.6	16.2	3.6	19.7	81.5	4.9
1967	36.2	25.2	64.5	14.8	1.9	16.6	3.9	20.6	85.0	4.8
1968	37.2	25.7	65.8	15.0	1.8	16.8	3.7	20.5	86.4	5.0
1969	37.5	25.0	65.1	15.8	1.6	17.5	3.8	21.2	86.3	5.1
1970	38.4	25.3	66.2	16.6	1.7	18.3	3.7	22.0	88.2	5.4
1971	38.1	27.5	67.9	16.5	1.7	18.3	3.8	22.1	90.0	5.2
1972	38.7	24.8	65.7	17.3	1.6	18.9	4.1	23.0	88.8	5.7
1973	36.5	22.2	60.5	16.7	1.5	18.1	3.8	21.9	82.4	5.8
1974	38.9	23.9	64.6	16.6	1.4	18.0	3.9	22.0	86.6	5.5
1975	40.0	19.5	61.9	16.5	1.2	17.7	3.8	21.5	83.5	5.5
1976	42.9	20.7	65.7	17.9	1.2	17.8	4.0	23.2	88.9	5.9
1977	41.7	21.3	65.1	18.3	1.2	19.4	4.0	23.4	88.6	5.8
1978	39.6	21.3	62.7	19.3	1.1	20.4	3.9	24.3	87.0	6.1
1979	35.5	24.4	61.2	20.9	1.0	21.9	4.2	26.2	87.3	5.9
1980	34.8	26.0	62.1	20.8	1.0	21.8	4.7	26.5	88.6	5.7
1981	35.1	24.8	61.3	21.3	1.1	22.4	4.8	27.2	88.5	5.8
1982	35.0	22.3	58.7	21.3	1.2	22.5	4.8	27.3	86.0	5.7
1983	35.7	23.5	60.7	21.5	1.1	22.6	5.0	27.6	88.3	6.1
1984	35.6	23.4	60.5	22.3	1.1	23.4	5.0	28.4	88.9	6.4
1985	36.0	23.6	61.0	23.2	1.0	24.1	5.3	29.4	90.4	6.9
1986	35.8	22.2	59.5	23.6	1.0	24.7	5.9	30.5	90.0	7.0
1987	33.6	22.3	57.2	25.0	1.0	26.1	6.7	32.7	89.9	7.4
1988	33.1	23.8	58.1	25.1	1.0	26.1	7.1	33.2	91.3	7.2
1989	31.3	23.6	56.1	26.0	1.0	26.9	7.5	34.5	90.6	7.1
1990	30.8	22.6	54.5	27.0	0.9	27.9	8.0	35.9	90.4	6.8
1991	30.2	22.8	54.3	28.2	0.9	29.1	8.1	37.2	91.4	6.8
1992	30.2	24.1	55.3	29.9	0.9	30.8	8.1	38.9	94.3	6.7
1993	29.6	23.7	54.3	31.1	0.8	31.9	8.0	40.0	94.3	6.8
1994	30.4	24.1	55.4	31.6	0.7	32.3	8.1	40.4	95.8	6.9
1995	30.6	23.8	55.4	31.2	0.7	32.0	8.1	40.1	95.5	6.8
1996	31.0	22.3	54.3	32.0	0.4	32.4	8.4	40.8	95.1	6.7
1997	30.4	22.1	53.4	32.6	0.2	32.9	8.0	40.9	94.3	6.6
1998	30.9	23.8	55.6	32.9	0.2	33.1	8.2	41.3	96.9	6.8
1999	31.3	24.4	56.7	35.0	0.2	35.2	8.1	43.4	100.1	6.8
2000	28.5	23.2	54.8	34.9	0.5	35.4	7.9	43.3	98.1	7.1
2001	30.1	22.8	53.6	34.7	0.5	35.2	7.9	43.2	96.8	7.0

[a]Includes beef/pork/veal, and mutton/lamb, but excludes edible offal.

Note: All products on a retail weight basis, except 'other chicken' and 'turkey' which are reported by USDA on a carcass-weight basis.

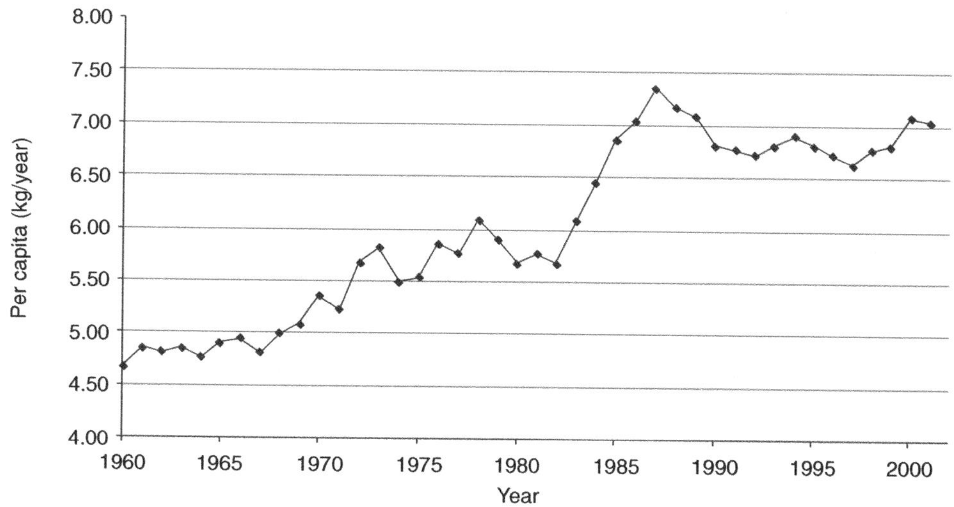

Fig. 9.1. Per capita (kg) consumption of fish and shellfish.

Table 9.6. Value by species and geographic area for US wild catch.

	Quantity captured (× 1000 t)		Value (US$/kg)	
Species	New England	USA	New England	USA
Flounder	17.7	38.5	2.29	1.59
Pollock	4.1	1447.6	1.50	0.16
Haddock	5.8	5.8	2.49	2.49
Cod	15.0	229.2	2.11	0.66
Tilapia	3.2	1.3	2.75	1.01
All species	287.5	4316.2	2.22	0.75
Average			2.23	1.11

Table 9.7. Comparison of production costs (US$/kg) for net pen salmon (current and most efficient operations), large-scale RAS-produced tilapia and commercial broiler production.

	Cost (US$/kg)			
	Tilapia	Salmon	Efficient salmon	Broilers
Cost of operations				
Direct labour and benefits	0.17	0.20		
Feed	0.46	1.26		
Oxygen	0.11	0.00		
Other operating costs	0.04	0.31		
Utilities – heat	0.22	0.00		
Utilities – electric	0.09	0.00		
Fingerlings	0.18	0.35		
Insurance	0.00	0.11		
Health treatments	0.00	0.02		
Total cost of operations (US$/kg fish produced)	1.27	2.25	1.76	0.66

produce, but the yield fillet disadvantage pushes tilapia fillet prices above current salmon prices. Tilapia for RAS have not been genetically improved to any significant degree to date. The productivity increases obtained by the broiler industry (Figs 9.2–9.4) can be mathematically described as follows:

Feed to gain ratio: $FG = 7.68E + 11\exp(-1.34E{-}02x)$

Weight at harvest (kg): $WT = 6.77E{-}08\exp(-8.64E{-}03x)$

Market age: $AGE = 9.88E + 11\exp(-1.19E{-}02x)$

where x is calendar year (valid from 1930 to 2001).

Combining the effects of reduced market age and increased harvest weight, there was a 19% gain in productivity expressed as rate of gain over a 10-year period from 1950 to 1960. The improved feed efficiency of the bird due to genetic selection and nutrition was 13% in this same period. It would be reasonable to assume that similar improvements, or even better, could be made in the tilapia industry, particularly in the fillet yield aspect, since there are tilapia that are currently producing skinless fillet yields exceeding 40%. Given industry-wide improvement in fillet yield, as is indicated by history in the broiler industry, it would place tilapia at a lower freight on board (FOB) fillet price than

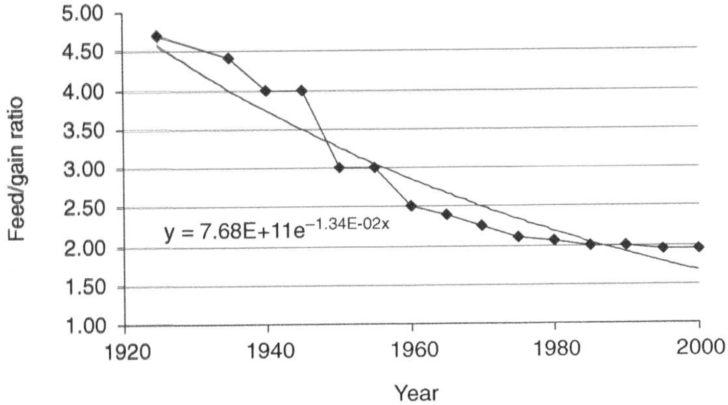

Fig. 9.2. Change in feed to gain ratio for US broiler chicken.

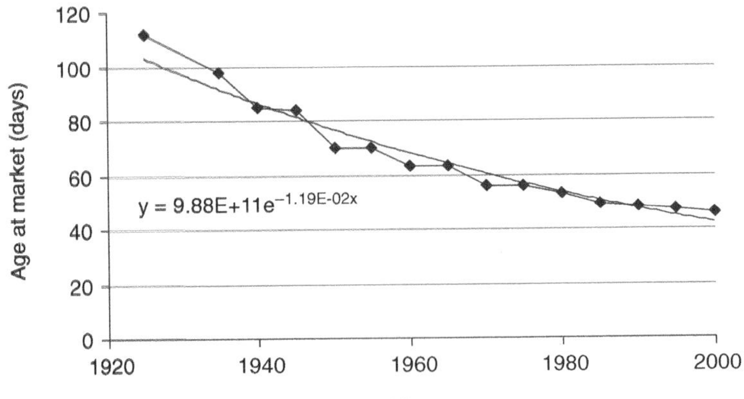

Fig. 9.3. Change in age at market for US broiler chicken.

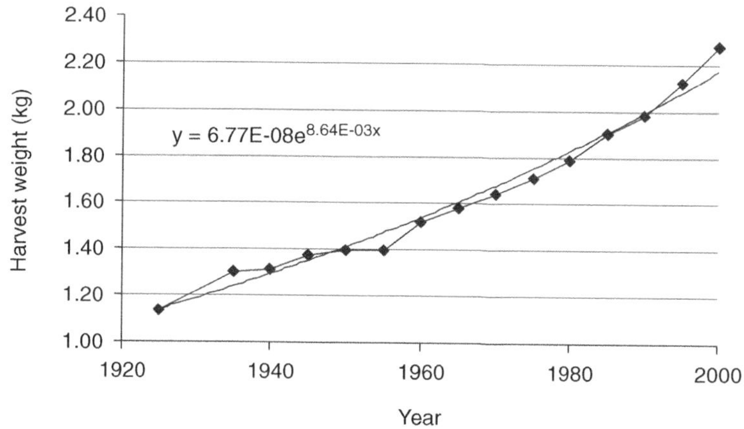

Fig. 9.4. Change in harvest weight (kg) for US broiler chicken over time.

salmon. Let it be noted that there is ample opportunity for a large number of companies to begin to address the market need for competitively priced fish fillets. In the broiler industry, there are 40 companies that are producing more than 23×10^6 kg/year of product, even though the top five companies make up 55% of the total production (Table 9.8).

Although broiler costs of production are relatively low compared to other fish products, the percentage of breast meat from a whole chicken is only 15% (Table 9.10), and this is the high value of the total chicken. Using 15% breast meat yield and similar other prices for packaging etc., the data in Table 9.9 can be converted to show FOB costs for the premium flesh pieces of an animal carcass.

The relatively low percentage yield for broiler breast meat has been a distinct disadvantage for the broiler industry. This has been addressed by steadily progressing to a further processed product (Table 9.11). It should be no surprise to the seafood industry that marketing whole fish products has no future if the broiler industry can be used as any guide to consumer preference, and as a means to add value to carcass parts that are not readily marketed.

The broiler industry's ability to adapt their product to the needs of the marketplace are very evident in Table 9.11. The commercial turkey industry is probably a more poignant example, where genetic selection improved breast yield from 10% to 32% over a 40-year period.

Being competitive long term in the commodity meat market will depend to a large degree on the cost of feed used to grow the animals and the associated efficiency of converting feed energy into meat flesh. Again, the broiler industry has been quite successful at improving conversion efficiencies and other production characteristics (Figs 9.2–9.4 and Table 9.12).

Both feed price ingredients and feed conversion efficiency will impact the ultimate competitiveness of a commodity meat. Table 9.13 shows current ingredient prices and cost per kcal provided for ration mixes for hogs, broilers, three types of tilapia and a salmon diet. Feed to gain ratios or FG (dry weight of feed to wet weight of animal gain) are roughly 2.5, 2.0 and 1.8, and 1.2 for hogs, poultry, salmon and tilapia. Forster (1999) points out though that when recruitment is considered, feed used for maintaining breeding stock and reproduction, the conversion efficiency for broilers increases about 25%, or to a FG of 2.5. This is a large advantage for tilapia production as ultimately the costs of feed will dominate the cost of production, since all other cost inputs can be reduced by increasing scale and obtaining associated production

Table 9.8. 2001 Top broiler companies and weekly production numbers.

Ranking	Company	Average weekly production (million kg, ready-to-cook weight basis)	Market share (%)
1	Tyson Foods, Inc.	66.48	22.6
2	Gold Kist, Inc.	27.37	9.3
3	Pilgrim's Pride Corporation	25.74	8.7
4	ConAgra Poultry Cos.	21.96	7.5
5	Perdue Farms, Inc.	21.10	7.2
6	Wayne Farms, Inc.	12.48	4.2
7	Sanderson Farms Inc.	10.36	3.5
8	Cagle's, Inc.	9.93	3.4
9	Foster Farms	9.03	3.1
10	Mountaire Farms, Inc.	8.45	2.9
11	O.K. Foods, Inc.	6.35	2.2
12	George's, Inc	5.97	2.00
13	Fieldale Farms Corporation	5.83	2.00
14	Peco Foods, Inc.	5.31	1.8
15	House of Raeford Farms, Inc.	5.28	1.8
16	Choctaw Maid Farms, Inc.	4.72	1.6
17	Townsends, Inc.	4.46	1.5
18	Allen Family Foods, Inc.	4.31	1.5
19	Simmons Foods, Inc.	4.30	1.5
20	Case Foods, Inc.	3.15	1.1
21	Marshall Durbin Companies	2.68	0.9
22	Koch Foods, Inc.	2.66	0.9
23	Mar-Jac, Inc.	2.59	0.9
24	B.C. Rogers Poultry, Inc.	2.50	0.9
25	Claxton Poultry Farms, Inc.	2.42	0.8
26	Gold n Plump Poultry, Inc.	2.35	0.8
27	Peterson Farms, Inc.	2.27	0.8
28	Sylvest Farms, Inc.	2.03	0.7
29	Amick Farms, Inc.	1.79	0.6
30	Golden-Rod Broilers, Inc.	1.58	0.5
30	Harrison Poultry, Inc.	1.58	0.5
31	Charoen Pokphand USA, Inc.	1.56	0.5
32	Farmers Pride, Inc.	1.11	0.4
33	Draper Valley Farms, Inc.	0.79	0.3
34	Empire Kosher Poultry, Inc.	0.68	0.2
35	Holmes Foods	0.57	0.2
36	Pennfield Farms	0.52	0.2
37	Park Farms, Inc.	0.49	0.2
38	Lady Forest Farms	0.48	0.2
39	Gentry Poultry Co., Inc.	0.45	0.2
40	MBA Poultry, LLC	0.40	0.1
41	College Hill Poultry, Inc.	0.36	0.1
Total		294.47	

efficiencies. So feed conversion efficiency coupled with productivity per unit area for indoor RAS-produced tilapia should allow tilapia to command a strong long-term advantage in having the lowest possible production costs. Productivity for broiler chickens is around 76 kg/m^2 of building space per year, and tilapia using 2.4 m deep tanks will produce around 290 kg/m^2 of building floor area (assuming tank space is 40% of floor coverage).

Table 9.9. Premium FOB prices (US$/kg) for fish fillets and broiler breast based upon costs of production; assumes fillet yield of 31% for tilapia, 50% for salmon and 15% for broiler breast meat.

	Cost US$/kg			
Component cost	Tilapia	Salmon	Efficient salmon	Broilers
Finance operations	0.07	0.07	0.07	0.07
Finance capital	0.04	0.04	0.04	0.04
Depreciation	0.09	0.07	0.07	0.07
Harvesting/processing/packaging	0.55	0.55	0.55	0.55
Total cost of production (US$/kg)	2.02	2.97	2.49	1.39
Cost/kg dressed weight (83% yield)	2.43	3.58	3.00	1.67
Fillet cost/kg (31% tilapia and 50% salmon)	6.52	5.95	4.98	9.25

Table 9.10. Broiler part percentages of whole carcass (Denson, 1997).

Chicken part	Percent live weight
Inedible processing loss	22.0
Boneless breast meat	12.0
Tender (small breast muscle)	2.8
Thighs, bone in	13.3
Drumsticks	9.9
Wings, includes drumette	7.8
Back	9.4
Breast skin	4.2
Frame/rack	8.1
Giblets (liver, neck, gizzard, heart)	8.6
Fat	0.7
Shrink	1.2
Total	100.0

Table 9.11. Product form for marketing broiler products.

Year	Whole bird	Cut-up	Further processed
1962	83	15	2
1965	78	19	3
1970	70	26	4
1975	61	32	7
1980	50	40	10
1985	29	53	17
1990	18	56	26
1995	11	53	36
2000	9	46	45

Urban Aquaculture and the Northeast Competitive Advantage

The Northeast US competitive advantage is the ability to grow the highest possible quality fish product on the doorstep of the consuming market. In particular, tilapia offers unique opportunity for the Northeast to create a major new industry and also one that does not steal jobs from neighbouring regions or competing commodity products. Tilapia, a mild white fish fillet, is increasing in US consumption by over 30% per year, and is currently estimated at 135 × 10^6 kg/year (Seafood Business, 2002). Tilapia is now the tenth largest contributor to the US per capita consumption of seafood (Table 9.14).

Nearly all fresh tilapia fillets being sold in the US today are imported from Latin America, the Caribbean or the Far East. These importers face considerable transport costs and higher feed costs, which Northeast producers will avoid due to their 'home field' advantage and the abundance of grains and grain by-products in the USA. Already, US commercial-scale tilapia farms are producing tilapia for less than US$2.20/kg on a whole weight basis. If these same technologies are implemented on a large scale, or as a collection of farms that supply fish to a central processing facility, the FOB price of tilapia fillets becomes competitive with Central

Table 9.12. US broiler performance from 1925 to 2000 (National Chicken Council, 2002).

Year	Market age (average days)	Market weight (kg, live weight)	Feed to meat gain (kg of feed to kg of broiler, live weight)	Mortality (%)
1925	112	1.14	4.70	18
1935	98	1.30	4.40	14
1940	85	1.31	4.00	12
1945	84	1.38	4.00	10
1950	70	1.40	3.00	10
1955	70	1.39	3.00	7
1960	63	1.52	2.50	6
1965	63	1.58	2.40	6
1970	56	1.64	2.25	5
1975	56	1.71	2.10	5
1980	53	1.78	2.05	5
1985	49	1.90	2.00	5
1990	48	1.98	2.00	5
1995	47	2.12	1.95	5
2000	46	2.28	1.95	5

Table 9.13. Relative cost of feed for various commodity animals.

Component	Cost (US$/t)	Hog	Broiler	Tilapia	Salmon
Protein		16%	21%	36%	55%
ME of diet (kcal/kg)		3465	3300	2800	4400
Fat (bulk)	260	6%	6%		
Maize	112	70%	62%	15%	
Soy (48%)	187	23%	21%	40%	20%
Wheat	153			20%	
Fish meal (62% protein)	550		2.5%	10%	50%
Fish oil	508			2%	12%
Blended ingredient cost (US$/kcal)		0.044	0.046	0.074	0.093
Blended ingredient cost (US$/t)		137	138	187	373

ME=metabolizable energy.

American imported products, e.g. US$7.70/kg of fillet. As the normal increased efficiencies associated with a developing industry are obtained, tilapia fillets produced using recirculating aquaculture system technology will be produced at costs less than Central American competition. And at this point, tilapia fillets will be competitive with the other premium forms of poultry and red meat (collectively a 96.7 kg per capita per year, Table 9.5) so that a 1–2 kg per capita tilapia consumption level becomes an achievable and realistic goal. Such an industry could be achieved over a 10-year period with strong support from states within the Northeast region. Such a per capita consumption, when expressed on a whole fish basis for the US, translates to 1.4×10^9 kg/year production industry (assuming a 30% yield from whole fish to fillets). A realistic goal would be to aim for an immediate production goal of 0.5×10^9 kg/year. The question is how can such a commercial enterprise be implemented? We could use the salmon industry and its growth

Table 9.14. Current levels of seafood consumption (kg per capita) by fish species (Seafood Business, 2002).

	2001		1990	
Species	Rank	kg per capita	Rank	kg per capita
Shrimp	1	1.54	2	1.00
Tuna	2	1.32	1	1.68
Salmon	3	0.92	5	0.33
Pollock	4	0.55	4	0.58
Catfish	5	0.52	6	0.32
Cod	6	0.25	3	0.63
Clams	7	0.21	10	0.28
Crabs	8	0.20	7	0.13
Flatfish	9	0.18	8	0.26
Tilapia	10	0.16		
Scallops	11	0.12	9	0.14
Total		5.97		5.33

over that last 20 years as an indicator of the potential for explosive growth once a technology is refined, e.g. net pens for salmon and RAS for tilapia (Table 9.15). Similarly, the growth of the broiler industry and the early explosive growth once a growing technology is perfected is shown in Fig. 9.5 and Table 9.16.

Ultimately, Northeast US producers should prevail over producers from overseas and traditional forms of aquaculture due to the following cost advantages:

- Lower transportation costs due to proximity to the market.
- Cost-effective and sustainable environmental treatment of fish waste compared to open net pens and ponds.
- Automated seafood processing techniques to minimize disadvantages of high labour costs.
- Lower US feed costs.
- More efficient feed conversion ratios in comparison to outdoor aquaculture, due to RAS technology.

Feed represents the single largest component of fish production costs. Because of lower domestic feed costs compared to overseas producers that must import much of their grains, there is a production price floor protecting the USA from domination by imported product. Low US feed costs coupled with high productivity per unit worker enabled by continually improving RAS technology will make Northeast producers cost competitive with overseas producers. These simple facts of commerce combined with the overseas producers' higher shipping costs, position Northeast producers to prevail long term against overseas suppliers of farmed seafood in much the same way the US poultry and hog industries have dominated the US marketplace.

Over 70% of fresh fish fillets are sold to restaurants as the final purchasing unit, generally through some type of seafood wholesaler. Since the restaurant buyer is paying near retail prices for fresh seafood, growers in urban areas could focus primarily on this customer base as *the primary marketing target* for the small fish-farming enterprise. For these markets, costs of pro-

Table 9.15. Farmed salmon growth.

Year	Million kg/year	Growth % last 5 years
1980	20	
1985	50	20.1
1990	250	38.0
1995	550	17.1
2000	1100	14.9

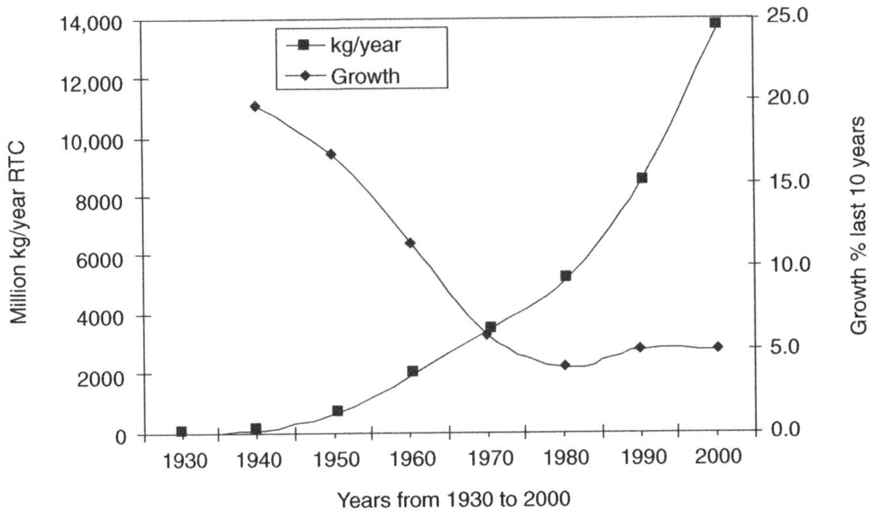

Fig. 9.5. Growth of US broiler production from 1930 to 2000 in million kg ready to cook (RTC) weight (left ordinate axis) and compounded growth rate for preceding last 10 years (right ordinate axis).

Table 9.16. Historical increase in US broiler production (USDA Economic Research Service).

Year	Production (million kg ready-to-cook basis)	Change from year before (%)	Year	Production (million kg ready-to-cook basis)	Change from year before (%)
1950	627	+23.9	1976	4,091	+12.4
1951	790	+25.9	1977	4,213	+3.0
1952	858	+8.6	1978	4,496	+6.7
1953	949	+10.7	1979	4,960	+10.3
1954	1,072	+13.0	1980	5,147	+3.8
1955	1,095	+2.1	1981	5,388	+4.7
1956	1,396	+27.4	1982	5,446	+1.1
1957	1,531	+9.7	1983	5,596	+2.7
1958	1,775	+16	1984	5,866	+4.8
1959	1,884	+6.1	1985	6,138	+4.6
1960	1,968	+4.5	1986	6,438	+4.9
1961	2,245	+14.1	1987	6,998	+8.7
1962	2,269	+1.1	1988	7,267	+3.9
1963	2,392	+5.4	1989	7,821	+7.6
1964	2,472	+3.3	1990	8,367	+7.0
1965	2,668	+7.9	1991	8,894	+6.3
1966	2,922	+9.6	1992	9,490	+6.7
1967	2,975	+1.8	1993	9,995	+5.3
1968	3,020	+1.5	1994	10,744	+7.5
1969	3,257	+7.8	1995	11,271	+4.9
1970	3,490	+7.1	1996	11,860	+5.2
1971	3,507	+0.5	1997	12,277	+3.5
1972	3,699	+5.5	1998	12,536	+2.1
1973	3,615	−2.3	1999	13,378	+6.7
1974	3,647	+0.9	2000	13,713	+2.5
1975	3,641	−0.2	2001	14,046	+2.4

duction could be considerably higher, and a reasonable profit still be made by the producer. Costs still need to be at or around US$2.00/kg whole fish basis. This can currently be achieved using today's RAS technology with expert management. RAS technology today is dramatically different than that of just a few years ago.

Urban aquaculture, to survive and grow, faces the same supply and demand market realities as any other business. Consumers are looking for quality, value and, most of all, convenience. American buyers have shown a long history of paying for convenience. Perhaps the most glaring example is the US$1 billion market that developed in the last 10 years for ready-to-eat salad mixes from the grocery. Historically, heads of lettuce, carrots, tomatoes, onions and other greens were purchased and then cut up and made into a salad. So even though the consumer is paying in the order of three times the raw ingredient costs that make up a mixed green salad, the convenience of this product has developed a strong demand and resulting supply.

Countless times we hear about the need for marketing programmes so that aquaculturalists can receive the prices they deserve for their products. This is another way of saying that we need to convince consumers to pay more for aquacultured products than alternative wild-caught products or other meat choices. This approach will not work. It has never worked and it never will work. The consumer will buy for value. Aquaculturalists must recognize this simple fact. The key to success will be to produce fish on an economically competitive basis.

Aquaculture Requirements for Infrastructure and Capitalization

An aquaculture development project will require significant infrastructure in terms of water, waste disposal capacity, enclosed building space, electrical energy and load demand supply, and transportation logistics. While each site considered will require a thorough engineering analysis, there are approximate minimal site requirements (Table 9.17).

The urban aquaculturalist must compete against the other commodity meats and large-scale fish farming such as currently being practised by the net pen salmon industry or the US catfish industry (Fig. 9.6).

Urban Setting Opportunities

The potential for an urban eco-park site could be viewed as a closed ecosystem where use, processing, transportation and consumption of resources flow as a 'closed loop' with feedback. Throughout, wastes are minimized and by-products are recycled as recovered resources. Energy is minimized by improvements in efficiency and increasingly focusing on opportunities to utilize non-fossil fuel sources. Bioprocessing of agricultural materials, including waste biomass, is a source for bioproducts and energy. Biological processes are a preferred path for processing agricultural resources due to higher reaction specificity and fewer toxic by-products. These characteristics are consistent with the goal of developing industrial processes and systems that are environmentally friendly.

Table 9.17. Approximate infrastructure and utility requirements.

	Trout	Tilapia
Production per year (kg)	454,000	454,000
Footprint of buildings (m^2)	5,600	3,700
Water required and discharge per day per m^3	3,000	300
Heating requirements (MJ per day)	20,000	40,000
Electrical requirements (kWh per day)	6,000	4,000
Liquid oxygen (m^3 per day)	1,000	1,000

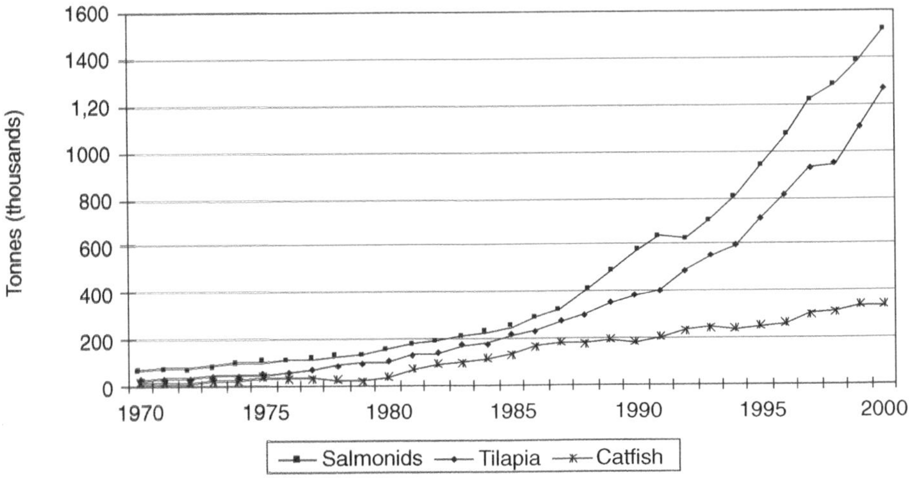

Fig. 9.6. World yearly production of salmonids, tilapia and catfish.

A challenge for ultimate agricultural production efficiency is to take advantage of urban characteristics that may allow the creation of a sustainable system that integrates energy, environmental, agricultural and industrial innovations. Fuel cell technology is a new technology that is approaching economic viability in certain applications. Fuel cells can convert biogas directly into electrical and thermal energies for both on-farm uses and to supply community energy needs. The advantage of an aquaculture operation, particularly for the warm water species, is that the waste heat from a generator/fuel cell can be used throughout the year.

Future success may depend upon fish production systems being viewed as a series of integrated farming systems that has a system of materials and energy flows. The fish farm concept could be expanded beyond a contemporary system of just producing fish flesh, to one that produces energy which can drive more integrated food and fibre production systems, as well as generate energy for enterprises on or near the fish farm or for energy needs of the surrounding community.

Current costs of fuel cell systems are roughly (Minott, 2002), with current capitalization costs, at US$4000/kW of capacity. Costs of fuel cell systems are expected to drop to the US$1000/kW by 2010 (N.R. Scott, pers. comm.). As a relative example, a 0.5×10^6 kg/year tilapia farm would require a 200 kW fuel cell system designed into the system. At current prices, this is not economically feasible, but at the expected capital cost in 10 years, this becomes a realistic option. A depiction of how a fuel cell driven energy system might fit in to a greater industrial or ecology park is shown in Fig. 9.7.

Industry Vertical Integration

The success of the poultry industry was based upon two simple concepts:

- Vertical integration – the success of the broiler and catfish industries are attributed to their vertical integration of breeding, growing, processing and distribution operations under a single business structure.
- Branding – value in chicken became real when a generic bird became a named product, e.g. Perdue, Holly Farms and Country Pride. Ownership of a product by name became associated with value and quality and hence, added value.

More recently than the broiler industry, the catfish industry developed a 270×10^6 kg/year production base over the last 20

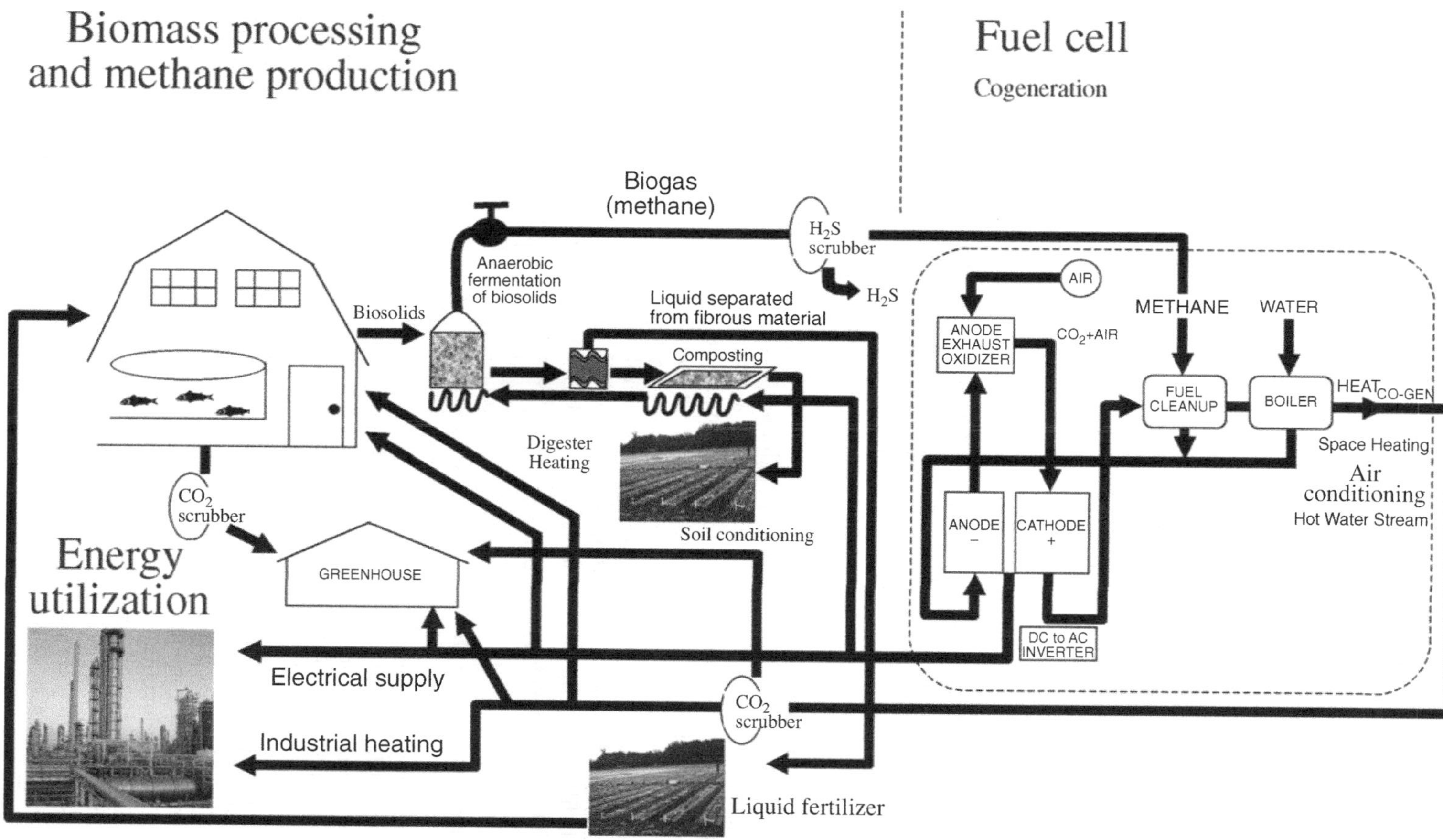

Fig. 9.7. Integrated fuel-cell driven energy system for a large-scale fish farm in an urban setting.

years, adding nearly 50 $\times$ 10^6 kg of production in the last 2 years! This whole industry was created as a concerted effort to transform poorly performing cropland into productive fishponds. A collection of bankers, farmers and entrepreneurs created this industry. To a large degree, the catfish industry was patterned after the chicken broiler industry. In 50 years the broiler industry grew from a few million kg per year to a 14 $\times$ 10^9 kg/year industry. The Atlantic salmon industry has used large-scale net pen operations to become a 0.5 $\times$ 10^9 kg/year force in the US market place. Salmon production has been adding 45 million kg/year of production on a yearly basis since the industry reached a critical production mass of around 200 $\times$ 10^6 kg/year. The tilapia industry is rapidly approaching this level. Others have already done the hard part of creating market awareness and acceptance. A consumer, given a choice between a fresh fillet from Central America or a fresh fillet from upstate New York or Hartford, CT, will choose the locally produced product and even pay a slight premium for their choice. Further, the general consensus and perception that fresh fish are fragile and spoil quickly will only enhance the Northeast producer's competitive advantage of producing fresh fillets within the region targeted to regional consumers.

So, How Can It Be Done?

The US is currently consuming nearly 136 $\times$ 10^6 kg/year of tilapia, but there is only 7 $\times$ 10^6 kg/year being produced in the US. The US production is almost exclusively targeted to the live Oriental markets. Farm prices for tilapia targeted for the US live markets are from US$2.75 to US$3.3/kg, depending upon distance to the marketplace and whether or not the market is currently over or undersupplied. During periods of minor over-supply, farm prices have dipped to the mid US$1.75/kg level, resulting in small farms dropping out of the competition. Currently, only a very few producers have production costs sufficiently low to withstand farm prices below US$0.45/kg at the farm. None appears on the verge of being able to compete in the fillet market. China is a wild-card country that currently produces 500,000 t/year that could disrupt the US marketplace even for fresh fillet product.

China already dominates the frozen whole tilapia market and the individually quick frozen (IQF) markets. Based upon landed frozen whole fish prices of US$1.10/kg, China conceivably has production costs in the mid US$0.50s range per kilogram of whole fish. Quality in general of the China product is suspect, and the costs of shipping fresh product to the US and Europe is the primary disadvantage of the Chinese producers.

It is without question that the only real market for expansion in the US is the fillet market. It is also accepted that the market sector that can most easily be competed for is the fresh fillet market. The general consumer perception of freshness of a locally (domestic US) produced fillet compared with a product produced overseas is that US-produced products are of higher quality and are safer to eat and this is a key advantage for US producers.

A simple exercise can establish the farm price of tilapia necessary to allow a company to compete in the fillet market (Table 9.18). Current tilapia fillet yields are around 28% for automated fillet machines and custom processing (such systems have been demonstrated in Ecuador). Selected tilapia animals are known to give yields as high as 42%, so it is reasonable to assume over the next 10 years that genetic improvements will result in fillet yields comparable to catfish processing yields, which are around 42%. Assuming an integrated farm with processing plant that is selling product through the normal distribution system, FOB cost for the integrator can be estimated as shown in Table 9.19. Comparing Tables 9.18 and 9.19 indicates the producer must be able to purchase fish in the round and transport the fish (live or on ice) to the processing plant for US$1.20–$1.32/kg with fillet yields at 28%. The effect of yield and allowable farm price is shown in Fig. 9.8.

Table 9.18. Freight on board costs (US$/kg FOB) for various farm gate prices paid for tilapia in the round and for various fillet yields.

	FOB cost (US$/kg)		
Farm price (US$/kg)	Yield 0.28	0.33	0.42
0.60	2.91	2.59	2.20
0.70	3.27	2.89	2.44
0.80	3.63	3.19	2.67
0.90	3.98	3.50	2.91
1.00	4.34	3.80	3.15
1.10	4.70	4.10	3.39
1.20	5.06	4.41	3.63
1.30	5.41	4.71	3.87
1.40	5.77	5.01	4.10
1.50	6.13	5.32	4.34
1.60	6.48	5.62	4.58

Assumes a processing cost of US$0.77/kg fillet basis.

Table 9.19. FOB price as affected by distributor and integrator margins and sales retail price.

Sales retail price	US$3.50
Distributor margin	15.0%
Integrator margin	20.0%
FOB price	US$2.38

Integrator and Contract Farming Approach

Contract farming has worked from a financial perspective in the broiler and hog industries, and has resulted in high efficiencies of production and low priced inputs for feed and processing. These industries required many years to develop their own successful methods. Developing a similar infrastructure and scale of operation will not happen overnight. It should also be painfully obvious that to compete in the fish fillet markets, the overall costs of production must be low and that any loss of efficiency in the production operation that results in higher feeding costs or other input costs will result in a non-competitively produced product. Thus, for contract tilapia farming to be successful, everything from the hatchery to the processing plant delivery must be extremely efficient. The advantage brought by the contract farming approach is that economies of scale allow reductions in price for the raw commodity inputs into the equation, e.g. feed and oxygen, and the scale of operations also allows specialization of labour activities, e.g. fingerling production and delivery to farms, feed delivery, harvesting, surveillance, maintenance and fish health management. These specialized functions could be provided by

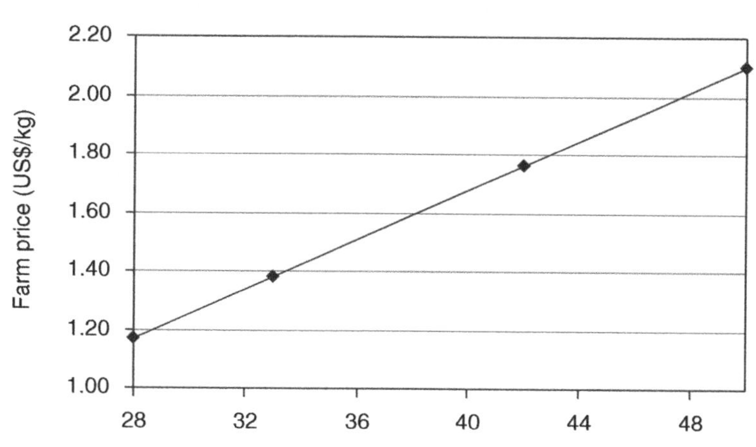

Fig. 9.8. Allowable farm price as affected by fillet yield for tilapia for a target FOB price of US$5.00/kg (assumes a US$0.77/kg processing and packaging cost).

the contract farmer on an as needed basis (repair) or on a scheduled basis repetitively over the growout period, e.g. water quality assessment or fish condition factors to judge effectiveness of the feeding programme. It is also these same specialty functions that a small independent farmer will lack with a normal consequence of poor production or catastrophic failure.

A successful integrated contract farming approach should only be built upon an already successful large tilapia farming and processing operation. Currently available automated processing equipment will easily process 30 fish per minute. The minimum farm year capacity can be estimated as 2.7×10^6 kg per year of whole fish, using the following assumptions:

- Minimum size fillet of 170 g (140–200 g is a premium size classification).
- Processing yield of 28%.
- Processing line speed of 30 fish per minute.
- Six active hours per 8 hour shift.
- Four days per week of processing.
- Fifty-two weeks per year of market delivery.

The above assumptions and resulting estimate of farm output predicts that a farm should start with 2.7×10^6 kg/year of production to support an automated processing facility. An automated processing facility is required, or else the hand labour costs for filleting will drive the cost of a finished fillet product above market competitive price. Once the farm is successfully managing the first 2.7×10^6 kg/year of production and processing, and market demand exists for additional product, then and only then, should the parent farm (integrator) consider adding additional production through a contract farming approach to generate additional raw product for the processing line and market demands.

At a base level of production of 2.7×10^6 kg/year, and by locating contract farmers within 10–15 miles of the processing plant, staff labour used for the large central farm could probably absorb the responsibilities of an additional 0.5×10^6 kg/year of production spread over four contract farmers for such activities as fingerling placement, stock movement between tanks (if any) and harvesting. The parent farm would provide continuous monitoring of contract farmers' water quality and provide proactive recommendations on alterations of a feeding schedule or water exchange or other management activities to control fish growth performance. Such support from the parent farm is not conceivable until the parent farm has these same activities under extremely efficient operation. Once the skills and personnel are developed in-house, expanding production efficiently should be a fairly simple task.

Why the contract farming approach has not worked to date is because there are only two or three farms in the entire US that have refined their management procedures and efficiencies to the point that they could support a contract farming approach. It should also be mandatory that each contract farmer be given a hands-on training period at the parent farm to become familiar with their responsibilities of managing fish and water quality prior to having fish on their farm. Again, this has never been done. To a large degree and without exception, every RAS-based farming operation has always underestimated the sheer difficulty of raising fish from a biological perspective on a consistent basis throughout the calendar year. It is not easy to raise fish under any conditions, and it is more challenging initially trying to do this in an RAS because there is an additional biological system that must be mastered, i.e. the RAS itself. The differentiation of skills and raw inputs are needed in RAS production and the responsibility for each are shown in Table 9.20.

Conclusions

Aquaculture will need to supply 49 billion kg of product (16 billion kg above the 1999 levels) to maintain the current per capita consumption for a projected world population of 6.8 billion people in 2010 for a total demand of 142 billion kg. Growth of aqua-

Table 9.20. A suggested differentiation in responsibility for inputs and management between farm and integrator.

	Supplied by	
Input	Farmer	Integrator
System capitalization	x	
Feed	x	
Heating	x	
Oxygen	x	
Electric	x	
Water and sewer	x	
Daily management labour	x	
Fingerlings		x
Surveillance labour		x
Repairs and preventive management		x
Health and engineering management		x
Harvesting		x
Marketing of product		x

culture in the US has remained constrained for a variety of reasons, with environmental concerns and permitting processes being dominant. These constraints naturally raise the advantages of recirculating aquaculture system (RAS) technology, as RAS can eliminate potentially any negative environmental impact from the production operation. RAS also conserves water and land, eliminates escapement of cultured animals and is basically site independent. The recycling nature of RAS also permits culture of marine or freshwater species and allows the farms to be located primarily to the benefit of market proximity, as opposed to being sited based upon the availability of natural resources, such as high volume water or open ocean sites.

There are no commercial recirculating aquaculture operations in the Northeast (or US) that have sufficient scales of production or processing to compete with large-scale aquaculture or off-shore commercial fishing operations in the food service or wholesale markets. Analysis supports the premise that if RAS production is scaled to comparable levels as salmon net pen aquaculture or other commercial forms of meat farming, then RAS-produced seafood can be economically competitive. Scales of production for RAS-produced tilapia would need a minimum scale of 2.3×10^6 kg/year annual production (to justify an automated processing plant) and direct costs of production of approximately US$1.10/kg.

References

Delmarva Poultry Industry, Inc. (2002) 16686 County Seat Highway, Georgetown, Delaware.

Denson, W.D. (1997) Product costing: how to allocate meat cost. *Broiler Industry* October, 24–27.

Food and Agriculture Organization (2002) *World Review of Fisheries and Agriculture.* FAO, Rome.

Forster, J. (1999) Aquaculture, chickens, salmon – a case study. *World Aquaculture* 30(3), 33–38, 40, 69, 70.

Hicks, B. and Holder, J.L. (2001) Net pen farming. *Proceedings 2001 AES Issues Forum,* Shepherdstown, WV, 11–14 November 2001, NRAES-157 Publication. Cornell University, Ithaca, New York.

Macinko, S. and Bromley, D.W. (2002) *Who Owns America's Fisheries?* Report provided and sponsored by the Pew Charitable Trusts, Washington, DC.

Minott, S.J. (2002) Feasibility of fuel cells for energy conversion on the dairy farm. MSc thesis, Cornell University, Ithaca, New York.

National Chicken Council (2002) http://www.nationalchickencouncil.com

National Marine Fisheries Service http://www.nmfs.noaa.gov/Budget-html/IT&2001.html

Seafood Business (2002) *September 2002 Buyers Guide – Tilapia.* Seafood Business, Portland, Maine.

Timmons, M.B., Ebeling, J.M., Wheaton, F., Summerfelt, S.T. and Vinci, B.J. (2002) *Recirculating Aquaculture Systems,* 2nd edn. Cayuga Aqua Ventures, Ithaca, New York.

United States Department of Agriculture Economic and Research Service. http://www.ers.usda.gov

United States Department of Commerce. http://www.nmfs.noaa.gov/trade/DOCAQpolicy.htm#top

This page intentionally left blank

10 Commercially Feasible Urban Recirculating Aquaculture: Addressing the Marine Sector

Yonathan Zohar, Yossi Tal, Harold J. Schreier, Colin R. Steven, John Stubblefield and Allen R. Place
Center of Marine Biotechnology, University of Maryland Biotechnology Institute, Baltimore, USA

Abstract

With the global collapse of marine fisheries and the environmental issues associated with net-pen aquaculture practices, there is a pressing need to develop fully contained and environmentally sustainable approaches to producing seafood. Recirculating aquaculture systems (RAS), which provide this approach, have been widely used over the last two decades to farm freshwater species. However, the cost associated with the technology, along with strong competition from pond culture and low market prices for the freshwater product, resulted in numerous economic failures. This chapter describes the application and optimization of the RAS technology in the marine sector, in particular the development of urban recirculating mariculture for high-value marine fish. The system's performance and economic feasibility were tested in a pilot urban mariculture programme in the City of Baltimore, studying the Mediterranean gilthead seabream (*Sparus aurata*) as its candidate species. This fish, a non-native species in North America, commands a local retail price of up to US$20/kg. The Baltimore Urban Recirculating Mariculture System was designed to produce high-value marine fishes that cannot be farmed in net-pens or ponds, to use municipal pre-existing infrastructure and services, to have the ability to locate anywhere and to maximize the re-use of water. The life support system consisted of a particle removal microscreen drum filter, a moving bed nitrifying reactor, an ozone-based protein skimmer and a low head oxygenation unit. Conditioned artificial seawater was automatically delivered to provide the desired salinity and temperature. pH, ozone levels and photoperiod were continuously monitored and adjusted. Strict biosecurity was achieved by disinfecting all waste effluents before their discharge to the municipal sanitary sewer. Using this system, gilthead seabream of two strains were grown from 0.5 to 400 g commercial size in 268 days (first strain) and to 410 g in 232 days (second strain). Survival rates exceeded 90% and food conversion rates varied from 0.87 to 1.89, depending on fish growth. Growing densities ranged from 44 to 47 kg/m^3 at 7–10% daily water exchange rates. Total ammonia and nitrite levels remained significantly below stressful concentrations. To increase the economic feasibility of the system, we studied microbial communities associated with biofiltration in an effort to improve nitrogen removal and thus maximize re-use of the saltwater. New bacterial-mediated nitrogen removal processes are described herein and addition of an anaerobic denitrification unit was also studied, both of which enhanced our ability to minimize saltwater discharge. The environmentally compatible recirculating mariculture pilot system described here can be scaled up to cost-effectively produce high-value marine fish in an urban setting.

Introduction

Global marine fisheries have exhibited continuous declines in recent years. According to Food and Agricultural Organization (FAO) statistics, 74% of the world's commercially fished species are either depleted (9%), overfished (18%) or fully fished (47%) (FAO, 2002; Schiermeier, 2002). Fisheries scientists are in agreement that the oceans have attained their maximal sustainable yield, and if current trends continue, world fisheries could collapse within a few decades (Watson and Pauly, 2001; Pauly *et al.*, 2002). At the same time, consumer demand for seafood has steadily increased, with a 6% rise in per-capita seafood consumption from 1996 to 2001 (FAO, 2002). It is very clear that in order to ease pressures on wild fisheries stocks, and to satisfy the growing global consumption of seafood, marine species must be produced through aquaculture. Responding to that situation, aquaculture has been growing steadily during the last decade, and has in fact been the fastest growing sector in both US and world agricultural industries. During the last 10 years, global aquaculture production has expanded at a rate of over 10% per year, reaching a total production in 2000 of 45.7 million metric tonnes and a value of US$56.5 billion (FAO, 2002). Despite the industry's overall growth, the culture of freshwater species still comprises about two-thirds of global aquaculture production. The farming of marine species must be accelerated to meet the challenge of providing seafood to the growing world population, especially in the face of dwindling marine landings.

One of the major obstacles to the development of marine aquaculture is its interaction with the environment. The potential adverse effect of net-pen mariculture on the environment has been publicized widely (Naylor *et al.*, 1998, 2000; Goldburg *et al.*, 2001) and has become a source of unending controversy between environmentalists and mariculturists. At the centre of the controversy is the chemical pollution generated by net-pens and pond farming, as well as 'biological' pollution – the result of farmed fish mixing with wild stocks. However, the potential negative effect of mariculture on the marine environment is only one side of the problem. In net-pen and marine pond practices, the environment also exerts an adverse effect on farmed organisms. Fish reared in net-pens or ponds are exposed to water-borne contaminants and pathogens, algal toxins and sub-optimal environmental conditions. Recirculating, fully contained marine aquaculture, in either urban or rural communities, is one very promising avenue in the development of an environmentally sustainable and economically feasible marine aquaculture industry. This chapter discusses the rationale for the development of urban recirculating mariculture, and provides basic and applied findings generated by our urban mariculture programme in the city of Baltimore (Maryland, USA).

The Promise of Recirculating Mariculture

The most significant benefit of recirculating mariculture is that it alleviates the potential deleterious effect of fish farming on the environment. Our recirculating systems can grow high densities of marine fish with 98% containment of effluents, or better. The nominal amount of wastewater is disinfected and can easily be handled by a city or municipal sewer system – there is no direct waste release to the environment. Moreover, the high percentage of recirculation efficiency, together with the disinfection of effluents, provides an increased level of 'biosecurity.' The risk of escape of farmed organisms to the environment and, in turn, biological pollution, is significantly reduced.

The total containment and biosecurity of recirculating systems make them the only aquaculture practice that can safely farm non-indigenous species. A major advantage of aquaculture over the farming of other livestock is the ability to easily diversify to multiple high-value species. Being able to choose and alternate the species to be farmed based on their economic viability, as opposed to their geographic origin, would be highly beneficial for the industry. This is what drove our efforts to develop a Baltimore-based urban mariculture approach using the high value Mediterranean gilthead seabream (*Sparus aurata*).

Moreover, the full biosecurity provided by recirculating aquaculture systems make them the only aquaculture practice expected to receive regulatory clearance to grow genetically modified shellfish and finfish. Although the issue of transgenic fish is highly controversial, it is clear that genetically modified fish may offer a huge advantage to the industry (Devlin *et al.*, 1994; Hew *et al.*, 1995; Hew and Fletcher, 1997, 2001). Aquaculture scientists and regulators must be prepared for the potential future integration of genetically modified fish to the industry, and recirculating systems provide the only current solution to handle this developing scenario.

In addition to having minimal environmental impact, recirculating marine aquaculture systems have a number of advantages, making them the optimal approach for efficiently growing seafood and for future urban aquaculture expansion:

- Recirculating systems, unlike net-pens or ponds, usually use municipal or well water (and, in the case of marine systems, use municipal water for artificial seawater preparation) and are thus disease free. Moreover, water is continuously disinfected and potential pathogens eliminated.
- In recirculating systems, again unlike in ponds or net-pens, fish are grown in depurated water with no contaminants, toxins or off-flavour sources. Consequently, the produced fish are very 'clean' and can be marketed as such.
- Environmental conditions (water temperature, salinity, etc.) can be fully tailored to optimally fit the requirements of the fish of interest, thus ensuring optimal performances and the fastest growth rates to market size. As such, recirculating systems are generic, and can be modified to accommodate the species for which the market feasibility is highest.
- Because of the excellent water quality, recirculating systems can grow fish at very high densities. As is discussed later for our marine species, those densities are much higher than the densities practised in the net-pen industry.
- Recirculating aquaculture facilities are fully contained and thus can be located almost anywhere; they can be developed in rural or urban areas, in inner cities and in most warehouses. Site selection is not dictated by proximity to a natural source of water (e.g. lake, ocean), but rather by the business opportunity.

Economic Feasibility of Recirculating Mariculture Systems

The concept of recirculating aquaculture systems (RAS) was introduced relatively recently, in the 1970s (for review see Timmons *et al.*, 2002). The history and advancement of RAS technologies are well documented in two excellent books (Timmons and Losordo, 1994; Timmons *et al.*, 2002) and in the collection of proceedings from the 1996, 1998, 2000 and 2002 International Conferences on Recirculating Aquaculture (Virginia Polytechnic Institute and Virginia Sea Grant Extension Program, 2002). Although much progress has been made during the last 30 years in developing and refining RAS, it is clear that there is much work to be done to reduce the production cost of fish in such systems. A brief case history provided by Timmons *et al.* (2002) shows multiple examples of financial failures in the recent use of the RAS technology. Currently, initial investment in the facilities and the operational costs associated with RAS are relatively high. This led Timmons *et al.* (2002) to caution potential investors in RAS by reminding them of the following rule of thumb – 'Only invest what you can afford to lose'.

In almost all RAS business failures the problem was cost effectiveness. The vast majority of previous and current RAS facilities focus on freshwater or low salinity species, mainly tilapia and hybrid striped bass. With very low ex-farm selling prices (~US$3.30/kg for tilapia and US$5.00/kg for hybrid striped bass) it is very difficult for RAS production of these species to be profitable. Addressing the marine sector promises to be more economically feasible. Some of the high-value marine species, such as black or grey grouper, red and yellowtail snapper and summer flounder, command wholesale prices of US$7.50–10.00/kg for fresh fish on ice and US$11.00–13.50/kg for live fish (NOAA/NMFS, 2001). This has driven our efforts to develop and optimize a marine RAS and test its performance using a high-value marine species, such as the gilthead seabream (*Sparus aurata*).

Gilthead seabream is a highly valued farmed fish in the Mediterranean area, with an annual production of over 70×10^6 kg and a total value of about US$540 million (Kissil *et al.*, 2000; Theodorou, 2002). It is also a close relative of the red seabream (*Pagrus major*), a primary sushi fish in Japan. The gilthead seabream is imported to the US on a regular basis from Europe, with an estimated annual import volume of about 0.5×10^6 kg. The fish is relatively well known in North America, where it is considered a seafood delicacy, retailing for approximately US$20.00/kg. Our market research indicated that it could be sold ex-farm for US$12.00/kg live and US$9.00/kg fresh on ice. The fact that seabream is a non-indigenous species in North America means that it will be allowed to be grown only in fully contained and biosecure systems. Consequently, future production of this species in North America using RAS will not face competition from pond or net-pen production. Such competition was the main reason that the selling price of fish, such as hybrid striped bass, was driven down, which in turn led to the failure of several RAS producers of this species.

An additional sector of the industry where the RAS technology stands a higher probability of being profitable is that of hatchery and nursery production. Intensive and cost-efficient aquaculture requires a reliable production of 'seeds' (eggs and juveniles) year-round. This in turn means full control and phase shifting of all environmental parameters to which broodstock are exposed to enable year-round spawning. Optimal out of season larval and fry rearing also requires manipulating temperatures, salinities, day length, etc. Obviously, this is best and most economically achieved in fully contained RAS. Moreover, RAS-based hatcheries and nurseries are best situated to produce disease-free seeds, which is becoming an essential requirement for the growing aquaculture industry. Finally, good quality and disease-free seeds of aquacultured species are a high value-added product, especially when available year-round (Bromage and Roberts, 1995). These considerations make hatcheries and nurseries perfect candidates for economically feasible RAS operations.

In the remainder of this chapter, we will describe research efforts at the Center of Marine Biotechnology, University of Maryland Biotechnology Institute, to develop an efficient recirculating marine system for the cost-efficient production of high-value marine fish, focusing on the gilthead seabream model (Zohar *et al.*, 2002).

The Baltimore Urban Recirculating Mariculture System (BURMS)

During the design of our recirculating mariculture system, we kept several criteria in mind:

- We would be growing only fish (e.g. seabream) that had high market value and could not be grown in ponds or pens because of federal and state restrictions on non-indigenous species.
- We would make use of the pre-existing physical infrastructure in an urban environment (i.e. steam, city water and sewer, electricity).
- We would locate near the buyers to keep transportation costs low.
- We would attempt to grow fish at minimal discharge levels because of the cost associated with producing artificial seawater.

Given these criteria, we then heeded some sage advice accumulated during 30 years of recirculating aquaculture research and development (Timmons *et al.*, 2002) to choose components we deemed most important for growing high-value marine species. Our entire experimental marine recirculating facility is built in the basement of our building and occupies 1700 m^2. It contains hatchery, nursery and growout components and tanks ranging in volume from 350 l to 20 m^3. The total tank water volume of the facility is 205 m^3. For the experiments described in this chapter, designed and implemented to develop an intensive and high-density recirculating mariculture system, we used four tanks of 2.44 m diameter, with two life support systems, each controlling two tanks. Figure 10.1 presents a schematic of that experimental system with its major life support components. The rationale for the design of this system and the detailed description of its components are as follows.

Component 1 – Tank design

Our experimental unit included four 2.44 m diameter fibreglass tanks, each holding 3.8 m^3 of water. We chose circular, sloped-bottom tanks because of their self-cleaning operation and natural tendency to develop currents, which many marine species require. The tanks were covered, with a light source provided under the lid, to enable us to manipulate the photoperiod (e.g. provide long days) and maximize feeding. The tank covers also reduce heat loss and water loss due to evaporation. The only disadvantage associated with circular tanks involves harvesting and grading, which we overcame with radial crowders.

Component 2 – Particle removal

Particle removal (next to oxygen delivery) is the most important component in maintaining water quality. We used Hydrotech microscreen drum filters, which are mechanical, self-cleaning filters designed to achieve high performance in systems where prevention of particle fragmentation is essential. Solid-laden water flows, via gravity, through the filter, and solids are retained on a 60 μm screen. We decided to use drum microscreen removal of particles because it scales to larger tank volumes better than other solid removal systems, and is self-cleaning, thereby reducing maintenance needs and costs.

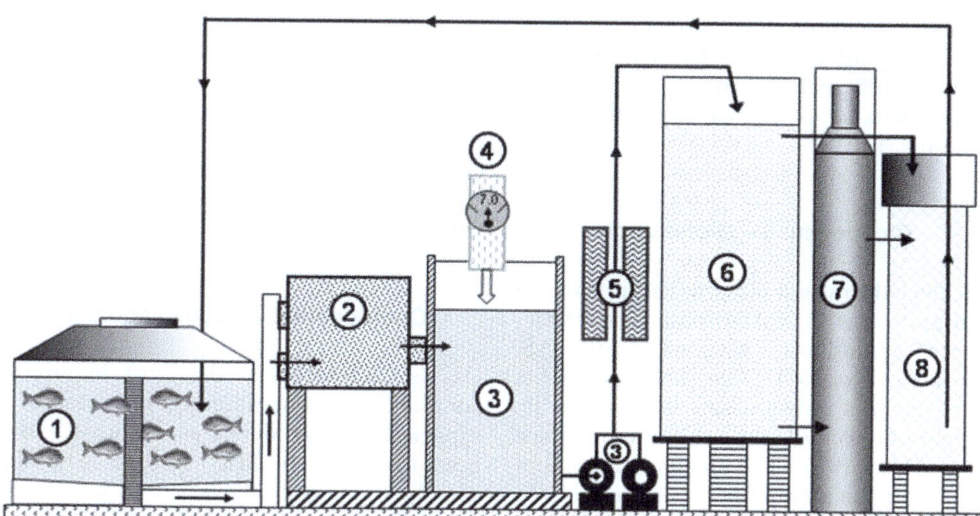

Fig. 10.1. A schematic drawing of the experimental Baltimore Recirculating Mariculture System. Numbered system components are: (1) fish tank, (2) particle removal, (3) sump [left] and pump [right], (4) pH doser, (5) temperature control; (6) biofiltration, (7) protein skimmer, (8) oxygen delivery.

Component 3 – Sump and pumping

Water from the drum screens flowed to a sump basin where water chemistry probes were located, pH adjustments were made and new water was added. It was essential for optimal performance of our systems to maintain two exchanges of the tank volume per hour. In order to achieve this, we used two pumps running at half capacity. With this redundancy, we ensured that water flow would not be impacted by the loss of one pump.

Component 4 – pH doser

pH was maintained at the desired level by an automatic doser that added 10% sodium hydroxide to the water in the sump basin if the pH dropped below the programmed level.

Component 5 – Temperature control

Being an experimental unit, we designed our systems to provide water temperature in the broad range of 10–32°C. In order to maintain constant water temperatures, we used our building's heating and cooling systems. Titanium heat exchangers were used to avoid corrosion by seawater and provide sufficient control to maintain the desired tank temperatures within the specified ranges.

Component 6 – Biofiltration

We elected to use a moving bed process of biofiltration with KMT-Biofilm carrier elements (Kaldnes Miljoteknoloi AS, Tonsberg, Norway). The KMT elements were made of polyethylene with a specific weight of 0.95 kg/l and a minimum surface area of 500 m^2/m^3. The elements were kept in continuous motion by air delivered from aerators located at the bottom of the reactor. The carrier elements were about 7 mm long and 9 mm in diameter and were shaped in such a way as to provide a large protected surface area for the microbial biofilm. The main advantages of the moving bed reactor (MBR) were that it was a compact biofilter with exceptional ability to accept high organic load, and that it was able to support a robust and very stable microbial biofilm production. Additionally, because of the constant water movement in the MBR, the KMT carrier elements remained clean, therefore clogging and sludge production and accumulation were eliminated. We used a single 2 m^3 cylindrical biofilter providing approximately 1050 m^2 of bioreactive surface area to handle two culture tanks.

The above-described biofilter was fully aerobic and, as is shown later, provided excellent nitrification. In order to enhance the efficiency and capacity of nitrogen removal, and in view of minimizing the seawater discharge from our system, we have initiated a comprehensive programme aimed at characterizing the microbial communities associated with biofiltration and at adding an anaerobic denitrification unit.

Component 7 – Ozone addition

In order to remove the high dissolved organic load and help flocculate the bacterial load, we used protein skimmers with ozone supplementation. Ozone was generated on-site from house air using an ozone generator (Pacific Ozone Technology) and piped into the protein skimmer venturi. Ozone was added on demand to maintain an oxidation–reduction potential (ORP) value of about 300 mV.

Component 8 – Oxygen addition

Oxygen delivery was the most important component in maintaining fish viability, especially at high densities. We used a low head oxygenation (LHO) unit to maintain dissolved oxygen above 4.5 ppm at all times. The overflow from the protein skimmer was added to the feed water of the LHO providing additional oxygen from the breakdown of the ozone.

Additional Controls

Photoperiod

All of our experimental tanks were covered by cone-shaped lids, under which were installed incandescent (natural light) greenhouse bulbs. Through computerized software, the light intensity and the photoperiod (hours of light and dark per day, and their gradual change-over time) were continuously controlled. An electronic system provided light dimming, enabling the simulation of dawn and dusk. Control of photoperiod is important to enable year-round spawning and juvenile production in hatcheries, and to enable optimal feed consumption at the growout stage.

Salinity and seawater make-up system

Our main objective was to develop recirculating marine aquaculture systems based on the use of artificial seawater. For experimental purposes, we developed a flexible system able to provide salinity levels ranging from fresh to full seawater (0–35 ppt) on a continuous basis. Salt water was prepared from Baltimore city water in a water-makeup system that delivered the desired salinity to all tanks. A salt storage tank contained 18,000 kg of technical grade sodium chloride. City water was pumped through charcoal filters and then through the salt storage tank where a saturated brine solution of sodium chloride was produced. This solution was then moved into a 9.5 m^3 reservoir where it was manually mixed with essential chemical ingredients found in seawater to provide a 35 ppt saltwater solution. Making our own salt mix from individual ingredients reduced the cost of saltwater by 30% as compared to the use of ready-made, commercially available salt mixes. A second reservoir of 9.5 m^3 held charcoal-treated freshwater. Water at the desired salinity was provided automatically and on-demand to each of the culture tanks.

Disinfecting wastewater

Our entire facility was designed to house non-indigenous species. As such, it was built with safety measures to ensure strict biosecurity. One such measure was the discharge of all waste effluents to a 9.5 m^3 reservoir, where the water was automatically chlorinated prior to its release to the municipal sanitary sewer.

Control of environmental and operational parameters

All the above-mentioned environmental and operational parameters of the system were continuously tracked through a real-time monitoring system connected to specific probes located in the tanks or the sumps. The monitored information was fed to a desktop computer that responded to readings outside the allowed ranges by triggering an appropriate warning. Actual control of these parameters was handled at each tank (e.g. oxygen delivery, pH buffering, etc.).

Operational Performance: Growing a High-value Marine Fish – Gilthead Seabream

Experimental conditions

Gilthead seabream (*Sparus aurata*) fry weighing 0.25–0.6 g were purchased from a Greek producer (referred to as Greek strain). They were air-shipped to Baltimore in plastic bags and grown to market size of 400–425 g under intensive conditions in the above-described tank system (Experiment 1). In Experiment 2, fry from our Spanish strain seabream broodstock were grown under identical conditions. Effluent from the culture tanks was gravity-fed to the 60 μm drum screen filter to remove solids, after which the water was collected in a sump basin. From the sump, the water was pumped through the titanium heat exchanger, into the nitrifying mixed bed bioreactor. Upon leaving the

MBB, water was gravity-fed into the LHO for carbon dioxide 'stripping' and supplementation of liquid oxygen, before returning to the fish tanks. A side stream fed water from the MBB to the ozone protein skimmer and then into the LHO. Additional oxygen was supplied to the fish via a pair of diffusers placed directly in the fish tanks. This latter addition was done to ensure that the tanks would be oxygenated even if flow stopped. The average flow rate in the system was 10 m^3 per hour. New water was added to the system at a rate of 7–10% of the total tank volume per day to make up for water lost to evaporation and during the self-cleaning cycle of the drum filter. pH levels were maintained at 7.5 as described above. Several water quality parameters, including temperature, O_2, CO_2, pH and ORP were monitored in real-time via computer, while other water chemistry parameters (NH_3, NO_3, NO_2, alkalinity and phosphate) were measured daily by staff.

Water temperature was maintained at 26°C. Photoperiod in the tanks was regulated via computer and maintained at 18:6 h light/dark cycle. The fish were grown at salinities ranging from 15 to 25 ppt, depending on the stage of their growth cycle. They were fed a commercially available red seabream extruded diet (EWOS Seabream Omega LE) containing 14% fat, 47% protein, 2% fibre and 28% carbohydrate. Food was delivered twice every hour via two automatic feeders per tank. Feeding rate was determined by body weight (BW). The fingerlings were fed at an initial rate of 6% BW/day. As they grew, this percentage was gradually reduced to 3.2% BW/day by the time they reached an average size of 40 g after which they were fed 1.5% BW/day. Growth rates were tracked by determining the average weight of 40 fish every 2–4 weeks. The fish were graded by size three times over the duration of the experiment.

Results

The results of two seabream growout experiments are illustrated in Fig. 10.2A and B. Figure 10.2A provides the full growth curve for the first experiment, in which seabream fingerlings of a Greek strain grew from 0.5 to 400 g commercial size in 268 days (under 9 months) and to 450 g in 300 days (10 months). Figure 10.2B presents the growth curve for the Spanish strain fingerlings, which grew from 0.5 to 410 g in 232 days (7.7 months). Overall survival rate in both experiments was over 90%. The above growth rates are very fast for seabream compared to the best growth performances in net-pens, reported at 387 days (12.9 months) and 420 days (14 months) from 1 to 400 g (Lupatsch and Kissil, 1998; and personal communications with industry, respectively). Commercial standards for growing seabream from 1 to 400 g in the Mediterranean net-pen indus-

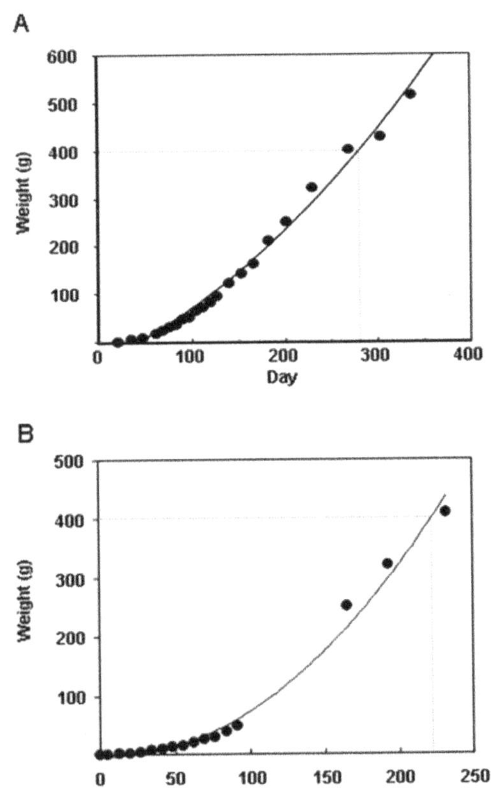

Fig. 10.2. Growth rate of seabream in the recirculating mariculture system. (A) Greek strain, 0.5–516 g. (B) Spanish strain, 0.5–410 g. The dotted lines indicate the time (days) needed to grow fish to 400 g. Note difference in x-axis (time) scales.

try are in the range of 13–17 months, depending on locations and strain used (Sahin, 1996; Kissil *et al.*, 2000). We attribute the excellent growth performance in our system to the fact that all environmental parameters (temperature, salinity, photoperiod, etc.) were tailored to optimally meet the specific physiological requirements of seabream at its different stages of growth and that feeding was on a high frequency schedule. Additionally, the continuous long days to which the fish were exposed delayed the onset of sexual maturation and gonadal growth, resulting in more energy diverted to muscular growth (Kissil *et al.*, 2001).

Table 10.1 summarizes some of the growth performance parameters of gilthead seabream in our system. As can be seen, the above-discussed growth rates were obtained at densities of 44–47 kg/m^3, much higher than the average 20 kg/m^3 standard densities in commercial seabream net-pens. In more recent experiments, we have been able to grow seabream at stocking densities exceeding 60 kg/m^3. Moreover, as can be seen in Table 10.1, at least for the first stages of growout to 76 g, we obtained excellent food conversion rates (FCR) of under 1 (0.9–40.8 g and 0.87–87 g). These values are much better than the 1.31–1.58 FCRs predicted for those weight ranges in the net-pen industry (Lupatsch and Kissil, 1998). We suggest that our better FCR values reflect the fact that fish in our tanks, in addition to all the enhanced environmental parameters noted above, have better access to food, compared to fish in floating net-pens. Our FCR value increased to 1.89 for fish in the 83–403 g size range. A similar increase in FCR, associated with the growth of the fish, has also been predicted by Lupatsch and Kissil (1998), who list theoretical FCR values of 1.58–1.79 for seabream in the size range of 100–400 g. We attribute the large FCR increase observed in our experiments to the fact that at a feeding rate of 1.5% of their body weight daily, we slightly overfed our large fish. Lupatsch and Kissil (1998) predict a 1.2–1.3% BW daily feed intake in seabream weighing 300–400 g. Nevertheless, overall the experimental FCR we observed for seabream grown in the marine recirculating systems is much better than the net-pen industry standard of 1.8 (Theodorou, 2002).

The water chemistry of the culture systems during our growout experiments is summarized in Table 10.2. Average total ammonia and nitrite levels were quite low, much below concentrations that are considered to be stressful to fish or to interfere with fish growth (Losordo and Westers, 1994; Timmons *et al.*, 2002). This is a reflection of the high efficiency of our nitrifying moving bed reactors. Surprisingly, relative to our earlier freshwater studies, the tank nitrate levels never built up. This is mainly the result of the 7–10% daily water exchange (which will have to be reduced, see below), but also due to the high heterotrophic bacterial growth in these organic rich systems. However, another factor that is probably responsible for the low nitrate level was revealed by a recent study, which characterized the microbial communities present in our biofilters (Tal *et al.*, 2003). This study demonstrated considerable denitrification and anaerobic ammonia oxidation (anammox) activities in our nitrifying filters. Anammox is a microbial process responsible for the direct conversion of ammonia to free nitrogen.

Table 10.1. Grow-out densities and FCR values for gilthead seabream grown in recirculating mariculture systems (Experiment 1).

Weight (g)	Duration (days)	Density (kg/m^3)	FCR
0.62–40.8	91	44	0.9
6.5–76	89	47	0.87
83–403	158	44	1.89

Table 10.2. Physical and chemical parameters (average where applicable) recorded during the seabream growth (Experiment 1).

Parameter	Mean ± SD
Water exchange	7–10% per day
Temperature	26 ± 2°C
pH	7.47 ± 0.11
Alkalinity	202 ± 36 mg/l ($CaCO_3$)
Total NH_3	0.50 ± 0.25 mg/l
NO_2	0.85 ± 0.45 mg/l
NO_3	50.4 ± 10.5 mg/l
PO_4	9.71 ± 2.6 mg/l

Increasing the Cost-efficiency of Marine Aquaculture Systems: Improving Nitrogen Removal Through Biofiltration

As explained earlier, the major goal of our studies is to develop the technology and protocols that will enable the industry to expand the production of marine finfish in urban recirculating systems. Cost-analysis of our technology has indicated that despite the excellent farm-gate value of seabream, its production cost is still relatively high. Dissecting the components of the production cost has shown that salt used for producing seawater accounts for about 25% of that cost. This figure is based on daily discharge and renewal of 10% of the total tank saltwater volume. As discussed above, our encouraging results and growth performance for seabream were obtained in systems in which we replaced 7–10% of the system's volume daily. In an effort to reduce the production cost in the marine RAS, we initiated a research programme aimed at decreasing the daily saltwater discharge from our tanks. Since the most important limiting factor to increasing the degree of recirculation is the build up of chemical waste, specifically of nitrogenous compounds, we set out to develop a better understanding of the biofiltration process as a basis for enhancing its biological nitrogen removal efficiency.

The process of biological nitrogen removal in recirculating aquaculture systems is carried out by microorganisms that colonize the biofilters (for review see Wheaton *et al.*, 1994; van Rijn, 1996; Timmons *et al.*, 2002). Despite the importance of nitrogen removal in recirculating aquaculture, there is a dearth of information about the identity of the microbial communities present in the biofilters, their biology, ecology and beneficial activities. Generally speaking, RAS biofilters are largely 'black boxes', and very little is known about the marine biological filtration process and the specific contribution of each microorganism to the process. Consequently, the main objectives of our study were to: (i) use modern tools of molecular biology to identify the microbial communities associated with our marine moving bed bioreactors (MBB); (ii) study the physiology of the identified microorganisms and their contribution to the nitrogen cycle; (iii) enhance the nitrogen removal activity of the MBBs by modifying and engineering its microbial consortia; and (iv) add a denitrifying, anaerobic component to our biological filtration process.

Our initial studies resulted in the identification of several unique autotrophic and heterotrophic bacteria associated with ammonia and nitrite oxidation (Tal *et al.*, 2003). In addition to nitrification activity, the microbial consortia associated with the MBRs were shown to carry out denitrification. More significantly, we have identified two unique marine *Planctomycetes* species capable of nitrogen removal via the anaerobic ammonia oxidation (anammox) process (Tal *et al.*, 2003). Anammox, which has only recently been shown to occur in the marine environment (Strous *et al.*, 1999), involves the reduction of ammonia to nitrogen gas using nitrite as an electron acceptor. Laboratory scale studies using the MBBs demonstrated that the anammox process could be induced in the system (Fig. 10.3). This complete nitrogen removal process is likely due to the participation of all nitrogen and carbon utilizers within the filter's microbial community.

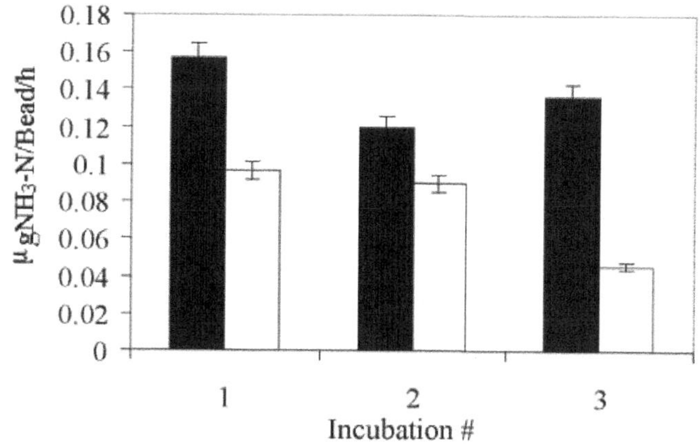

Fig. 10.3. Anammox potential of the bacterial consortia in MBRs. Ammonia removal rates of MBRs incubated with ammonia and nitrite (closed boxes) or ammonia alone (open boxes). Three separate incubations were performed under conditions that would stimulate anammox activity, i.e. anaerobic conditions with no added organic carbon source. The ammonia oxidation rate with nitrite as electron acceptor was between 0.29 and 0.33 mg NH_3–N/m^2/h compared to values of 0.092–0.20 mg NH_3–N/m^2/h during incubations without nitrite, suggesting the presence of anammox activity. For additional details see Tal *et al.* (2003).

Our innovative results demonstrated that the microbial consortia present in the MBRs have the potential to support different nitrogen transformation processes that enable closing the nitrogen cycle and releasing nitrogen back to the atmosphere. We are now in the process of studying the potential of using MBRs for complete nitrogen removal in our marine systems by combining nitrification with denitrification and the anammox process. We expect that these improvements will enable us to establish standardized marine nitrogen-removing consortia for use in inoculating MBR biological filters in recirculating mariculture systems.

Another avenue to achieving our goal of decreasing saltwater discharge from our system is the addition of an anaerobic denitrification unit (ADU) to our MBR nitrifying module, thus creating a two-stage nitrogen removal process. In our recent studies (Tal and Schreier, 2004), we supplemented our system (Fig. 10.1) with an ADU that was directly attached to the main biofiltration tank (label 6 in Fig. 10.1). The denitrification unit was found to reduce 90–100% of the daily nitrate production of the nitrifying filter resulting in minimal nitrate accumulation in the system. The two-stage biofiltration device was tested in our two 3.8 m^3 tanks in which adult seabream were grown at 40–50 kg/m^3 and fed daily at 1% of their body weight. During a 4-month experimental period, we were able to obtain daily water exchanges that averaged as low as 1% of the tank volume. This water exchange was significantly lower than the 7–10% used in our earlier study and is expected to reduce the costs associated with salt use in recirculating marine systems from as high as 25% to as little as 3% of the total production cost, thereby considerably enhancing the profitability of such systems. Presently, we are also engineering ways to recover salts lost through solids removal in order to gain additional savings in system costs.

Conclusions

This chapter describes the successful development and characterization of a small-scale pilot of environmentally compatible urban mariculture, based on the use of water recirculating technologies. Using a high-value marine species, the gilthead seabream, we demonstrated both the feasibility of this approach and excellent performances in terms of growth and food conversion rates, production densities and water quality. In additional studies not presented here, we closed the entire life cycle of seabream and other marine species in recirculating systems, including spawning and larval rearing. Based on our initial economic feasibility studies, we believe that, when properly scaled-up, our technology will support a profitable operation producing high-quality marine fish. The first such urban, commercial operation to be located in the city of Baltimore is now in the planning stages. While this technology is ready for scaling up, it is clear that much more research is required to strengthen its long-term outlook and success. Our current and future research efforts focus on diversifying the species to be farmed to include additional finfish, as well as shellfish species, and decreasing production cost associated with the loss of water and salt, through understanding and improving the efficiency of the biological filtration process.

Acknowledgements

We are grateful to Steven Rodgers, Eric Evans, James Frank and Jon Deeds for the technical help provided in the development and maintenance of our systems. Stanley Serfling and Dr Moti Harel also contributed to the design of the systems. Dan Gross contributed useful technical editing to the final text. This work was supported in part by a Maryland Industrial Partnership grant to ARP and a Binational Agricultural Research and Development Foundation award to HJS. This is contribution #05–112 from the Center of Marine Biotechnology.

References

Bromage, N.R. and Roberts, R.J. (1995) *Broodstock Management and Egg and Larval Quality.* Blackwell Scientific, Oxford.

Devlin, R.H., Yesaki, T.Y., Biagi, C.A., Donaldson, E.M., Swanson, P. and Chan, W.K. (1994) Extraordinary salmon growth. *Nature* 371, 209–210.

FAO (2002) *The State of World Fisheries and Aquaculture.* FAO Press, Rome, pp. 1–125.

Goldburg, R.J., Elliott, R.L. and Naylor, R.L. (2001) *Marine Aquaculture in the United States: Environmental Impacts and Policy Options.* Pew Oceans Commission, Arlington, Virginia, pp. 1–33.

Hew, C.L. and Fletcher, G.L. (1997) Transgenic fish for aquaculture. *Chemistry & Industry* 21, 311–314.

Hew, C.L. and Fletcher, G.L. (2001) The role of aquatic biotechnology in aquaculture. *Aquaculture* 197, 191–204.

Hew, C.L., Fletcher, G.L. and Davies, P.L. (1995) Transgenic salmon: tailoring the genome for food production. *Journal of Fish Biology* 47, 1–19.

Kissil, G.W., Tandler, A., Elizur, A., Colorni, A. and Zohar, Y. (2000) Gilthead seabream culture. In: Stickney, R.S. (ed.) *Encyclopedia of Aquaculture.* John Wiley & Sons, Chichester, pp. 408–416.

Kissil, G.W., Lupatsch, I., Elizur, A. and Zohar, Y. (2001) Long photoperiod delayed spawning and increased somatic growth in gilthead seabream (*Sparus aurata*). *Aquaculture* 200, 363–379.

Losordo, T.M. and Westers, H. (1994) System carrying capacity and flow estimation. In: Timmons, M.B. and Losordo, T.M. (eds) *Aquaculture Water Reuse Systems. Engineering Design and Management.* Elsevier Press, Amsterdam.

Lupatsch, I. and Kissil, G.W. (1998) Predicting aquaculture waste from gilthead seabream (*Sparus aurata*) culture using a nutritional approach. *Aquatic Living Resources* 11, 265–268.

Naylor, R.L., Goldburg, R.J., Mooney, H., Beveridge, M., Clay, J., Folke, C., Kautsky, N., Lubchenco, J., Primavera, J. and Williams, M. (1998) Nature's subsidies to shrimp and salmon farming. *Science* 282, 883–884.

Naylor, R.L., Goldburg, R.J., Primavera, J.H., Kautsky, N., Beveridge, M.C.M., Clay, J., Folke, C., Lubchenco, J., Mooney, H. and Troell, M. (2000) Effect of aquaculture on world fish supplies. *Nature* 405, 1017–1024.

NOAA/NMFS (2001) *Wholesale Fresh Prices at Fulton Fish Market – Annual Summary.* http://www.st.nmfs.gov/st1/market_news/
Pauly, D., Christensen, V., Guenette, S., Pitcher, T.J., Sumaila, U.R., Walters, C.J., Watson, R. and Zeller, D. (2002) Towards sustainability in world fisheries. *Nature* 418, 689–695.
Sahin, M. (1996) *Seabass and Bream in Floating Cages in Turkey.* Cahiers Options Mediterraneennes (CIHEAM). CIHEAM-IAMZ, Zaragoza, pp. 57–63, v. 14.
Schiermeier, Q. (2002) How many more fish in the sea? *Nature* 419, 662–665.
Strous, M., Fuerst, J.H., Kramer, E.H.M., Logemann, S., Muyzer, G., van de Psa-Schoonen, K.T., Webb, R., Kuenen, J.G. and Jetten, M.S.M. (1999) Missing lithotroph identified as new planctomycete. *Nature* 400, 446–449.
Tal, Y. and Schreier, H.J. (2004) Dissimilatory sulfate reduction as a process to promote denitrification in marine recirculating aquaculture systems. *Proceedings of the 5th International Conference on Recirculating Aquaculture.* Roanoke, Virginia.
Tal, Y., Watts, J.E.M., Schreier, S.B., Sowers, K.R. and Schreier, H.J. (2003) Characterization of the microbial community and nitrogen transformation processes associated with moving bed bioreactors in a closed recirculating mariculture system. *Aquaculture* 215, 187–202.
Theodorou, J.A. (2002) Current and future technological trends of European seabass–seabream culture. *Reviews in Fisheries Science* 10, 529–543.
Timmons, M.B. and Losordo, T.M. (1994) *Aquaculture Water Reuse Systems. Engineering Design and Management.* Elsevier Press, Amsterdam.
Timmons, M.B., Ebeling, J.M., Wheaton, F.W., Summerfelt, S.T. and Vinci, B.J. (2002) *Recirculating Aquaculture Systems.* USDA Northeastern Regional Aquaculture Center, Publication no. 01–002.
van Rijn, J. (1996) The potential for integrated biological treatment systems in recirculating fish culture – a review. *Aquaculture* 139, 181–201.
Virginia Polytechnic Institute and Virginia Sea Grant Extension Program (2002) *The Collection of Proceedings from the 1996, 1998, 2000 and 2002 International Conferences on Recirculating Aquaculture.* Virginia Polytechnic Institute and Virginia Sea Grant Extension Program.
Watson, R. and Pauly, D. (2001) Systemic distortions in world fisheries catch trends. *Nature* 414, 534–536.
Wheaton, F.W., Hochheimer, J.N., Kaiser, G.E., Krones, M.J., Libey, G.S. and Easter, C. (1994) Nitrification filter principles. In: Timmons, M.B. and Losordo, T.M. (eds) *Aquaculture Water Reuse Systems. Engineering Design and Management.* Elsevier Press, Amsterdam.
Zohar, Y., Serfling, S., Stubblefield, J., Place, A. and Harel, M. (2002) *Recirculating Marine Aquaculture Process.* US Process Patent No. 6,443,097.

This page intentionally left blank

11 Shrimp Culture in Urban, Super-intensive Closed Systems

Craig L. Browdy[1] and Shaun M. Moss[2]
[1]*South Carolina Department of Natural Resources, Charleston, USA;*
[2]*The Oceanic Institute, Waimanalo, USA*

Abstract

The development and expansion of a viable marine shrimp-farming industry in the USA has been a goal of the US Marine Shrimp Farming Program research consortium for many years. Production research has focused on the development of systems for open pond culture, and a small but stable industry has developed based on these technologies. New advances have facilitated the intensification of production, reduction or elimination of water exchange, application of biosecurity protocols and stocking of high-health and genetically improved shrimp, all of which currently serve as the foundation for the US shrimp-farming industry. Based on these research advances, consortium scientists have shifted resources over the past few years to re-evaluate super-intensive raceway production technologies. At the Waddell Mariculture Center (WMC) in South Carolina, and the Oceanic Institute (OI) in Hawaii, prototype and pilot-scale systems have been operated over the past 3 years. In recent WMC trials, production of up to 3 kg/m^2 was demonstrated with survival ranging from 55 to 71%, and harvest sizes from 14.6 to 17.1 g in 137 days. At harvest, production results for the latest trial based on the stocking of nursed 1 g juveniles were: survival = 91%; mean weight = 16.6 g; FCR = 1.54; mean growth/week = 1.44 g; and yield = 4.50 kg/m^2. At OI, three trials have been conducted with increasing stocking densities ranging from 100 to 300/m^2. In the most recent 85-day trial, juvenile shrimp (~2 g) grew to a harvest weight of 19.9 g with a mean growth rate of 1.47 g/week. Productivity of the system reached 5.2 kg/m^2 and survival was 86.3%. All of these trials are based on high-output, minimal water usage, enclosed raceway designs that assure biosecurity and provide excellent potential for application in non-traditional, urban and/or contaminated environments. The demonstration trials described in this chapter, along with supporting research on breeding, microbial dynamics, feeds, engineering, financial feasibility and marketing, have brought these technologies to the point of commercialization. Urban and peri-urban application of these technologies in the USA could offer opportunities for the development of integrated marketing initiatives to improve the outlook for financial viability.

Introduction

Over the past two decades, increasing efforts have been made to intensify pond production of cultured penaeid shrimp in the United States. With a limited growing season in the continental US, coupled with high labour, energy and land costs, and a strong technology and infrastructure base, achievement of high production per unit

(eds B. Costa-Pierce, A. Desbonnet, P. Edwards and D. Baker)

area has been an important strategy to allow domestic shrimp farmers to compete with foreign producers (Sandifer *et al.*, 1991a,b; Wyban and Sweeney, 1991). Research advances in the application of aeration technologies and management of intensive ponds have improved the environmental sustainability of intensive pond culture while increasing production (Hopkins *et al.*, 1992; Avnimelech, 2001; Delgado *et al.*, 2003). One of the most important areas in which management improvements have been made is in the reduction and elimination of water exchange (Hopkins *et al.*, 1995a; Browdy *et al.*, 2001a). These technologies have allowed growers to stabilize the pond environment, improve control over pond microbial communities, and reduce overall pollutant discharge to the environment. Concurrently, research and development efforts in the US have resulted in the establishment and widespread use of specific pathogen free (SPF) shrimp stocks along with other comprehensive biosecurity measures, which reduce the risk of excludable pathogen contamination of cultured stocks (Lotz, 1997; Jahncke *et al.*, 2002). Together with the advances made over the past decade in genetic selection and breeding, US growers now have commercially available, genetically improved shrimp stocks that can further improve the outlook for financial viability in a competitive global shrimp commodity marketplace (Moss *et al.*, 2001, 2003; Wyban *et al.*, 2003). Pond culture has undergone a relatively slow but stable expansion in the US over the past two decades, and recent application of shrimp culture in inland regions from Florida to Arizona will likely support continued expansion in the future. Nevertheless, pond production in the US will always be limited by competing land uses and financial pressures from foreign producers.

Further intensification of shrimp culture in the US beyond earthen pond systems has been achieved by several research groups. For example, beginning in the 1970s, a research effort by the Environmental Research Laboratory at the University of Arizona led to attempts to commercialize raceway production of *Litopenaeus stylirostris* in Hawaii (Moore and Brand, 1993). Although successes in reaching almost 5 kg/m^2 were achieved, financial viability was ultimately doomed by production issues and recurring infections with the IHHN virus (one of the first examples illustrating the importance of controlling or excluding pathogens for the sustainability of intensive shrimp culture systems). Researchers at the University of Texas also developed systems that facilitated biomass loadings as high as 14 kg/m^2 (Reid and Arnold, 1992; Davis and Arnold, 1998). Again, significant investments were made aimed at commercializing these technologies, but efforts to scale-up production did not yield the consistency necessary for financial viability. A third example of efforts to develop super-intensive raceway systems for shrimp production is illustrated by the work at the Harbor Branch Oceanographic Institution (Scarpa, 1998; Van Wyk, 1999, 2000, 2001). Researchers successfully developed a model for a three-phase raceway production system, and research and development efforts aimed at evaluating raceway designs, filtration systems and varying management protocols yielded promising results which demonstrated feasibility of super-intensive production in essentially fresh water. A small commercial operation based on these technologies exists, and shrimp sales into niche markets is ongoing. Over the past few years, HBOI shrimp research efforts have been privatized and commercialization efforts are continuing. Although these and other examples have demonstrated some successes, a commercial-scale, financially viable application of super-intensive raceway production technologies has been an elusive goal.

The US Marine Shrimp Farming Program consortium recently has focused its research efforts on developing advanced, controlled production systems for shrimp based on successful US agricultural models, including poultry and swine. With the introduction of pathogenic agents to the US in the mid 1990s, increasing research efforts have been devoted to improving the biosecurity of domestic shrimp production systems (Moss, 1998,

2000b; Bullis and Pruder, 1999). The consortium is integrating advanced technologies and applying expertise from each of the participating institutions. These include, but are not limited to: (i) intensive nursery raceway system development at the Texas Agriculture Experimental Station of Texas A&M University (Samocha *et al.*, 1993, 2002); (ii) development of SPF shrimp quarantined at the University of Arizona, the University of Southern Mississippi's Gulf Coast Research Laboratory, and the Oceanic Institute (Wyban *et al.*, 1992; Lotz and Lightner, 2000) and bred at the Oceanic Institute for disease resistance and fast growth (Moss *et al.*, 1998a; Argue *et al.*, 2002); and (iii) development and evaluation of intensive production systems at the Waddell Mariculture Center in South Carolina based on little or no water exchange, maximizing contributions of natural productivity and nutrient conversion efficiencies by managing autotrophic and heterotrophic components of the microbial community in the system (Hopkins *et al.*, 1995a, 1996; Browdy *et al.*, 2001b).

This chapter describes recent advances made at the Waddell Mariculture Center (WMC) in South Carolina and the Oceanic Institute (OI) in Hawaii in the development and demonstration of prototype and pilot-scale super-intensive raceway systems for the culture of the Pacific white shrimp, *Litopenaeus vannamei*. Because these systems are enclosed and managed with minimal water exchange, biosecurity procedures may be implemented easily to prevent losses due to infectious diseases (Moss *et al.*, 1998b; Ogle and Lotz, 1998; Bratvold and Browdy, 1999; Leung and Moss, 1999; Moss, 1998; Lotz and Lightner, 2000), year round production can be achieved and the environmental impacts due to discharges are reduced or eliminated (Browdy *et al.*, 2001b). The enclosed minimal exchange systems use considerably less land than that required for open pond culture, can serve as an excellent model for sustainability in aquaculture and provide the opportunity for urban and peri-urban culture operations.

South Carolina

The initial WMC pilot-scale system, constructed in 1999, consists of two 55 m^2 HDPE-lined, sediment-free raceways (18.3 $\times$ 3 $\times$ 1 m) enclosed in a greenhouse. A second commercial-scale prototype system, consisting of one 282 m^2 greenhouse-enclosed raceway (38.7 $\times$ 7.3 $\times$ 1 m), was completed in 2001. Each raceway contains PVC-suspended AquaMats™ for vertical surface enhancement (about 1 m^2 mat material per m^3 raceway water) and each has a centre partition to facilitate water movement. Aeration and water movement are supplied by blown air, Aire-O2® propeller-aspirator aeration units and/or paddlewheel aerators placed at both ends (deep and shallow) of each raceway. During the most recent trial, oxygen injection replaced the surface-mounted aeration, and a 0.7 m^3 bead filter was operated during the latter half of the cycle for solids removal. An auxiliary blower is available in the event of interruption of electrical power. Automatic switching, generator back-up, and alarm systems are of great importance as aeration failure under high oxygen demands can quickly lead to catastrophic crop loss. Supplemental propane-based heating is supplied via heat exchange units. Raceway bottoms are sloped to facilitate drainage and harvest. In the commercial-scale prototype, a 3.75 kW water pump system recirculates water within the systems with outlets situated at 1 m intervals along the bottom of the raceway periphery. A 65% shade cloth structure is used to cover each greenhouse during summer months to prevent elevated water temperatures. In addition, each system is equipped with fans for temperature control. To reduce shrimp losses due to jumping, 1.3 cm mesh plastic netting encircles the periphery of each raceway.

During the first phase of technology development (1999–2002), five shrimp production trials were completed using the pilot-scale system and one trial was completed using the commercial-scale system. With the exception of stocking density, salinity and origin/age of culture water, operational and experimental procedures employed during each of these trials has remained constant. The duration of each trial was about 3 months.

For each trial, raceways were filled and fertilized as necessary to ensure establishment and maintenance of microbial communities. After *L. vannamei* (PL 14–20) were stocked, #1–3 starter diets (40% protein, 8% lipid, 5% squid meal; Rangen, Inc., Buhl, Idaho, USA) were fed until shrimp attained a size ranging from 0.5 to 1.0 g. At this time, a sinking pelleted feed (35% protein, 8% lipid, 2.5% squid meal; Rangen, Inc.) was offered, and sampling of shrimp weight initiated. About 25% of the daily feed ration was distributed on feed trays located at the deep and shallow ends of each raceway; the remaining amount was dispersed evenly throughout the culture unit. Shrimp were sampled every 10 days. Feeding rates were adjusted based on estimated weight and monitoring of feed tray consumption.

Water temperature, dissolved oxygen (DO) concentration, salinity and pH were measured daily between 0800 and 0900 hours. Total ammonia–nitrogen (TAN), nitrite–nitrogen (NO_2–N), nitrate–nitrogen (NO_3–N) and total alkalinity were measured weekly. To maintain pH levels > 7.0, sodium bicarbonate was added periodically according to the method of Loyless and Malone (1997) to maintain total alkalinity levels > 120 mg/l as $CaCO_3$. Fresh groundwater was periodically added to offset evaporative losses and to maintain salinity levels. Surface sludge and flocculants were removed manually as needed to prevent excessive accumulation.

At harvest, water was drained to concentrate shrimp in the deep end of each raceway. Shrimp movement was aided via water flow. Shrimp were collected, and total harvest weight was used to determine yield, percentage survival and FCR. Further details on design and operation parameters may be found in Weirich *et al.* (2002).

Four trials were completed in the pilot-scale system. Year-round production and reuse of culture water up to three times was demonstrated. Yields of 2.3 and 2.4 kg/m^2 were achieved for Trial 2 at a stocking density of 200/m^2. Mean weight at harvest of 19.3 and 18.9 g and survival of 60.1 and 63.9% were recorded for pilot Raceways 1 and 2, respectively. Lower survival rates were recorded in Trials 1, 3 and 4 due to problems with mechanical aeration failure, and issues related to husbandry at stocking densities of 300 and 400 shrimp/m^2. Further details on production results for these trials were reported in Weirich *et al.* (2002).

Trial 5 was conducted from September 2001 to January 2002 (duration = 137 days) using the original pilot-scale system, as well as the recently constructed commercial-scale system. Water used in prototype RW 1 during Trial 4 was discarded and this culture unit was refilled with filtered seawater and fresh groundwater to achieve a nominal salinity of 15 g/l. Water from Trial 4 was reused in RW 2, thereby allowing a comparison of 'new' versus 'old' culture water. The commercial-scale system (RW 3) was filled with seawater (bag-filtered to 150 μm) from an adjacent outdoor earthen pond, and fresh groundwater, to achieve a nominal salinity of 15 g/l. Mean growth rate of shrimp from PL to harvest was 0.8, 0.9 and 1.0 g/week in RW 1, 2 and 3, respectively (Table 11.1). Growth remained at or near the target rate of 1.3 g/week until December (growth also declined during the latter stages of Trials 1–4). Notwithstanding reduced growth observed during the latter phase of Trial 5, final weight of shrimp reared in both prototype systems was of acceptable market size (Table 11.1). In addition, FCR and yield were lower and higher, respectively, than that observed for previous trials. Regarding

Table 11.1. Waddell Mariculture Center Trial 5 super-intensive raceway *L. vannamei* production characteristics.

Parameter	RW 1	RW 2	RW 3
Weight (g)	14.6	15.4	17.1
Survival (%)	70.5	71.5	55.2
FCR	1.8	2.0	1.9
Yield (kg/m^2)	3.1	3.3	2.8

the comparison of 'new' (RW 1) and 'old' (RW 2) water, no distinct differences with respect to production characteristics were observed.

Values (means and ranges) of selected water quality parameters measured during Trial 5 are shown in Table 11.2. With the exception of a three-week period during the middle portion of the trial in which NO_2–N levels were elevated in RW 3, all water quality parameters were maintained within acceptable ranges for culture of *L. vannamei* (Van Wyk and Scarpa, 1999). As with production characteristics, there were no obvious differences regarding water quality parameters between 'new' (RW 1) and 'old' (RW 2) water sources, suggesting that water can be reused over the course of multiple production cycles in minimal exchange systems. Moreover, reuse of water may not only be beneficial with respect to environmental and biosecurity issues, but this practice may provide a more stable production environment. For example, TAN and NO_2–N levels of the original pilot-scale system culture units were fairly unstable during Trials 1 and 2, and NO_2–N levels of the commercial-scale system exhibited a moderate spike during Trial 5. However, because TAN and NO_2–N levels of RW 1 (new water) remained low throughout Trial 5, it is possible that microbial and algal communities attached to AquaMats™ (and surviving between harvest and restocking) may play a significant role in system stability. Although the effects of vertical surface addition on shrimp production systems have not been fully evaluated, Bratvold and Browdy (2001) found significant production advantages in systems containing high densities of AquaMats™. The authors observed that pH levels were more stable, and that water column nitrification rates of tank systems containing AquaMats™ were enhanced relative to that of tanks with no vertical surface additions, suggesting that nutrient recycling rates may be improved via the use of this culture technology.

As described above, raceways were operated with no external filtration or removal of organic material during the production cycle. One of the distinguishing characteristics of this type of minimal exchange shrimp production system is the eventual development of suspended and surface flocculants (Browdy *et al.*, 2001a; Chamberlain *et al.*, 2001; McIntosh, 2000, 2001). These flocculants develop typically within 6–7 weeks after production is initiated (Browdy *et al.*, 2001a; McIntosh, 2001). It has been suggested that flocculants and associated organic particles contribute to shrimp nutrition and therefore nutrient recycling (Moss, 1995, 2000a; Moss and Pruder, 1995; McIntosh, 2000, 2001). Although the contribution of flocculants to the nutrition of shrimp reared in WMC systems has not been evaluated on a qualitative or quantitative basis, shrimp are routinely observed near the surface, presumably feeding directly on flocculant material.

Table 11.2. Means and ranges (in parentheses) of water quality parameters measured during Waddell Mariculture Center Trial 5 super-intensive raceway growout of *L. vannamei*.

Parameter	RW 1	RW 2	RW 3
Temperature (°C)			
AM	28.5 (25.7–30.3)	28.8 (25.7–30.5)	28.5 (24.2–30.6)
PM	29.5 (26.2–31.1)	29.8 (25.9–31.1)	29.7 (25.0–32.7)
Dissolved oxygen (mg/l)	6.3 (5.0–7.9)	6.1 (4.9–7.7)	5.8 (4.3–7.2)
Salinity (g/l)	15.0 (14.3–15.6)	14.9 (14.1–16.1)	15.1 (14.6–15.7)
pH	7.7 (7.1–8.3)	7.7 (7.1–8.2)	7.6 (7.1–8.3)
Alkalinity (mg/l as $CaCO_3$)	180 (140–240)	229 (160–340)	144 (120–220)
Total ammonia–nitrogen (mg/l)	0.01 (0–0.10)	0.01 (0–0.10)	0.15 (0–1.80)
Nitrite–nitrogen (mg/l)	0.13 (0.03–1.40)	0.18 (0.03–0.50)	0.94 (0–7.60)
Nitrate–nitrogen (mg/l)	22 (1–38)	29 (10–40)	19 (1–40)

Because water quality parameters were within acceptable ranges (see Table 11.2), reduced growth simply may have been due to late-season density/water effects inherent to these heavily aerated super-intensive culture systems when operated with no removal of organic material. It is important to note that a large percentage of shrimp reared in the commercial-scale system began to exhibit clinical signs indicative of black spot disease (Main and Laramore, 1999) in mid-December, characterized by melanization of wounds located dorsally on abdominal segments 1–3. The outbreak of this condition corresponded with an aeration malfunction event during which shrimp were observed jumping within the system followed by mass moulting. It is speculated that shrimp became injured as they moulted, and came into contact with other shrimp, aerators and/or raceway structures. However, no mortalities were observed. The condition persisted until harvest, although the severity of the condition decreased over time.

Phase one results suggested that satisfactory production of shrimp up to about 3 kg/m^2 can be achieved without external filtration using only supplemental aeration. However, problems with reduced growth and associated shell disease suggested that the system was nearing maximum carrying capacity under these conditions. Therefore, several modifications in operational protocols were made before the subsequent trial stocked in January 2003. A 0.7 m^3 bead filter was operated during the latter half of the cycle, effectively cropping solids from the system. Backwash was collected in an external reservoir and water was returned to the system following sludge sedimentation. Also, in place of the paddlewheel and aspirator units used in previous trials, oxygen supplementation was provided via injection by an Air Products Specialty VSA oxygen generator (Model No. A-150 L). With the addition of airlift units, raceway flows were reduced and a more even water movement was achieved without turbulence associated with paddlewheel and aspirator aerators. The system was stocked at 300 shrimp/m^2 with 1 g juveniles. Other operational parameters were similar to previous trials.

Harvest production results demonstrated an improvement over previous trials. A survival of 91% was observed. Survival of nursed shrimp in growout can be higher and more predictable when compared to direct stocked PL (Samocha *et al.*, 1993). Growth rates improved with a mean of 1.44 g/week. Growth increased significantly above target levels of 1.3 g/week when oxygen supplementation and particulate removal were initiated (Fig. 11.1). With improved growth and stocking of nursed juveniles, growout time was reduced to 11 weeks. Mean weight at harvest was 16.6 g, resulting in a final production rate of 4.50 kg/m^2. Preliminary economic feasibility analyses suggest that with appropriate economies of scale, production results achieved in this trial of the prototype system may provide a basis for profitable shrimp production in South Carolina.

Hawaii

The recirculating system used for the production of market sized *L. vannamei* at the Oceanic Institute has been described by Moss *et al.* (2002) and Otoshi *et al.* (2002) as operated during three growout trials from 1999 to 2001. Briefly, the system consisted of a concrete 58 m^2 raceway that was filled with seawater (34 ppt) from an underground aquifer to a depth of about 65 cm. A 2 HP, aspirator-type aerator was used to provide aeration and to move water in a circular pattern around a central baffle. Water flow produced a scouring velocity to keep solids in suspension. For filtration, a 0.7 m^3 propeller-washed bead filter was used for solids removal and biological filtration (Malone *et al.*, 1998). The bead filter allowed a sufficient amount of microalgae and other suspended particles to pass through so a 'green water' environment was maintained. This is important because shrimp reared in water with high concentrations of microalgae and microbial/detrital aggregates grow better than shrimp reared in clean, filtered seawater (Moss *et al.*, 1992; Moss and Pruder, 1995). Clear,

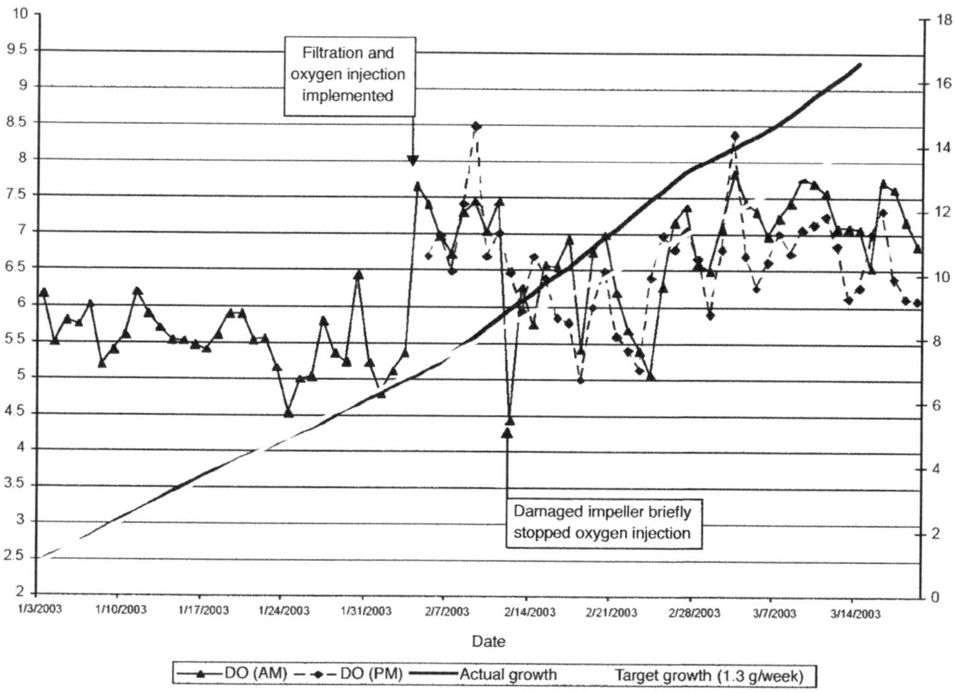

Fig. 11.1. Growth of *Litopenaeus vannamei* in a super-intensive raceway system at the Waddell Mariculture Center operated with mechanical filtration and oxygen supplementation.

plastic sheeting (0.1524 mm) was used to cover the raceway as a biosecurity feature to reduce pathogen introduction by airborne vectors. The cover also served as an effective thermal insulator to maintain desirable water temperatures.

Shrimp harvest results are summarized for three production trials conducted between 1999 and 2001 in Table 11.3. The first trial was stocked with nursed juveniles at a density of 100 shrimp/m^2 (mean weight = 1.87 g). Shrimp were fed a commercial, 30% protein feed (Rangen Inc., Buhl, Idaho, USA) five times daily. After 16 weeks, shrimp were harvested from the raceway at a mean weight of 22.99 g. Final survival was 88.2%, mean growth rate was 1.32 g/week, final production was about 2.0 kg/m^2 and 483 l of water were used to produce 1 kg of shrimp. The second trial was stocked with juveniles (mean weight = 2.15 g) at an initial density of 200 shrimp/m^2. Shrimp were fed a commercial, 35% protein feed five times daily. After 16 weeks, shrimp were harvested from the raceway at a mean weight of 23.24 g. Final survival was 85.0%, mean growth rate was 1.32 g/week, final production was about

Table 11.3. Shrimp performance during three growout trials for *L. vannamei* in super-intensive raceway systems at the Oceanic Institute.

	Stocking density (shrimp/m^2)	Duration (weeks)	Survival (%)	Harvest weight (g)	Growth rate (g/week)	Production (kg/m^2)
Trial 1	100	16.0	88.2	22.99	1.32	2.0
Trial 2	200	16.0	85.0	23.24	1.32	4.0
Trial 3	300	12.2	86.3	19.89	1.47	5.2

4.0 kg/m^2 and 370 l of water were used to produce 1 kg of shrimp. The third trial was stocked with juvenile shrimp (mean weight = 2.01 g), this time at an initial density of 300 shrimp/m^2. Shrimp were fed a commercial, 35% protein feed five times daily. After about 12 weeks, shrimp were harvested from the raceway at a mean weight of 19.89 g. Final survival was 86.3%, mean growth rate was 1.47 g/week, final production was 5.2 kg/m^2 and 352 l of water were used to produce 1 kg of shrimp.

These results indicate that stocking density did not negatively affect shrimp growth or survival in the recirculating raceway. In fact, shrimp stocked at 300/m^2 exhibited a mean growth rate that was 11% greater than the growth rate for shrimp stocked at 100/m^2. In addition, survival was similar at these two stocking densities. Improved shrimp performance in Trial 3 may be attributed to several factors, including the benefits associated with two generations of selective breeding for growth under intensive culture conditions, a 5% increase in protein level in the diet and a better understanding of the culture system gained through the experience of managing two additional growout trials. In contrast to the results indicated above, several reports in the literature point to an inverse linear relationship between stocking density and growth for marine shrimp in intensive systems (Williams *et al.*, 1996; Davis and Arnold, 1998; Tseng *et al.*, 1998).

The importance of selective breeding as a tool to attain rapid shrimp growth and high survival under intensive culture conditions cannot be overstated. In each of the three raceway growout trials described above, shrimp were tagged to evaluate family performance for a selective breeding programme at OI. The fastest growing families from each of the three growout trials exhibited mean growth rates greater than 1.72 g/week. In contrast, the slowest growing families from each of the three trials were controls that exhibited mean growth rates less than 1.05 g/week (Moss *et al.*, 2002). Data for shrimp families selected for growth at harvest in these raceway trials were compared to unselected controls and results indicate that significant improvement in growth has been achieved through selective breeding. For example, in Trial 1, selected families exhibited a mean weight gain (±SD) of 22.9 ± 2.1 g after 16 weeks, whereas control families exhibited a mean weight gain (±SD) of 18.9 ± 2.7 g (Argue *et al.*, 2002). Thus, there was a 21.2% difference in mean weight gain between selected and unselected shrimp, and this difference would significantly impact production and profitability for the shrimp farmer.

Water quality parameters for these production trials are reported in detail in Moss *et al.* (2002). Temperature was maintained at optimal levels (averaging 29.0, 28.7 and 29.5°C in Trials 1–3, respectively) due to the low water use and plastic cover over the raceway. With few exceptions, DO concentrations were maintained above 5.0 mg/l, 3.0 mg/l and 4.0 mg/l throughout Trials 1–3, respectively. TAN and NO_2–N concentrations in Trials 1–3 are shown in Table 11.4. At times, TAN and NO_2–N concentrations exceed levels suggested to be toxic to penaeid shrimp (Chen and Tu, 1991; Chien, 1992). However, in these systems shrimp continue to grow, suggesting system conditions or shrimp stocks in the present studies enhance tolerance to high TAN and NO_2–N concentrations.

Conclusions

Water use and eutrophication of receiving waters have been identified as important factors constraining environmental sustain-

Table 11.4. Mean and range (in parentheses) of total ammonia–nitrogen (TAN) and NO2–N concentrations during three growout trials for *L. vannamei* in super-intensive raceway systems at the Oceanic Institute.

	TAN (mg/l)	NO2–N (mg/l)
Trial 1	0.33 (0–2.56)	2.11 (0.52–6.34)
Trial 2	1.31 (0.01–5.13)	4.39 (0.04–14.79)
Trial 3	4.57 (0.09–15.58)	7.21 (0.05–20.46)

ability of marine shrimp-farming (Boyd and Musig, 1992; Hopkins *et al.*, 1995b; Boyd and Clay, 1998; Moss *et al.*, 2001). An important metric for assessing the efficacy of water conservation in shrimp aquaculture is the amount of water used to produce shrimp biomass. The super-intensive raceway technologies described in the present paper do not rely on water exchange. Water addition is only used to make up for losses in filter backwash cycles and evaporation. Thus, seawater use reported here is less than 500 l/kg of market-sized shrimp produced. The present results, coupled with advanced system designs aimed at facilitating dewatering of solids removed from the system and reuse of water between crops, will eliminate the need for building these production systems in proximity to coastal waters. This immediately improves the outlook for control of facility biosecurity, while greatly expanding the range of possible sites. The elimination of reliance on natural seawater is one of the most important factors contributing to the facilitation of urban or peri-urban application of super-intensive shrimp-production technologies.

Although results reported here demonstrate the biological feasibility of these advanced production technologies, the assurance of financial feasibility is a prerequisite to successful large-scale application of these technologies. As an initial starting point for preliminary economic feasibility analyses, a 'base case' production system, along with production and economic assumptions, is being hypothesized based upon the experimental work described above and from other information sources. A hypothetical base-case production unit in the initial analysis is comprised of five modules, each of which is comprised of six, 750 m^2 raceways covered by biosecure interlinked greenhouses. It was assumed that seawater would be pumped in and treated from a nearby source (e.g. an estuarine creek) and this treated seawater would be heated via a heat-exchanger system during colder months. Major preliminary base-case production assumptions included direct stocking of 1 g shrimp at 300/m^2, overall survival to harvest size of 85%, mean harvest size of 17 g, growth rate of 1.3 g/week allowing four crops per raceway over a 12 month period and an FCR of 1.50. Major economic assumptions included juvenile stocking costs of US$11.50/1000 and feed costs of US$0.50/kg. The projected total estimated cost of production per kg is US$4.36 with variable costs making up US$3.27/kg. This preliminary analysis indicates that one of the most important parameters relative to reducing production costs is growth rate. For example, maintaining a mean growth rate of 1.44 g/week (Table 11.3), rather than 1.3 g/week in the base case scenario, would reduce projected total production costs by US$0.21/kg. Consequently, it appears that cost-effective techniques that can improve growth rates, such as the selection of fast-growing shrimp strains, could significantly improve the financial performance of indoor recirculating biosecure systems.

Although the initial production cost projections appear to be high compared to those which may be achieved in existing pond-based tropical shrimp farms in developing countries, urban and peri-urban production of shrimp in proximity to major markets could provide significant marketing advantages and opportunities. A number of studies recently carried out in the southeastern US suggest some potential for product differentiation, as well as opportunities for directly accessing restaurant and consumer markets for farm-raised shrimp (Wirth and Davis, 2001a,b). Based on consumer preferences and initial positive responses to the quality of the shrimp produced from these systems, significant opportunities for product differentiation from imported frozen commodity shrimp may exist. Thus, development of these advanced systems in peri-urban areas will likely involve integrated specialized marketing initiatives designed to utilize the consistent output of fresh, high quality shrimp. This becomes particularly attractive in large urban areas away from traditional sources of seafood. Energy efficient low water use systems could be developed in proximity to virtually any major urban metropolitan area.

Despite encouraging preliminary results, there are a number of critical areas where further research is needed, including: (i) establishment of selective breeding programmes that integrate molecular-based techniques to promote rapid shrimp growth and high survival in super-intensive, biosecure production systems; (ii) development of aquatic feeds appropriate for biosecure systems that rely on minimal water exchange; (iii) understanding the microbial ecology of these systems to enhance shrimp performance and improve water quality; (iv) identification of efficient and cost-effective system design and engineering to accommodate scale-up to commercially viable facility configurations, temperature control in temperate climates, water filtration, water reuse, aeration, application of oxygen supplementation and CO_2 degassing; (v) management of waste materials from solids removal and shrimp processing; (vi) cost-effective formulations for dissolved solids to facilitate application of artificial seawater for inland applications; (vii) marketing efforts to differentiate the environmentally friendly, locally produced, fresh, high-quality shrimp products produced in these systems; and (viii) economic feasibility models that incorporate sensitivity analyses to identify areas where further research may contribute to reducing production costs and increased competitiveness.

Acknowledgements

This work was funded by the US Marine Shrimp Farming Program, USDA-CSREES. Reference to trade names does not imply endorsement. The present research would not have been possible without the hard work and dedication of Clete Otoshi and Brad McAbee in operating the raceway systems. The authors wish to express their appreciation to S. Arce, M. Brown, C. Bruce, J. Bruce, L. Bush, A. Cammarano, D. Deloach, G. Hagino, C. Hamilton, K. Houston, E. Matsuda, S. Miller, T. Mongomery, D. Moss, G. Pruder, J. Richardson, A. Segars, A. Stokes and C. Weirich for their assistance during the course of this research. This is contribution number 538 from the SCDNR Marine Resources Division.

References

Argue, B.A., Arce, S.M, Lotz, J.M. and Moss, S.M. (2002) Selective breeding of Pacific white shrimp (*Litopenaeus vannamei*) for growth and resistance to Taura Syndrome Virus. *Aquaculture* 204, 447–460.

Avnimelech, Y. (2001) Aeration, mixing, and sludge control in shrimp ponds. *Global Aquaculture Advocate* 4(4), 51–53.

Boyd, C.E. and Clay, J.W. (1998) Shrimp aquaculture and the environment. *Scientific American* 278(6), 58–65.

Boyd, C.E. and Musig, Y. (1992) Shrimp pond effluents: observations on the nature of the problem on commercial farms. In: Wyban, J. (ed.) *Proceedings of the Special Session on Shrimp Farming*. The World Aquaculture Society, Baton Rouge, Louisiana, pp. 195–197.

Bratvold, D. and Browdy, C.L. (1999) Disinfection, community establishment and production in a prototype biosecure shrimp pond. *Journal of the World Aquaculture Society* 30, 422–432.

Bratvold, D. and Browdy, C.L. (2001) Effects of sand sediment and vertical surfaces (AquaMats™) on production, water quality, and microbial ecology in an intensive *Litopenaeus vannamei* culture system. *Aquaculture* 195, 81–94.

Browdy, C.L., Bratvold, D., Stokes, A.D. and McIntosh, R.P. (2001a) Perspectives on the application of closed shrimp culture systems. In: Browdy, C.L. and Jory, D.E. (eds) *The New Wave: Proceedings of the Special Session on Sustainable Shrimp Culture, Aquaculture 2001*. The World Aquaculture Society, Baton Rouge, Louisiana, pp. 20–34.

Browdy, C.L., Bratvold, D., Hopkins, J.S., Stokes, A.D. and Sandifer, P.A. (2001b) Emerging technologies for the mitigation of environmental impacts associated with shrimp aquaculture pond effluents. *Asian Fisheries Science* 134, 255–267.

Bullis, R.A. and Pruder, G.D. (1999) *Controlled and Biosecure Production Systems. Evolution and Integration of Shrimp and Chicken Models.* Oceanic Institute, Honolulu, Hawaii.

Chamberlain, G., Avnimelech, Y., McIntosh, R.P. and Velasco, M. (2001) Advantages of aerated microbial reuse systems with balanced C:N: II. Composition and nutritional value of organic detritus. *Global Aquaculture Advocate* 4(4), 22–24.

Chen, J.-C. and Tu, C.-C. (1991) Influence of ammonia on growth of *Penaeus monodon* Fabricius post-larvae. *Aquaculture and Fisheries Management* 22, 457–462.

Chien, Y.H. (1992) Water quality requirements and management for marine shrimp culture. In: Wyban, J.M. (ed.) *Proceedings of the Special Session on Shrimp Farming.* The World Aquaculture Society, Baton Rouge, Louisiana, pp. 145–156.

Davis, D.A. and Arnold, C.R. (1998) The design, management, and production of a recirculating raceway system for the production of marine shrimp. *Aquaculture Engineering* 17, 193–211.

Delgado, P.C., Avnimelech, Y., McNeil, R., Bratvold, D., Browdy, C.L. and Sandifer, P.A. (2003) Physical, chemical and biological characteristics of radially aerated shrimp ponds. *Aquaculture Engineering* 217, 235–248.

Hopkins, J.S., Stokes, A.D., Browdy, C.L. and Sandifer, P.A. (1992) The relationship between feeding rate, paddlewheel aeration rate and expected dawn dissolved oxygen in intensive shrimp ponds. *Aquacultural Engineering* 10, 281–290.

Hopkins, J.S., Sandifer, P.A. and Browdy, C.L. (1995a) A review of water management regimes which abate the environmental impacts of shrimp farming. In: Browdy, C.L. and Hopkins, J.S. (eds) *Swimming Through Troubled Water, Proceedings of the Special Session on Shrimp Farming, Aquaculture '95.* World Aquaculture Society, Baton Rouge, Louisiana, pp. 66–75.

Hopkins, J.S., DeVoe, M.R., Holland, A.F., Browdy, C.L. and Stokes, A.D. (1995b) Environmental impacts of shrimp farming with special reference to the situation in the continental United States. *Estuaries* 18, 25–42.

Hopkins, J.S., Sandifer, P.A., Browdy, C.L. and Holloway, J.D. (1996) Comparison of exchange and no-exchange water management for the intensive culture of marine shrimp. *Journal of Shellfish Research* 13, 441–445.

Jahncke, M.L., Browdy, C.L., Schwarz, M.H., Segars, A., Silva, J.L., Smith, D.C. and Stokes, A.D. (2002) *Application of Hazard Analysis Critical Control Point (HACCP) Principles as a Risk Management Tool to Control Viral Pathogens at Shrimp Production Facilities.* Publication Number VSG 02 10. Virginia Sea Grant, Charlottesville.

Leung, P.S. and Moss, S.M. (1999) Economic assessment of a prototype biosecure shrimp growout facility. In: Bullis, R.A. and Pruder, G.D. (eds) *Controlled and Biosecure Production Systems: Evolution and Integration of Shrimp and Chicken Models. Proceedings of a Special Session of the World Aquaculture Society,* Sydney, Australia. The Oceanic Institute, Honolulu, Hawaii, pp. 97–106.

Lotz, J.M. (1997) Disease control and pathogen status assurance in an SPF-based shrimp aquaculture industry, with particular reference to the United States. In: Flegel, T.W. and MacRae, I.H. (eds) *Diseases in Asian Aquaculture III.* Fish Health Section, Asian Fisheries Society, Manila, pp. 243–254.

Lotz, J.M. and Lightner, D.V. (2000) Shrimp biosecurity: pathogens and pathogen exclusion. In: Bullis, R.A. and Pruder, G.D. (eds) *Controlled and Biosecure Production Systems. Proceedings of a Special Session – Integration of Shrimp and Chicken Models.* The Oceanic Institute, Waimanalo, Hawaii, pp. 67–74.

Loyless, J.C. and Malone, R.F. (1997) A sodium bicarbonate dosing methodology for pH management in freshwater-recirculating aquaculture systems. *Progressive Fish-Culturist* 59, 198–205.

Main, K.L. and Laramore, R. (1999) Shrimp health management. In: Van Wyk, P., Davis-Hodgkins, M., Laramore, R., Main, K.L. and Scarpa, J. (eds) *Farming Marine Shrimp in Recirculating Freshwater Systems.* Florida Department of Agriculture and Consumer Services, Tallahassee, Florida, pp. 139–146.

Malone, R.F., Beecher, L.E. and DeLosReyes, A.A. (1998) Sizing and management of floating bead bioclarifiers. In: Libey, G.S. and Timmons, M.B. (eds) *Proceedings of the 2nd International Conference on Recirculating Aquaculture.* Virginia Polytechnic Institute and State University, Roanoke, Virginia, pp. 319–341.

McIntosh, R.P. (2000) Changing paradigms in shrimp farming: V. Establishment of heterotrophic bacterial communities. *Global Aquaculture Advocate* 3(6), 52–54.

McIntosh, R.P. (2001) High rate bacterial systems for culturing shrimp. In: Summerfelt, S.T., Watten, B.J. and Timmons, M.B. (eds) *Proceedings from the Aquacultural Engineering Society's 2001 Issues Forum.* Aquaculture Engineering Society, Shepherdstown, West Virginia, pp. 117–129.

Moore, D.W. and Brand, C.W. (1993) The culture of marine shrimp in controlled environment superintensive systems. In: McVey, J.P. (ed.) *CRC Handbook of Mariculture.* CRC Press, Boca Raton, Florida, pp. 315–348.

Moss, S.M. (1995) Production of growth-enhancing particles in a plastic-lined shrimp pond. *Aquaculture* 132, 253–260.

Moss, S.M. (1998) *U.S. Marine Shrimp Farming Program Biosecurity Workshop.* The Oceanic Institute, Honolulu, Hawaii.

Moss, S.M. (2000a) Benefits of a microbially dominated intensive shrimp production system: a review of pond water studies at the Oceanic Institute. *Global Aquaculture Advocate* 3(2), 53–55.

Moss, S.M. (2000b) Biosecure shrimp production: emerging technologies for a maturing industry. *Global Aquaculture Advocate* 2(4/5), 50–52.

Moss, S.M. and Pruder, G.D. (1995) Characterization of organic particles associated with rapid growth in juvenile white shrimp, *Penaeus vannamei* Boone, reared under intensive culture conditions. *Journal of Experimental Marine Biology and Ecology* 187, 175–191.

Moss, S.M., Pruder, G.D., Leber, K.M. and Wyban, J.A. (1992) Relative enhancement of *Penaeus vannamei* growth by selected fractions of shrimp pond water. *Aquaculture* 101, 229–239.

Moss, S., Arce, S., Calderon, F., Otoshi, C., Moss, D., Lotz, J., Lightner, D., Argue, B. and Pruder, G. (1998a) Breeding for disease resistance in penaeid shrimp: experiences from the US Marine Shrimp Farming Program. In: Jory, D.E. (ed.) *Proceedings of the 1st Latin American Shrimp Farming Congress.* Panama City, Panama.

Moss, S.M., Reynolds, W.J. and Mahler, L.E. (1998b) Design and economic analysis of a prototype biosecure shrimp growout facility. In: Moss, S.M. (ed.) *US Marine Shrimp Farming Program Biosecurity Workshop.* The Oceanic Institute, Honolulu, Hawaii, pp. 67–74.

Moss, S.M., Arce, S.M., Argue, B.J., Otoshi, C.A., Calderon, F.R.O. and Tacon, A.G.J. (2001) Greening of the blue revolution: efforts toward environmentally responsible shrimp culture. In: Browdy, C.L. and Jory, D.E. (eds) *The New Wave: Proceedings of the Special Session on Sustainable Shrimp Culture.* World Aquaculture Society, Baton Rouge, Louisiana, pp. 1–19.

Moss, S.M., Otoshi, C.A., Montomery, A.D. and Matsuda, E.A. (2002) Recirculating aquaculture systems for the production of market-sized shrimp. In: *Proceedings of The Fourth International Conference on Recirculating Aquaculture.* Virginia Tech University, Blacksburg, Virginia, pp. 245–254.

Moss, S.M., Arce, S.M., Moss, D.R., Hwang, Y.S., White-Noble, B. and Lighter, D.V. (2003) Selectively bred shrimp survive varied TSV exposure. *Global Aquaculture Advocate* 6(4), 16–17.

Ogle, J.T. and Lotz, J.M. (1998) Preliminary design of a closed, biosecure shrimp growout system. In: Moss, S.M. (ed.) *US Marine Shrimp Farming Program Biosecurity Workshop.* The Oceanic Institute, Honolulu, Hawaii, pp. 39–48.

Otoshi, C.A., Arce, S.M. and Moss, S.M. (2002) Use of recirculating systems for the production of broodstock shrimp. In: *Proceedings of The Fourth International Conference on Recirculating Aquaculture.* Virginia Tech University, Blacksburg, Virginia, pp. 271–279.

Reid, B. and Arnold. C.R. (1992) The intensive culture of penaeid shrimp *Penaeus vannamei* Boone in a recirculating raceway system. *Journal of the World Aquaculture Society* 23, 146–153.

Sandifer, P.A., Hopkins, J.S., Stokes, A.D. and Pruder, G.D. (1991a) Technological advances in intensive pond culture of shrimp in the United States. In: DeLoach, P., Dougherty W.J. and Davidson, M.A. (eds) *Frontiers in Shrimp Research.* Elsevier Science, Amsterdam, pp. 241–256.

Sandifer, P.A., Stokes, A.D., Hopkins, J.S. and Smiley, R.A. (1991b) Further intensification of pond shrimp culture in South Carolina. In: Sandifer, P.A. (ed.) *Shrimp Culture in North America and the Caribbean* World Aquaculture Society, Baton Rouge, Louisiana, pp. 84–95.

Samocha, T.M., Hamper, L., Emberson, C.R., Davis, A.D., McIntosh, D., Lawrence, A.L. and Van Wyk, P.M. (2002) Review of some recent developments in sustainable shrimp farming practices in Texas, Arizona and Florida. *Journal of Applied Aquaculture* 12, 1–42.

Samocha, T.M., Lawrence, A.L. and Bray, W.A. (1993) Design and operation of an intensive nursery raceway system for Penaeid shrimp. In: McVey, J.P. (ed.) *CRC Handbook of Mariculture.* CRC Press, Boca Raton, Florida, pp. 173–210.

Scarpa, J. (1998) Freshwater recirculating systems in Florida. In: Moss, S.M. (ed.) *Proceedings of the US Marine Shrimp Farming Program Biosecurity Workshop.* The Oceanic Institute, Honolulu, Hawaii, pp. 67–70.

Tseng, K.-F., Su, H.-M. and Su, M.-S. (1998) Culture of *Penaeus monodon* in a recirculating system. *Aquacultural Engineering* 17, 138–147.

Van Wyk, P. (1999) Harbor branch shrimp production systems. In: Van Wyk, P., Davis-Hodgkins, M., Laramore, R., Main, K.L. and Scarpa, J. (eds) *Farming Marine Shrimp in Recirculating Freshwater Systems.* Florida Department of Agriculture and Consumer Services, Tallahassee, Florida, pp. 99–114.

Van Wyk, P. (2000) Culture of *Penaeus vannamei* in Single-Phase and Three-Phase Recirculating Aquaculture Systems. *Global Aquaculture Advocate* 3(3), 41–43.

Van Wyk, P. (2001) Designing efficient indoor shrimp production systems: a bioeconomic approach. In: Browdy, C.L. and Jory, D.E. (eds) *The New Wave: Proceedings of the Special Session on Sustainable Shrimp Culture, Aquaculture 2001.* The World Aquaculture Society, Baton Rouge, Louisiana, pp. 44–56.

Van Wyk, P. and Scarpa, J. (1999) Water quality requirements and management. In: Van Wyk, P., Davis-Hodgkins, M., Laramore, R., Main, K.L. and Scarpa, J. (eds) *Farming Marine Shrimp in Recirculating Freshwater Systems.* Florida Department of Agriculture and Consumer Services, Tallahassee, Florida, pp. 128–138.

Weirich, C.R., Browdy, C.L., Bratvold, D., McAbee, B.J. and Stokes, A.D. (2002) Preliminary characterization of a prototype minimal exchange super-intensive shrimp production system. *Proceedings of the Fourth International Conference on Recirculating Aquaculture,* Virginia Tech University, Blacksburg, Virginia, pp. 255–270.

Williams, A.S., Davis, D.A. and Arnold, C.R. (1996) Density-dependent growth and survival of *Penaeus setiferus* and *Penaeus vannamei* in a semi-closed recirculating system. *Journal of the World Aquaculture Society* 27, 107–112.

Wirth, F.F. and Davis, K.J. (2001a) *Seafood Restaurant Shrimp Purchasing Behavior and Attitudes Toward Farm-Raised Shrimp. Staff Paper 01–19.* Food and Resource Economics Department, Institute of Food and Agricultural Sciences, University of Florida, Gainesville, Florida.

Wirth, F.F. and Davis, K.J. (2001b) *Assessing Potential Direct Consumer Markets for Farm-Raised Shrimp. Staff Paper 01–13.* Food and Resource Economics Department, Institute of Food and Agricultural Sciences, University of Florida, Gainesville, Florida.

Wyban, J.A. and Sweeney, J.N. (1991) *Intensive Shrimp Production Technology: The Oceanic Institute Shrimp Manual.* The Oceanic Institute, Honolulu, Hawaii.

Wyban, J.A., Swingle, J.S., Sweeney, J.N. and Pruder, G.D. (1992) Development and commercial performance of high health shrimp using specific pathogen free SPF broodstock *Penaeus vannamei.* In: Wyban, J. (ed.) *Proceedings of the Special Session on Shrimp Farming.* World Aquaculture Society, Baton Rouge, Louisiana, pp. 254–260.

Wyban, J., White-Noble, B. and Lightner, D.V. (2003) Breeding Taura virus resistant (TVR) *P. vannamei.* In: *Book of Abstracts, World Aquaculture 2003.* World Aquaculture Society, Baton Rouge, Louisiana, p. 842.

This page intentionally left blank

12 Aquaculture of the Florida Bay Scallop, *Argopecten irradians concentricus*, in Tampa Bay Florida, USA, an Urban Estuary

Norman J. Blake
University of South Florida, St Petersburg, USA

Abstract

Bay scallop (*Argopecten irradians*) populations along the east coast of the USA once supported both commercial and recreational fisheries. Since the 1950s, many of the local populations have collapsed. Both natural and anthropogenic factors, including hurricanes, red tide events, loss of habitat and overfishing, can result in large fluctuations in bay scallop populations. Urban estuaries that once supported bay scallop populations have been especially affected, and are slow to recover even if the water quality in the estuary is improved since natural linkages among local populations may be disrupted, for instance, preventing the transport of larvae. Tampa Bay is an example of an urban estuary that has lost its natural bay scallop population. Even though water quality conditions have greatly improved in recent years, bay scallops have not returned.

A bay scallop hatchery, which utilizes seawater from Tampa Bay, has been in operation since 1990. In order to maintain genetic variability, adult scallops nearing reproductive maturity are collected in the early autumn from a population approximately 160 km north of Tampa Bay. These scallops are allowed to mature in cages suspended in Tampa Bay and are spawned in a recirculating system. Larval mortality through settlement (8–10 days) usually ranges from 10% to 50%, similar to that of other hatcheries. The larvae are fed on algae grown in sterilized Tampa Bay water. Settled spat are maintained in the recirculating system for 30–45 days, then transferred to fine mesh bags which are suspended about 1.0 m below the surface of the bay. Mortality and growth are monitored bi-weekly and scallops transferred to larger mesh containers as they grow. Throughout the autumn and winter, cumulative mortality is about 20%, and growth is 1.5–2.0 mm/week. By early spring the scallops reach 25 mm, and these are placed in suspended cages at a density of 200–250 individuals/m^2. Mortality increases greatly in summer months due to high water temperatures and intense fouling. Nevertheless, more than 25% of the settled spat reach harvestable sizes of 45–50 mm by September. Surveys indicate that these scallops may be sold to processors dealing in the speciality market for US$0.25–0.30 each as a whole, in-shell product.

Introduction

Bay scallop (*Argopecten irradians*) populations along the east coast of the USA once supported profitable commercial and recreational fisheries. Since the 1950s, many of the local populations have collapsed. Both natural and anthropogenic factors, including

(eds B. Costa-Pierce, A. Desbonnet, P. Edwards and D. Baker)

hurricanes, red tide events, loss of habitat and overfishing, have resulted in large fluctuations in bay scallop populations. Bay scallop populations in urban estuaries have been especially affected, and have been slow to recover even if the water quality of the estuary is improved; natural linkages among local populations have been disrupted, in some cases preventing the transport of larvae and subsequent recolonization. Tampa Bay is an example of an urban estuary that has lost a viable natural bay scallop population, even though water quality conditions have improved greatly in recent years.

The bay scallop has a life span of 12–24 months (Gutsell, 1930; Marshall, 1963). The majority of bay scallops spawn once at an age of 1 year (Fig. 12.1), and spawning is largely catastrophic (Barber and Blake, 1991). Mass mortality occurs after their first spawn, and a small fraction survive to a second year (Belding, 1910; Gutsell, 1930; Barber and Blake, 1983). Increased mortality of Florida bay scallops occurs in the autumn after spawning (Barber and Blake, 1983).

The viability of a bay scallop population is determined primarily by the recruitment success of each consecutive reproductive season, which is determined by the success of gametogenesis of the adults and growth and survival of the young. One poor reproductive or recruitment season may reduce the population to such a low level that it may take years for the population to recover, since periods of poor recruitment are not buffered by the survival of adult stocks (Pollack, 1988). For example, recruitment failed to restock a bay scallop population in North Carolina in 1988, a year after a red tide caused a mass mortality (Summerson and Peterson, 1988). Bay scallop reproductive behaviour and environmental susceptibility may make the bay scallop one of the most vulnerable bivalve species to natural and human-induced environmental changes.

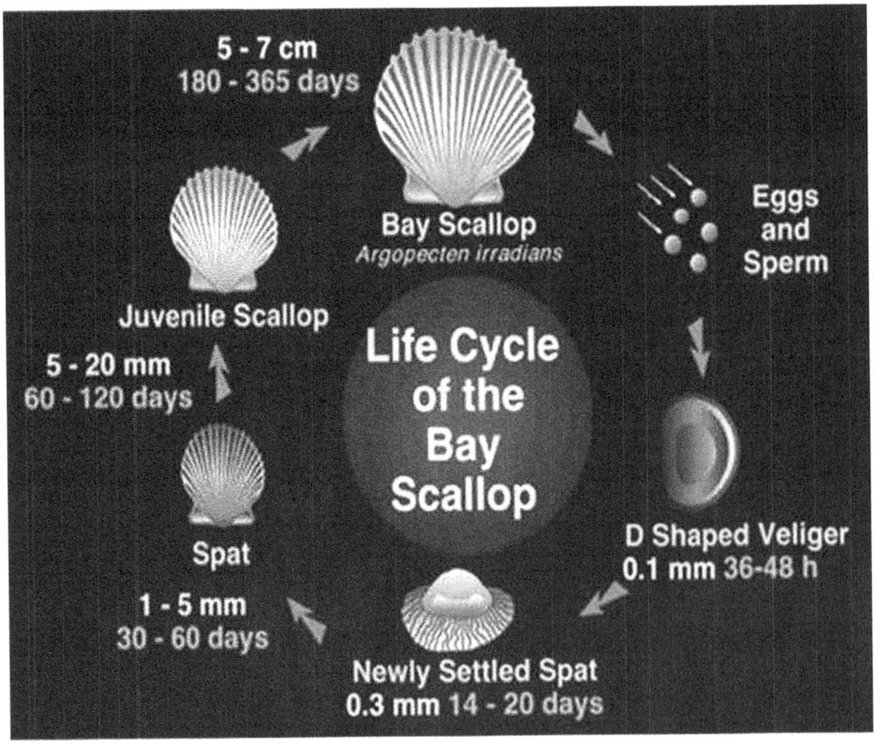

Fig. 12.1. Life cycle of the bay scallop, *Argopecten irradians*.

In many areas along the US east coast, such as the west coast of Florida, the distribution of the bay scallop is discontinuous, i.e. the population is largely self-sustaining with little larval exchange with other estuaries. Populations that disappear may not reappear even though the causes for the disappearance are removed, due to the discontinuous distribution and the reproductive strategy of, generally, one spawn per generation.

It is possible that recoveries of bay scallop populations in areas where habitats have been restored can be accelerated using hatchery-reared juveniles. Bay scallop stock enhancements have been carried out by planting juvenile scallops on potential scallop grounds in Connecticut (Morgan *et al.*, 1980) and in New York (Wenczel *et al.*, 1986). Spawning of the reseeded scallops was estimated to produce 25% of the scallops set in eastern Peconic Bays, New York, in 1989 (Tettelbach and Wenczel, 1993). Recent efforts in Connecticut (USA) have attempted to develop bay scallop aquaculture commercially.

This chapter reports on studies undertaken to determine if the waters of an urban estuary – Bayboro Harbor, Tampa Bay, Florida – could support a bay scallop hatchery, and if the scallops could be grown to maturity in nets suspended from marina docks.

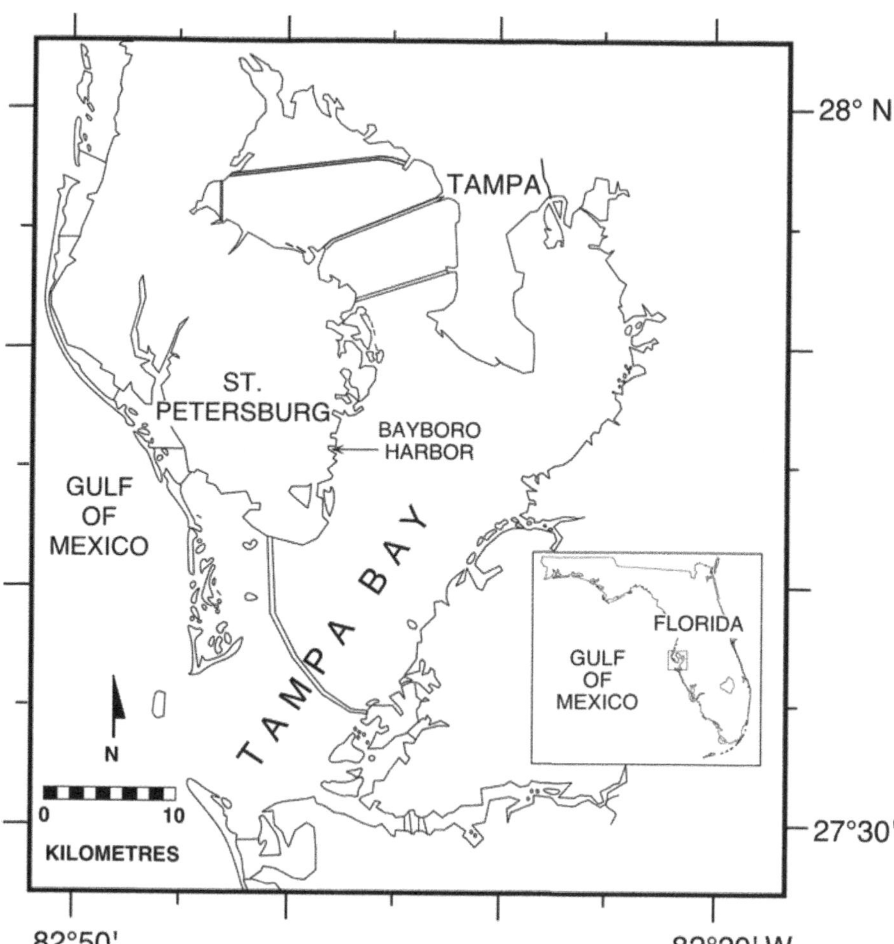

Fig. 12.2. Map of Tampa Bay, Florida.

Fig. 12.3. Bayboro Harbor, location of scallop aquaculture facility and associated marina.

Materials and Methods

Tampa Bay is a highly urbanized estuary with over two million people living in the vicinity. Bayboro Harbor contains several boat yards as well as a large marina (Fig. 12.2). In order to maintain genetic variability, bay scallops were collected from Homosassa, Florida, in October 1999 as they were nearing reproductive maturity. These scallops were allowed to mature in cages suspended from docks on Bayboro Harbor at the College of Marine Science at the University of South Florida (Fig. 12.3).

Prior to spawning, the scallops were kept in 500 l fibreglass tanks filled with seawater (25–28‰; 24–28°C) from Bayboro Harbor. One half of the water in each tank was replaced with fresh seawater from the same source each day. Scallops were fed an equal volume of cultured *Isochrysis galbana* (2×10^6 cells/ml) and *Tetraselmis* sp. (0.5×10^6 cells/ml) at 500 ml per individual per day until they were induced to spawn by temperature manipulation. Spawning was allowed to occur when gonads became ripe, and the fertilized eggs were allowed to hatch at a density of 10–20/ml. No antibiotics were used.

Larval culture

Fertilized eggs were allowed to develop for 20–30 h, when they became D-shaped larvae. They were then filtered onto a 35 μm screen and released into fresh seawater. Larvae were maintained at a density of 4–8/ml and fed with 10,000–30,000 cells/ml of *I. galbana* daily, with 1000–5000 cells/ml of *Tetraselmis* sp. added as a supplement. Water was replaced each day with Bayboro Harbor water in the amount of one-third of the total volume. Aeration was provided continuously. Each day a sample was taken from the culture, and shell length of 20 larvae was measured using a microscope fitted with an ocular micrometer. As soon as eyespots started to develop, pieces of black plastic *Thalassia* mimics, or plastic outdoor carpeting (Astroturf®), were added to the culture tanks to provide a settlement substrate for spat. Two to seven days were required for the eyed larvae to settle on the substrates and to complete metamorphosis. As soon as metamorphosis was observed, daily food ration was increased gradually from 30,000 cells/ml to 100,000 cells/ml of *I. galbana* and *Tetraselmis* sp.

Nursery culture

Once spat reached a mean shell length of >500 μm, a process that required 2–3 weeks, the substrate materials bearing byssate spat were placed into 300 μm nylon mesh bags (25 cm × 55 cm in size) at a density of approximately 10,000–30,000 spat per bag. Bags were suspended from a dock located on Bayboro Harbor. In 3–4 weeks the 300 μm bags were replaced with 800 μm nylon mesh bags. Spat density was reduced gradually to approximately 1000–2000 spat per bag during nursery culture. Bags were removed from the water briefly each week and cleaned with a brush to remove sediments and fouling organisms. Every 2–3 weeks, 50 spat were sampled from a marked bag, and shell height, percentage growth and survival determined. After measurements were completed, spat were returned to the same bag.

Growout

When juvenile scallops achieved a mean shell height of 7–10 mm, they were transferred from bags to small mesh lantern nets (mesh opening 5 mm) and returned to Bayboro Harbor. Lantern nets had four levels, with a 50 cm diameter and 20 cm height between levels. About 500–1000 scallops were placed in each level, making about 2000–4000 scallops per net. Small mesh lantern nets were replaced with growout lantern nets with 15 mm mesh openings at 4–6 weeks. Number of scallops was gradually reduced as they grew in size, until a final density of 50 scallops per net level was reached. Nets were cleaned every 2 or 3 weeks, at which time 50 scallops were sampled and measured in shell height with calipers to ±0.5 mm. Empty shells were counted and removed from nets.

Results and Discussion

A total of 14.6 × 10^6 eggs were spawned in 2000, producing 9.1 × 10^6 D-shaped larvae, resulting in a hatching rate of 62.3%. D-shaped larvae developed in 36–48 h, and eyed-larvae developed 7–9 days after fertilization. Larval growth rate was 15–20 μm/day. Metamorphosis (175–200 μm) occurred 8–12 days after fertilization. Larval mortality through settlement was 16%.

Castagna and Duggan (1971) found that larvae of the Virginia and North Carolina bay scallop begin settling in 10–19 days, with most occurring in 10–14 days at 20–28°C. They also found that from early post-setting to 2 mm in height, mortalities often reduced the number of live scallops by an estimated 50–80%. Despite the different food species and ratio used in the two studies, the rate of development and survival during the early stages for the Tampa Bay spawn were superior to the spawns of Castagna and Duggan in Virginia.

Spat were grown in the hatchery until they achieved a mean shell height of 0.8–0.9 mm after 30–35 days. They were then transferred to fine meshed bags and suspended about 1.0 m below the surface of the bay. Mean growth rate from pediveliger to this stage was 32.7–35.9 μm/day. Apart from the two best known critical periods limiting early survival, i.e. hatching of eggs and metamorphosis of larvae, nursery culture in mesh bags often represents another period of heavy spat mortality. Survival in nursery culture was around 26%. Nursery culture is often one of the key factors limiting production of scallop spat in commercial hatcheries. High mortality could be the result of crowding and low food flux, and oxygen levels associated with reduced circulation due to fouling of culture bags.

Scallop growth during intermediate culture in mesh bags and during growout in lantern nets (Fig. 12.4) was similar to that of a natural population in Anclote Estuary on the Gulf coast of Florida (Barber and Blake, 1983). Growth during January and

Fig. 12.4. Lantern nets used for scallop growout.

February was slow, then increased rapidly by early spring (Fig. 12.5) when scallops reached 20–25 mm, in accordance with the increase in water temperature and increased food supply. High growth persisted until June and July. After that, growth declined again.

Throughout the winter, cumulative mortality remained very low (Fig. 12.6), and growth was 1.5–2.0 mm/week. Mortality increased markedly in the summer months, resulting from high water temperatures and intense fouling (Fig. 12.7). Nevertheless, more than 35% of the settled spat reached harvestable size of 40–50 mm by September. In Tampa Bay, scallops grew from 9.0 mm to 50 mm in 8–9 months, a growth rate similar to that reported for Georgia where scallops grew from 9.8 mm to 49 mm in 8 months (Heffernan *et al.*, 1988).

Fouling suspension feeders reduce water circulation and compete for food resources with scallops (Belding, 1910; Broom, 1976), and this may be one of the main causes that prevent scallops from growing to a comparable maximum mean size of >60 mm as observed in the Anclote Estuary population (Barber and Blake, 1983). In culturing bay scallops in coastal waters of Georgia, Heffernan *et al.* (1988) suggested that the lower growth and survival of bay scallops reared in 'bottom' pearl nets may be related to varying interferences from fouling organisms.

Fouling represents a serious problem when using the caged scallops method for growout, especially in an estuary such as Tampa Bay. Main fouling organisms found on scallop shells of Tampa Bay scallops were barnacles, oyster spat, tunicates and polychaete worms (Fig. 12.7). The most extensive fouling starts in May, and extends throughout the whole summer. A scallop net of about 4.5 kg may be 13.5–22.5 kg in just one month, mainly due to the growth of tunicates on both scallops and nets. Scallop nets had to be cleaned or replaced every 3–4 weeks during summer. Scallops transferred from Tampa Bay in April and placed in bottom cages on seagrass beds in Homosassa grew to maturity without fouling (Fig. 12.8).

Fouling organisms affect scallops in several other ways than just competing for space and food. Tunicates grow very fast. It was often seen that one tunicate can 'cement' several scallops together, and some scallops were smothered. Oysters do the same; they outgrow scallops and often seal scallop valves, either smothering scallops or making it impossible for them to close their shells. Polychaete worms are common borers of scallop shells (Blake and Evans, 1973). They penetrate shells and produce black blisters in the inner shells of the scallops. In this study, 100% of the live scallops were bearing such worms and blisters in later summer, which was coincident with the period of high scallop mortality. The degree of fouling, and implementation of fouling reduction techniques, must be a primary consideration for any urban scallop aquaculture venture involving caged scallops in an estuary such as Tampa Bay.

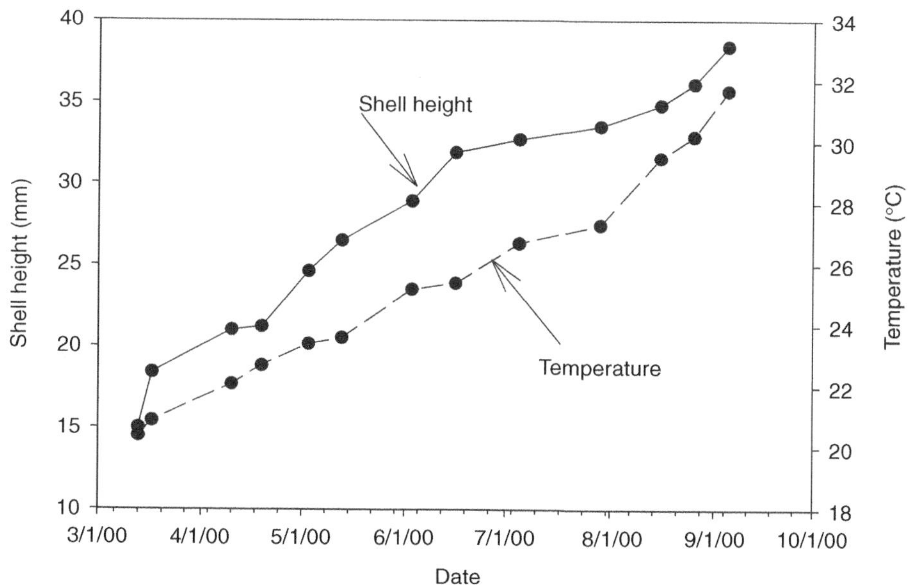

Fig. 12.5. Growth of bay scallops in Bayboro Harbor during growout.

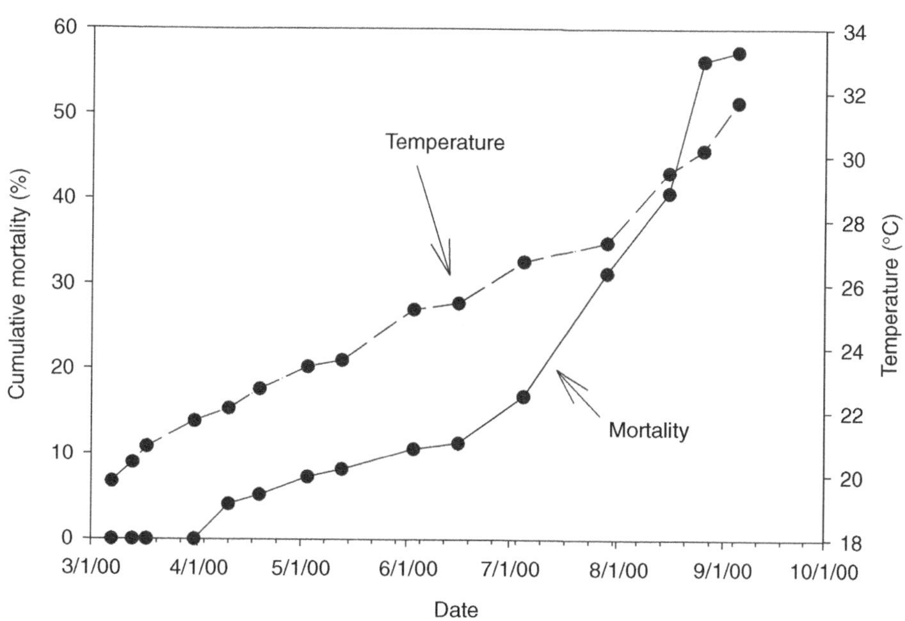

Fig. 12.6. Mortality of bay scallops in Bayboro Harbor during growout.

Conclusions

Bay scallops may be considered an ideal sustainable resource for urban aquaculture. Unlike most finfish aquaculture which may contribute to the organic load of the estuary and may face regulatory problems, scallops are filter feeders that can actually contribute to increasing water clarity and lowering organic and

Fig. 12.7. Fouling of bay scallops during growout in Bayboro Harbor.

Fig. 12.8. Scallops transferred from Bayboro Harbor to a seagrass bed in Homosassa, Florida in the early spring.

nutrient loading. In addition, the species grows to maturity very rapidly and can reach marketable size in 9–10 months, at least in Tampa Bay estuary. Although marketing the whole scallop has been demonstrated to be profitable (Adams *et al.*, 2001), current regulations prevent the sale of whole scallops from waters that do not meet specific water quality criteria. Given the current situation, Bayboro Harbor does not meet these criteria, nor would many other urban estuaries in the US. Growers would need to either relay product to approved waters for depuration purposes, or would have to compete in the existing market for scallop adductor muscle product. While perhaps not enough to completely remove the culture of scallops in urban estuaries from the realm of economic probability, it is a consideration that will have to be addressed for this market sector to move forward.

References

Adams, C., Sturmer, L., Sweat, D., Blake, N. and Degner, B. (2001) The economic feasibility of small-scale, commercial culture of the southern bay scallop (*Argopecten irradians concentricus*). *Aquaculture Economics and Management* 5(1/2), 80–98.

Barber, B.J. and Blake, N.J. (1983) Growth and reproduction of the bay scallop, *Argopecten irradians* (Lamarck) at its southern distributional limit. *Journal of Experimental Marine Biology and Ecology* 66, 247–256.

Barber, B.J. and Blake, N.J. (1991) Reproductive physiology. In: Shumway, S.E. (ed.) *Scallops: Biology, Ecology and Aquaculture.* Developments in Aquaculture and Fisheries Science, Vol. 21. Elsevier, New York, pp. 377–420.

Belding, D.L. (1910) A report upon the scallop fishery of Massachusetts, including the habits, life history of *Pecten irradians*, its rate of growth, and other factors of economic value. *Special Report of the Commonwealth of Massachusetts Commission on Fisheries and Game.* Boston, Massachusetts.

Blake, J.A. and Evans, J.W. (1973) *Polydora* and related genera as borers in mollusk shells and other calcareous substrates. *The Veliger* 15, 235–249.

Broom, M.J. (1976) Synopsis of biological data on scallops [*Chlamys (Aequipecten) opercularis* (Linnaeus) *Argopecten irradians* (Lamarck) *Argopecten gibbus* (Linnaeus)]. *FAO Fisheries Synopses* 114, 1–44.

Castagna, M. and Duggan, W. (1971) Rearing the bay scallop, *Aequipecten irradians. Proceedings of the National Shellfisheries Association* 61, 80–85.

Gutsell, J.S. (1930) Natural history of the bay scallop. *Bulletin of the US Bureau of Fisheries* 46, 569–632.

Heffernan, P.B., Walker, R.L. and Gillespie, D.M. (1988) Biological feasibility of growing the northern bay scallop, *Argopecten irradians irradians* (Lamarck, 1819), in coastal waters of Georgia. *Journal Shellfish Research* 1, 83–88.

Marshall, N. (1963) Mortality rates and the life span of the bay scallop, *Argopecten irradians. Proceedings of the National Shellfisheries Association* 54, 87–92.

Morgan, D.E., Goodsell, J., Matthiessen, G.C., Garey, J. and Jacobson, P. (1980) Release of hatchery-reared bay scallops (*Argopecten irradians*) onto a shallow coastal bottom in Waterford, Connecticut. *Proceedings of the World Mariculture Society* 11, 247–261.

Pollack, S. (1988) East coast bay scallops battle a variety of foes. *National Fisherman* October, 2–4, 40.

Summerson, H.C. and Peterson, C.H. (1988) Recruitment failure of the bay scallop, *Argopecten irradians concentricus*, during the first red tide, *Ptychodiscus brevis*, outbreak recorded in North Carolina. *Estuaries* 13(3), 322–331.

Tettelbach, S.T. and Wenczel, P. (1993) Reseeding efforts and the status of the bay scallop, *Argopecten irradians irradians* (Lamarck, 1819) populations in New York following the occurrence of 'brown tide' algal blooms. *Journal of Shellfish Research* 12(2), 423–431.

Wenczel, P., Scotti, J. and Smith, C. (1986) *Long Island Green Seal Committee Bay Scallop Rehabilitation Program: Plan of Work.* Report submitted to the Urban Development Corporation, New York.

13 Four Years of Recirculating Aquaculture in Urban Boston Harbor, USA

Clifford A. Goudey and Brandy M. Moran
Massachusetts Institute of Technology, Cambridge, USA

Abstract

Often urban ports hold a potential for aquaculture production due to the typical decline in more traditional maritime activities which can result in abandoned or underutilized waterfront warehouses, and a strong local market seeking fresh and quality products. A main issue surrounding siting an aquaculture facility in an urban area is the anthropogenic sources of contaminants that may impact the performance of the facility.

Boston Harbor exemplifies this situation as it has been subjected to decades of unregulated discharge through sewers, storm drains, air pollution and direct commercial and industrial discharges. A court-ordered clean-up has reversed that trend and remarkable progress has been made in harbour water quality. Although sediment quality problems remain, the question of the suitability of the harbour for aquaculture surfaced in 1996.

Massachusetts Institute of Technology (MIT) Sea Grant responded to this issue with the establishment of a demonstration marine finfish hatchery located in the Charlestown Navy Yard. The purpose of the facility has been to: (i) develop protocols for commercially important marine finfish species, (ii) track growth and environmental impacts, (iii) design and demonstrate recirculating system technologies, and (iv) conduct outreach and education about aquaculture.

In this chapter we present some results of our work with cod, haddock, winter flounder, red drum, tautog and black sea bass. We will discuss the environmental factors we have encountered during the larval rearing stages at our facility and explain some of the public and K-12 education programmes we have initiated. In particular, we have had great success in bringing aquaculture to college–preparatory high school students through linkages to science curricula already in place.

Introduction

There have been some dramatic changes in Boston Harbor's water quality as a result of a court-ordered plan to stop the fouling of beaches and waterways due to inadequate treatment of wastewaters from the metropolitan Boston area. Milestones in this upgrade process include the cessation of harbour sludge dumping in 1991, the closure of the Nut Island treatment plant in 1998 and the ending of all Deer Island Treatment Plant discharges into the harbour on 6 September 2000.

Today, a 9-mile-long outfall sends treated effluents into Massachusetts Bay, and the water quality in both the inner and outer harbours has significantly

(eds B. Costa-Pierce, A. Desbonnet, P. Edwards and D. Baker)

improved. Table 13.1 reveals some of the improvements as measured in spring 2002 (Massachusetts Water Resource Authority, 2002).

These changes have brought renewed biological activity to the harbour, and interest has been expressed in the possibility of harbour-based aquaculture activity (Table 13.2). This interest is based on the increased demand for high-quality seafood of a growing population and the emergence of economically viable methods of producing seafood in tank-based systems. Boston is viewed as a particularly suitable site due to its existing seafood infrastructure and buyers who appreciate fresh product. There also seems to be an ample supply of underutilized waterfront sites.

Table 13.1. Summary of Boston Harbor water quality improvements.

Sewage indicator bacteria	Down 30–70%
Nitrogen	Down 55%
Phosphorus	Down 31%
Chlorophyll	Down 49%
Water clarity	Up 12%
Dissolved oxygen	Up 3–7%

MIT Sea Grant Boston Harbor Aquaculture Initiative

Recognizing the opportunity for economic growth associated with this urban aquaculture opportunity, MIT Sea Grant established an experimental aquaculture facility in 1996. Called 'Aqualab', and located on Pier 2 in the Charlestown Navy Yard, we engaged in fingerling growout and public education.

This effort was expanded in May 1998 with the dedication of the Marine Finfish Hatchery nearby on Pier 3. In this laboratory we conducted research on species protocols (cod, haddock, tautog, winter flounder, black sea bass) and also developed innovative and cost-effective recirculating systems. We have also developed K-12 and undergraduate education through hatchery tours, classroom curricula and internships.

As is typical in the siting of such a facility, the location was one of convenience and not based on some optimization process. To understand the implications of our site, a look at its history is useful.

The Charlestown Navy Yard was constructed in 1801 as one of six sites in the United States for naval shipbuilding. It was also a major repair facility supply centre. It was closed in 1974 and 12 ha were designated as a National Historical Park. The 173 years of its operation, along with numerous other industrial operations surrounding the harbour, left its mark on the sediments of the harbour. Figure 13.1 charts the concentration of zinc, a typical

Table 13.2. Boston Harbor sites explored for aquaculture siting as shown in Fig. 13.3.

Map no.	Site	Owner
1	MIT hatchery	National Park Service
2	Charlestown Commerce Center	Private
3	Hess Petroleum	Private
4	Mystic Pier 1	Massachusetts Port Authority
5	Charlestown Pier 3	Boston Redevelopment Authority
6	East Boston Warehouse	Private
7	East Boston Pier 1	Massachusetts Port Authority
8	Immigration Building	Massachusetts Port Authority
9	S. Boston pier sheds	Massachusetts Port Authority
10	Reserve Channel parcel	Boston Redevelopment Authority
11	Black Falcon Cruise Terminal	Massachusetts Port Authority
12	Moon Island	City of Boston

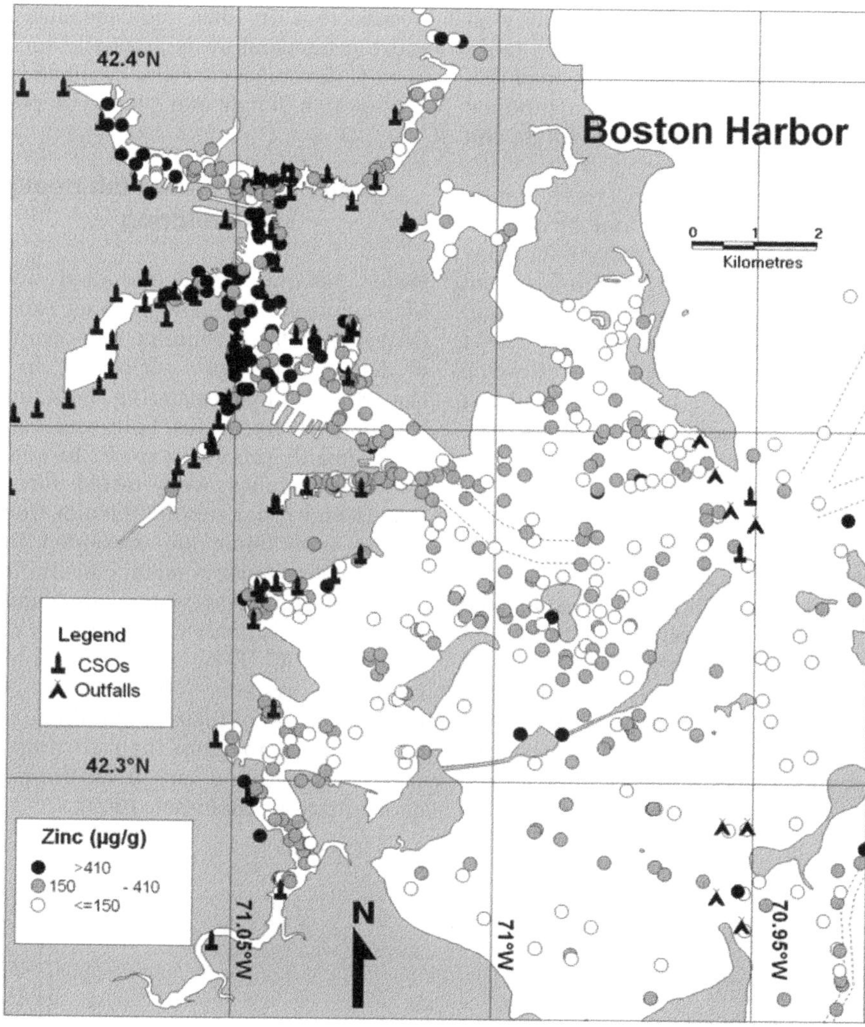

Fig. 13.1. Zinc concentrations in the Boston Harbor sediments.

heavy metal, in Boston Harbor. The inner harbour reveals particularly high levels of contamination (United States Geological Survey, 1999).

Though the objectives of the MIT Marine Finfish Hatchery were to establish culture protocols for marine finfish suitable for production in a New England urban setting, the presence of contaminated sediments and the ever-present risk of a pollution event in a busy harbour meant our operations evaluated the impact of water quality on recirculating finfish aquaculture.

Given the modest size of our facility and its high rate of recirculation, these challenges could have been obviated through the use of carbon filtration and other separation techniques appropriate to a research facility. However, our interest was in using water processing components appropriate for commercial aquaculture. We therefore used a combination of particle filtration and UV treatment for our incoming water.

Our intake pump is located at mid-depth and automated to pump only at high tide, flowing through 100 and 10 μm filters to storage tanks. Two 1100 l tanks are kept at ambient temperature and two 900 l tanks are kept heated for filling our live feed culture tanks. This seawater is then pumped on demand through 5 and 1 μm filters and a 40 W ultraviolet sterilizer.

There are two hatchery recirculating systems, one for hatching and larval culture and one for fingerling culture (Goudey *et al.*, 1999). The smaller hatching system includes six 380 l tanks and the larger growout system includes three 1900 l tanks. Both systems include the following full-flow components, in order: weir box settling tank, 30 μm particle filter bag, pump, protein skimmer, chiller, twin bioreactors, UV and degasser.

These systems have been used for a variety of marine finfish species including haddock, tautog, winter flounder and black sea bass. In the case of haddock, we adapted what is known about their culture to suit our systems and the conditions our water supply imposed (Moran and Goudey, 2000). These protocols include egg incubation at 7°C. Larvae are then cultured under 24-h lighting and fed rotifers beginning on day 3 and *Artemia* once they reach 8 mm. They are weaned to a dry diet upon metamorphosis and the temperature is gradually increased to 14°C. Juveniles are fed a commercial diet and see 16 h of light and 8 h of dark per day.

Water Quality and Fish Health Monitoring

Water testing was conducted by the Massachusetts Water Resources Authority (MWRA) every 6 months. The evaluation of fish health was performed by MIT Department of Comparative Medicine and by Micro Technologies, both routinely and when health problems arose. In addition, our feed batches were tested for heavy metals and other toxins. Results from our testing programme are presented in Fig. 13.2, where heavy metal levels in the intake and process waters are compared with EPA Critical Continuous Concentrations (CCC; EPA, 1999) and limits in intensive systems (Wedemeyer, 1996). Median values of these contaminate metals range from 4 to more than 20 times estimated background or pre-anthropogenic concentrations (Schubel, 1999).

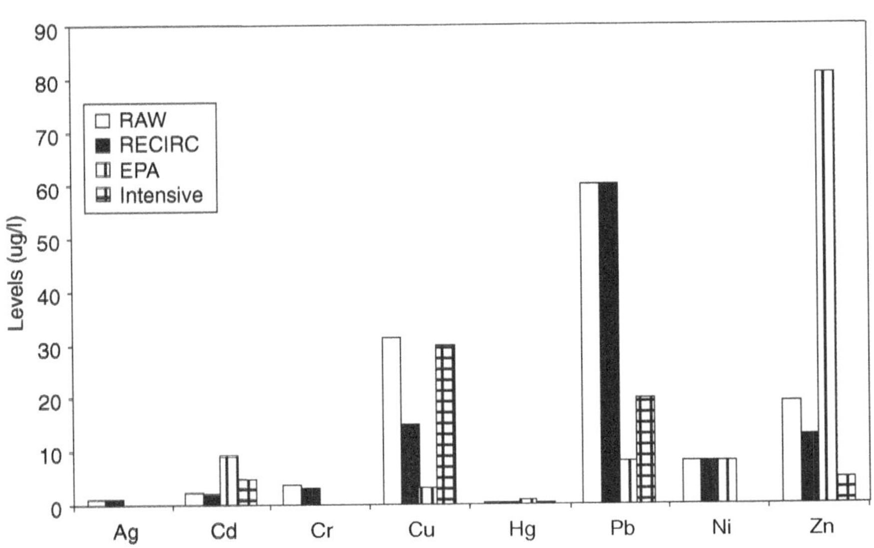

Fig. 13.2. Heavy metal levels in raw Boston Harbor water and post-treatment recirculating water compared with EPA CCC and limits in intensive systems.

During the culture of haddock at this facility we experienced mixed results. In June 1998, after metamorphosis, the juvenile fish began showing high mortalities. On Day 56, 50% of the fish were dead. Micro Technologies identified unicellular organisms on the gills of the fish. In July 1999 the haddock in the growout system from the 1998 spawn (Day 515) were showing chronic mortalities. Symptoms were caudal fin rot, haemorrhaging, scale loss and glaucoma in the eye. Reports from Micro Technologies concluded that *Vibrio fluvialis* or *V. splendidus* caused the mortalities. Harbour water quality was identified as the potential cause of the pathogens. In August 1999 fish from our growout tanks showed chronic mortalities (Day 540) and were diagnosed by Micro Technologies as having haddock vibriosis. Clinical signs were necrosis of heart tissue, kidney damage and eye glaucoma. Micro Technologies referred to a University of Maine case in 1997, reporting that the *Vibrio* and clinical signs were the same.

In April 2000 on Day 40, all fish were found dead in the larval tanks. This event was 5 days after a petroleum spill in the Mystic River, which is located less than 1 mile from the hatchery. The water was not tested during this event, therefore any conclusions are speculative.

In February 2001 haddock from the 2000 spawn were showing chronic mortality over a course of a week (Day 322). Micro Technologies showed that there were no bacterial or systemic causes of death. They questioned harbour water quality and if there were any materials in our intake line that would contain leachable metals. They concluded that there may be a 'toxic material' in the culture water causing compromised fish health.

In March 2001 there were very heavy rains in the Boston area. The hatchery is located next to the locks of the Charles River, and locks release a high volume of freshwater into the harbour under such conditions. This occurrence greatly affects the salinity and quality of the facility's intake water. We were forced to shut down our intake pump for seven days until our intake water reached acceptable levels. Once we began pumping water in for recirculating system water exchanges, the salinity in our system dropped from 29 to 24‰. Soon the larvae that were successfully feeding on *Artemia* were all found dead. Micro Technologies analysis yielded a high loading of a freshwater parasite in the system that stressed the larvae and resulted in the 100% mortality.

In June 2001 high mortality was observed in all tanks. Fish were weaned on to a Biokyowa diet at the end of the month, but heavy mortality continued to occur in July. Most of the mortalities were fish with spinal deformities. Laboratory results from MIT DCM veterinarians showed that there was an external fungus inhibiting the fish as well as lateral scoliosis.

The hatchery intake water was impacted by human and natural pollution since our operations began in May 1998. Although few studies investigate the impact of toxins on haddock, our clinical reports consistently show that harbour water impacts the health and survivorship of the fish being reared in our hatchery.

Human-induced impacts were exacerbated during the spring flood of 2001. Due to the influx of the vast amount of stormwater runoff and sewer overflow, the marine environment was saturated with freshwater and stimulated a bloom of freshwater parasites and poor water quality in the haddock facility. The parasites entered the intake water system of the hatchery and thrived due to the drop in salinity to 24‰. Lacking levels of UV sterilization above 50,000 MWS, we were unable to neutralize the freshwater parasites from the incoming water. Poor water quality accounted for the *Vibrio* diagnoses, since *V. fluvialis* is a human pathogen that can be found in harbour waters during high runoff periods.

The petroleum spill in the Mystic River impacted the harbour and apparently our facility. Spill notification was given to the hatchery 2 days after the spill. It is possible that petroleum was able to enter in our intake water as it moved under the influence of the currents and tides. We were not able to test the water for traces of petroleum, but trace amounts of petroleum in a water source detrimentally effect larval fish (Peterson, 2001). Negative effects of ingesting toxic levels of oil are poorly understood for many specific organisms, especially microorganisms (e.g. plankton), bottom-dwelling organisms and larval fish (Seymour and Geyer, 1992). Laboratory studies and field observations show that eggs and larvae are sensitive to oil exposure because of limited mobility to avoid contaminated areas, the undeveloped organs that help detoxify hydrocarbons in adults and larval stages of many species are concentrated at the water surface where exposure to oil spills is greater (Rice, 1999). During metamorphosis and in an intensive system, larval fish are highly sensitive to any toxic elements in the tank environment.

Contaminated sediments are prevalent in the Inner Harbor. Zinc, lead, chromium and copper have the highest concentrations in Boston Harbor sediment. Figure 13.1 shows that copper, lead and zinc are all found in hatchery water at levels higher than the EPA CCC and recommended limits of exposure for fish in an intensive system. Because fish are able to uptake and retain heavy metals dissolved in water through active or passive processes (Ay *et al.*, 1999), the heavy metals found in the raw and post-treated harbour water lead us to the conclusions that these chronic low levels of contaminants impact haddock quality. Threshold concentrations of toxic metal mixtures need not be high to produce toxic effects in codfish (Thurberg and Gould, 2004). Compression of the spine and jaw deformities are documented in cod from the Baltic Sea that were found to have elevated levels of cadmium in the liver and kidney (Lang and Dethlefsen, 1987).

Copper is documented to induce chloride cell necrosis (Bury *et al.*, 1998), interfere with Na and Cl regulation at the gills (Richards *et al.*, 1999) and can result in effects on growth rates and reproduction. Young fish are more sensitive to copper (Chakoumakos *et al.*, 1979; Mazon and Fernandes, 1999), which could account for our mortalities. Increased rates of necrosis and apoptosis are reported for skin and gills after exposure to several other toxic stressors, including acid water containing aluminium, cadmium and copper (Wendelaar *et al.*, 1990; Iger *et al.*, 1994). Four-week-old haddock larvae exposed to 500 μg/l copper for 18 h resulted in increased mortality and severe olfactory lesions (Thurberg and Gould, 2004). Pollutant exposure is an unnatural stress, and fish must reallocate energy expenditures to compensate (Brett and Groves, 1979; Kitchell, 1983; Beyers *et al.*, 1999). Energy reallocation results in lower growth rates and impacts osmoregulation.

Our investigations indicate that haddock, particularly in comparison with the other species cultured at our hatchery, are susceptible to toxins in the larval stages. The toxins induced disease and sudden mortality. Using similar larval-rearing protocols, tautog, winter flounder and black sea bass were reared successfully at the facility.

Siting of an aquaculture facility in an urban area can be very advantageous with respect to the marketing of live or superfresh fish if the facility is capable of taking appropriate measures to ensure proper water quality. Siting a facility in the outer Boston Harbor would likely result in successful rearing of fish under the same protocols due to its higher water quality.

Water pre-processing systems are key factors in the success of an urban area hatchery. Concentrated human populations result in highly developed areas and increased pollution presence. Facilities must take more precautions to assure water intake has acceptable water-quality levels for its housed activities. The higher sensitivity of larval stages indicated that additional filtration, such as carbon filters and high exposure levels of UV, are necessary. Additional filtration will aid in removing heavy metals and potential viruses, parasites and bacteria. Depending on the sensitivity of the species chosen, on-growing and live holding facilities would not be as susceptible to these problems and could require less stringent water filtration.

Urban aquaculture faces water-quality challenges that are simply not present at facilities located in pristine coastal areas. Stormwater runoff, sewer overflows and petroleum spills raise the risks in obvious ways. However, the re-suspension of contaminated sediments by natural and anthropogenic activities is an unavoidable risk factor, particularly in urban settings with an industrial past. Chronic and episodic water-quality problems make larval culture difficult without additional water treatment. However, even in such a compromised environment, site selection is important and will dictate the level of water pre-treatment needed and the amount of seawater storage needed to weather foreseeable pollution events.

Outreach to Industry

It is important that proponents of commercial aquaculture in urban settings such as Boston understand the complications associated with compromised water quality. To that end, the MIT Sea Grant Program has an active programme of outreach to the industry that includes the dissemination of culture protocols and providing assistance to ventures seeking to locate in Boston and other Massachusetts sites. To date, interest has come from both start-up aquaculture ventures and existing operations seeking to expand into Boston.

There are numerous sites in Boston Harbor that we have helped industry explore for the siting of commercial ventures. Figure 13.3 is a chart indicating locations that correspond to the sites and their owners. State, city governments or other authorities own most of these locations. In addition, most of these sites represent filled land and are therefore subject to the provisions of the Commonwealth's Chapter 91 regulations that require the accommodation of water-dependent activities. Unfortunately, pressure from residential and commercial developers is intense and, to date, commercial aquaculture has not been able to gain a foothold in Boston Harbor.

Conclusions

Boston Harbor and urban waterfronts in general may represent attractive locations from a market access standpoint. However, water-quality issues and the challenges of competing with development pressures on now desirable harbour access locations remain significant obstacles to commercial ventures. A combination of clever water management and political savvy will be needed to capture the economic possibilities.

Acknowledgements

We acknowledge the cooperation of the Massachusetts Water Resource Authority in conducting some of the work reported here. We also thank the NMFS Narragansett Laboratory, Heritage Salmon of New Brunswick, and the Department of Fisheries and Oceans in St Andrews, New Brunswick for help in providing haddock eggs to our hatchery. We also thank the National Park Service for providing the space for our hatchery and Aqualab.

References

Ay, O., Kalay, M., Tamer, L. and Canli, M. (1999) Copper and lead accumulation in tissues of a freshwater fish *Tilapia zillii* and its effects on the branchial Na, K-ATPase activity. *Bulletin of Environmental Contamination and Toxicology* 62, 160–168.

Beyers, D.W., Rice, J.A., Clements, W.H. and Henery, C.J. (1999) Estimating physiological cost of chemical exposure: integrating energetics and stress to quantify toxic effects in fish. *Canadian Journal of Fisheries and Aquatic Science* 56, 814–822.

Brett, J.R. and Groves, T.D.D. (1979) Physiological energetics. In: Hoar, W.S., Randall, D.J. and Brett, J.R. (eds) *Fish Physiology.* Vol. 8. Academic Press, New York, pp. 279–352.

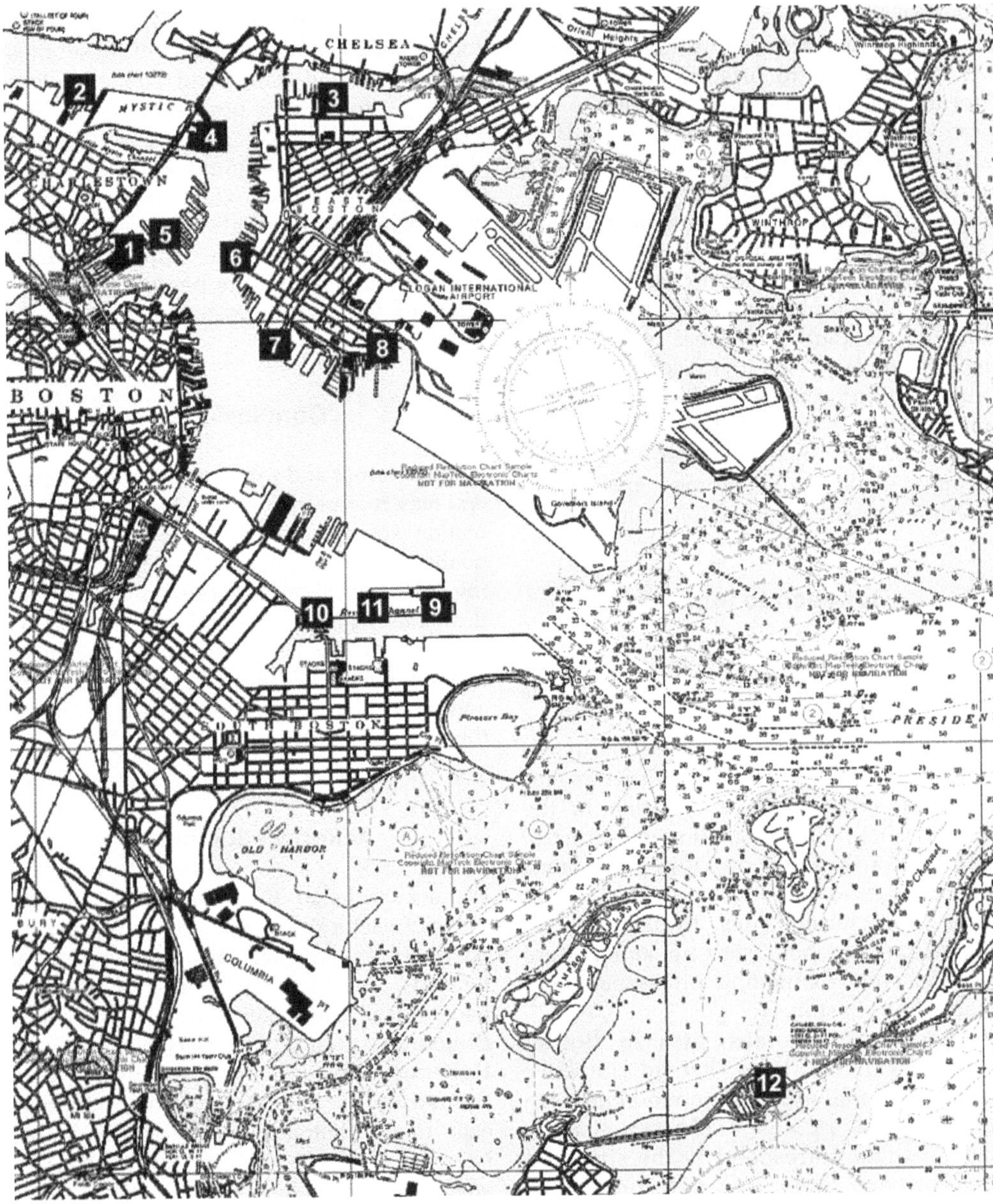

Fig. 13.3. Boston Harbor and potential aquaculture sites. Site details are given in Table 13.2.

Bury, N.R., Jie, L., Flik, G., Lock, R.A.C. and Bonga, W.S.E. (1998) Cortisol protects against copper induced necrosis and promotes apoptosis in fish gill chloride cells in vitro. *Aquatic Toxicology* 40, 193–202.

Chakoumakos, C., Russo, R.C. and Thurston, R.V. (1979) Toxicity of copper to cutthroat trout (*Salmo clarki*) under different conditions of alkalinity, pH and hardness. *American Chemical Society* 13, 213–219.

EPA (1999) *National Recommended Water Quality Criteria – Correction.* April. EPA 822-Z-99–001.

Goudey, C.A., Moran, B.M. and Mead, E. (1999) Larval culture in an urban setting: MIT's Boston Harbor Marine Finfish hatchery. *Recirculation Today* 1, 3.

Iger, Y., Balm, P.H.M. and Bonga, W.S.E. (1994) Cellular responses of the skin and changes in plasma cortisol levels of trout (*Oncorhynchus mykiss*) exposed to acidified water. *Cell Tissue Research* 278, 535–542.

Kitchell, J.F. (1983) Energetics. In: Webb, P.W. and Weihs, D. (eds) *Fish Biomechanics.* Praeger, New York, pp. 312–338.

Lang, T. and Dethlefsen, V. (1987) *Cadmium in Skeletally Deformed and Normally Developed Baltic Cod (*Gadus morhua *L.).* ICES CM-1987/E:30.

Massachusetts Water Resource Authority (2002) *Boston Harbor Update – Spring 2002.* http://www.mwra.state.ma.us/harbor/html/bhupdate.htm

Mazon, A.F. and Fernandes, M.N. (1999) Toxicity and differential tissue accumulation of copper in the tropical freshwater fish, *Prochilodus scrofa* (Prochilodontidae). *Bulletin of Environmental Contamination and Toxicology* 63, 797–804.

Moran, B.M. and Goudey, C. (2000) The culture of haddock, *Meanogrammus aeglefinus*, using a recirculating system in an urban setting. *Journal of Shellfish Research* 19(1), 563–584.

Peterson, C.H. (2001) The 'Exxon Valdez' oil spill in Alaska: acute, indirect and chronic effects on the ecosystem. *Advances in Marine Biology* 39, 1–103.

Rice, S. (1999) Pink salmon and oil: ten years later. Long-term studies have changed the way we think about oil pollution. EVOS Trustee Council, Anchorage, AK. Pink salmon synthesis 99329. US Dept. of Commerce, NOAA/NMFS. http://www.afsc.noaa.gov/abl/evos/index.htm

Richards, J.G., Burnison, B.K. and Playle, R.G. (1999) Natural and commercial dissolved organic matter protects against the physiological effects of a combined cadmium and copper exposure on rainbow trout (*Oncorhynchus mykiss*). *Canadian Journal of Fisheries and Aquatic Science* 56, 407–418.

Schubel, J.R. (1999) *Contaminated-Sediment Database Development and Assessment in Boston Harbor.* USGS Fact Sheet FS-078–99. http://pubs.usgs.gov/fs/fs78–99/

Seymour, R.J. and Geyer, R.A. (1992) Fates and effects of oil spills. In: Hollander, J.M. (ed.) *Annual Review of Energy and the Environment* 17, 261–83.

Thurberg, F.P. and Gould, E. (2004) Pollutant effects upon cod, haddock, pollock and flounders of the inshore fisheries of Massachusetts and Cape Cod Bays. In: Buchsbaum, R.N., Robinson, E.E. and Pederson, J. (eds) *The Decline of Fisheries Resources in New England: Evaluating the Impact of Overfishing, Contamination and Habitat Degradation.* Massachusetts Bay Programs, Massachusetts Institute of Technology Sea Grant, Cambridge, Massachusetts.

United States Geological Survey (1999) *Contaminated-sediment Database Development and Assessment in Boston Harbor.* http://pubs.usgs.gov/fs/fs78–99/

Wedemeyer, G.A. (1996) *Physiology of Fish in Intensive Culture Systems.* Chapman and Hall, New York, p. 62.

Wendelaar, B., Flik, S.E., Balm, P.H.M., and van der Meij, J.C.A. (1990) The ultrastructure of chloride cells in the gills of the teleost *Oreochromis mossambicus* during exposure to acidified water. *Cell Tissue Research* 259, 575–585.

This page intentionally left blank

14 Urban Aquaculture in Brooklyn, New York, USA

Martin P. Schreibman and Chester B. Zarnoch
Brooklyn College/CUNY, Brooklyn, USA

Abstract

If considered an independent city, Brooklyn would rank as the fourth largest in the United States. A case is made for Brooklyn, New York (US) as being an excellent site and a good template for urban aquaculture development. It typifies the characteristics needed for successful aquaculture, as well as reflecting the problems to be faced in moving urban aquaculture from concept, to practice and to ultimate success. The important role a university may play in the development and application of technology (e.g. Recirculating Aquaculture Systems – RAS), biological principles (induction of spawning, life cycles, nutrition) and economic feasibility (marketing, job training), each of which are important entities for successful urban aquaculture development, is also discussed. Diverse programmes which emphasize the use of RAS technology in an urban setting, including finfish, ornamental and bivalve aquaculture research; captive breeding of horseshoe crabs; educational programme development; use of aquatic organisms for biomedical research; and community outreach programmes are considered.

Introduction

Urban aquaculture, literally defined, means the rearing and breeding of aquatic organisms under controlled conditions for commercial or environmental enhancement purposes in or near a metropolitan area. Brooklyn, New York, is an excellent site and an excellent example for urban aquaculture development. It typifies the characteristics needed for successful aquaculture, as well as reflecting the many problems that are faced with the move towards urban aquaculture from concept, to practice and to ultimate success. Conducting aquaculture in an urban environment has many positive attributes. Primarily, there are a large number of multi-cultural consumers with a need for high quality, low cost, diversified food sources which are free from pollutants and chemical contamination. Urban centres can also serve as major distribution hubs, with an available labour force and diverse sources of investment capital.

New York City is an area ripe for development of urban aquaculture. The people of New York eat considerably more seafood than the United States norm (Timmons *et al.*, 2004), and it serves as a major distribution hub for fish and aquatic products from around the country and the

(eds B. Costa-Pierce, A. Desbonnet, P. Edwards and D. Baker)

world. The Fulton Fish Market in the borough of Manhattan accounts for about one-third of the value of all wholesale seafood activity in the entire state (Sea Grant, 2001). The waters in proximity to New York City are home to many species for potential aquaculture development and, according to leaders of local religious and ethnic organizations, a large multicultural population for consuming these new products exists in a wide variety of ethnic neighbourhoods that have distinctive seafood tastes and needs that are currently not being satisfied.

There are more than 933 km of land–water interface within Metropolitan New York that can provide a network of water passages, in addition to the extensive rail and highway routes for product distribution. New York City is central to a distribution network that stretches to the tip of Long Island, and into the large markets in New Jersey, Pennsylvania and Delaware – all within a few hours drive (Fig. 14.1). New York's superb interstate roadway system links suburban markets close to New York through a vast trucking and transportation industry. With JFK and Newark international airports and major freight businesses located on or near the airports, fresh seafood product can reach remote markets quickly. New York's port can also serve to transport properly packed cargo to other port cities.

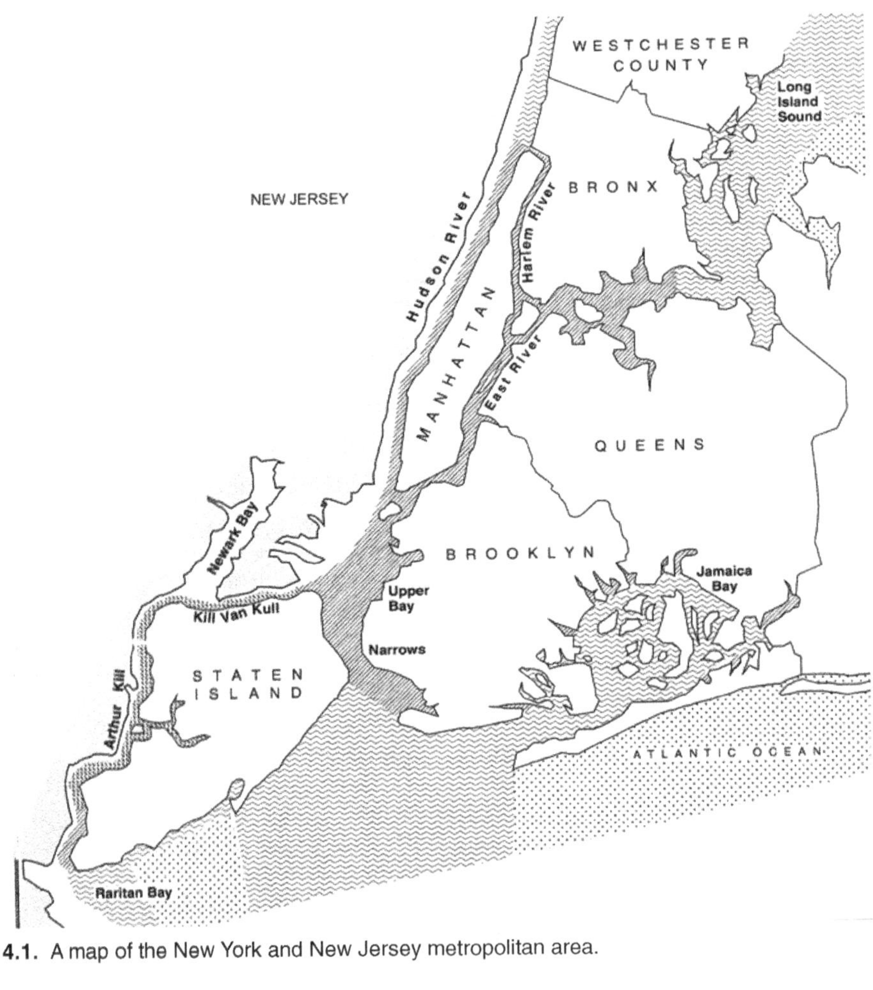

Fig. 14.1. A map of the New York and New Jersey metropolitan area.

The issue of limited water availability in New York City could be addressed, in part, as a by-product derived from water-treatment procedures used in aquaculture based on the new recirculation technology. New York City experiences, as many cities do, blatant needs to improve its economic climate by developing new industries, jobs and to provide training, especially for socially disadvantaged and unemployed citizens. A vibrant aquaculture industry could help address these social and economic issues.

For now, New York City relies heavily on aquaculture product imports from elsewhere in the nation and the world. In 1999, the largest source of fish and seafood purchased by the New York seafood industry were imports from outside the USA. The state's seafood industry, and others, purchased an estimated US$786 million worth of fish and seafood products from foreign sources (Sea Grant, 2001). Shrimp, almost all of which was frozen, accounted for 42% of the value of fish and seafood imported to New York in 1999 (Sea Grant, 2001). The New York seafood industry purchased an estimated US$535 million worth of fish and seafood products from sources in other states in 1999 (Sea Grant, 2001). This is in addition to purchases from other countries.

The ever-growing plethora of restaurants makes the greatest economic contribution from among the seafood-consuming industry segments. This contribution is attributable to the substantial value added by restaurants to the fish and seafood products they purchase, and from the great number of jobs generated in restaurants (Sea Grant, 2001).

Thirty per cent of all New Yorkers live in Brooklyn, making it the most populous of the city boroughs. Brooklyn is the leading destination for new emigrants in New York City – 200,000 in the last 10 years (McCall, 1999). The Asian population has quadrupled over the last 10 years, and the Hispanic population numbers almost 520,000 (McCall, 1999). It also has the third largest African-American population – estimated at 925,000 – of all the counties in the USA (McCall, 1999). As a result, Brooklyn is within the top 15 counties in the USA in terms of population concentration of many minority groups (Fig. 14.2; McCall, 1999; US Census Bureau, 2000). Central Brooklyn is proximate to, and can easily take advantage of, a potential market beyond the counties of New York City, once the business has solidified its base. Piers, warehouses, lofts and undeveloped geography are ready for aquaculture development. A Brooklyn urban aquaculture

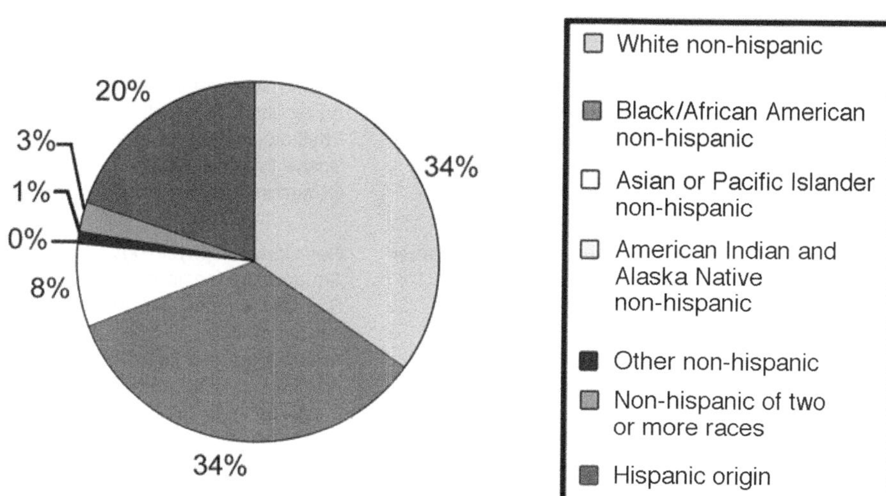

Fig. 14.2. A pie graph describing Brooklyn's ethnic diversity.

industry can be a timely, feasible and profitable undertaking.

The City University of New York (CUNY) is an essential building block for the creation of urban Brooklyn aquaculture. It provides research and development aspects that are essential for keeping the aquaculture industry vibrant, alert and up to date with new technology and candidate species for culture. Brooklyn College, one of the flagship colleges, is located in the heart of Brooklyn. Its two million dollar centre of excellence, the Aquatic Research and Environmental Assessment Center (AREAC), is located on the campus and is dedicated to basic and applied studies of aquatic organisms and the environments they inhabit. AREAC is an innovative leader known for its achievements over many years in biological and medical research, environmental assessment and restoration, educational programme development and aquaculture research and development.

Economic Development

AREAC is playing a major role in urban aquaculture programmes. This includes finfish for food production, ornamental species and aquaculture for research of animal reproduction and basic biological studies (Table 14.1). In many of these programmes, the AREAC facility utilizes state-of-the-art recirculating aquaculture systems (RAS) technology (Fig. 14.3). Recirculating aquaculture systems hold the key for the success of urban aquaculture, for they permit high-density growth of aquatic organisms in facilities that could be located in just about any part of town, including low real-estate value areas and brown fields. Essential to its utility is that it is environment friendly, adding only minimal amounts of water as a by-product of the process (Timmons *et al.*, 2001). The RAS at Brooklyn College permit the flexible design of aquatic ecosystems, allowing for the creation and execution of a number of diverse aquaculture and environmental assessment-restoration programmes.

For example, water temperature can be adjusted from near freezing to tropical temperatures, and environments can be 'constructed' that are salt water, fresh water or with salinities in between. A sophisticated lighting system provides flexibility in design of intensity, manipulation and control of calendar design, and photoperiod duration. An elegant water-quality monitor-

Table 14.1. A compendium of organisms studied at AREAC.

Common name	Scientific name	Research use
Freshwater species		
Tilapia	*Oreochromis niloticus*	Economic development and job training
Walleye	*Stizosledion vitreum*	Aquaculture feasibility
Platyfish	*Xiphophorus maculatus*	Physiological processes
Swordtails	*Xiphophorus helleri*	Space travel studies
Guppies	*Lebistes reticularis*	Ornamental aquaculture
Saltwater species		
Winter flounder	*Pseudopleuronectes americanus*	Reproductive physiology
Summer flounder	*Paralichthys dentatus*	Aquaculture feasibility
Horseshoe crabs	*Limulus polyphemus*	Scientific studies and environmental enhancement
Eastern oysters	*Crassostrea virginica*	Aquaculture and field studies
Hard clams	*Mercenaria mercenaria*	Aquaculture and physiology
Clownfish	*Amphiprion ocellaris*	Propagation/ornamental aquaculture
Peppermint shrimp	*Lysmata wurdemanni*	Propagation/ornamental aquaculture
Soft corals	Several species	Propagation and behaviour/ornamental aquaculture
Chambered nautilus	*Nautilus pompilius*	Behaviour

Fig. 14.3. Top: Panoramic view of AREAC RAS. Bottom: Close up of our '8' foot RAS. (A) biological filter, (B) chiller unit, (C) water parameter monitoring device, (D) in-line heater, (E) ozone generator, (F) mechanical screen filter and (G) foam fractionator.

ing system provides continuous, real-time gathering of water parameters, which include pH, dissolved oxygen, percent oxygen, conductivity and temperature. Mechanical screen filters, biological filters, ozone generators and foam fractionators facilitate the maintenance of high-quality water for aquatic organisms to thrive (Fig. 14.4). In AREAC, the RAS may take different forms. Animal holding tanks can range in capacity from a few litres to thousands of litres and biological and mechanical filters may take many forms depending on needs.

The use of RAS at AREAC demonstrates the unique marriage of water reuse systems and urban environments. AREAC is located on the Brooklyn College campus in the same building that houses offices and classrooms. This demonstrates that these systems can be located in a variety of diverse places from warehouses to skyscrapers, from green fields to brown fields, from close proximity to a body of water to one that is remote from it. AREAC is located approximately 3.2 km from the nearest major water source and, therefore, derives its water directly from the New York City water supply system. This is an ideal situation for urban aquaculture ventures in developed cities. In under-developed countries and cities that lack a consistent supply of high-quality water, the fact that RAS use

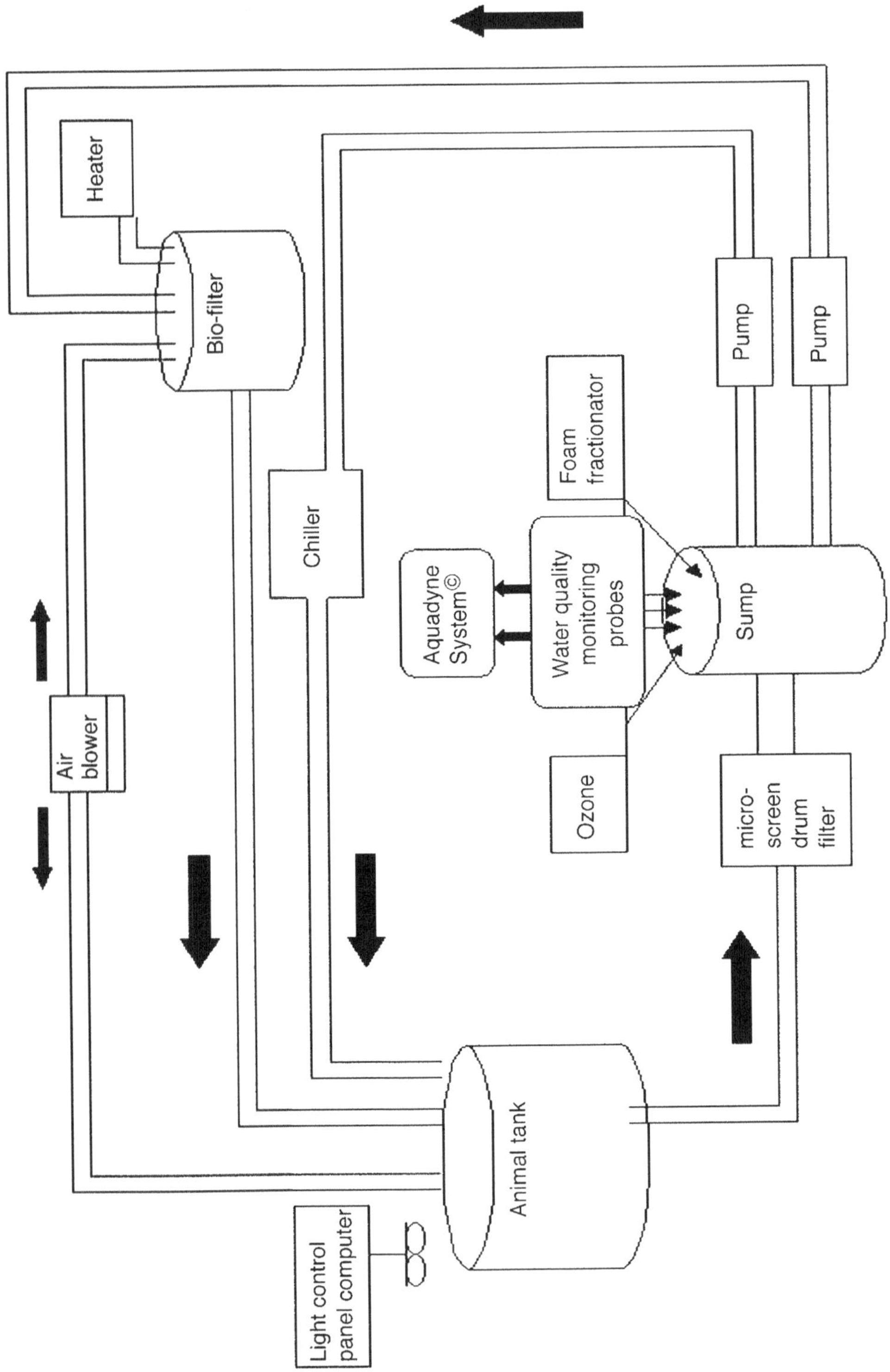

Fig. 14.4. A schematic of the connectedness of AREAC recirculating aquaculture systems.

90–99% less water than conventional aquaculture methods (Timmons *et al.*, 2001, 2004) eases the burden of water demand.

An additional benefit of RAS in urban ventures is that it produces a minimal amount of effluent. In New York City, effluent is processed through the city's water treatment system as is done likewise in many other major cities. Some ventures may even capture the effluent for other purposes, for example, in the production of fertilizer. Thus, the RAS is appropriate to both urban environments for the production of large volumes of protein in limited space and to countries whose water supply and water treatment systems are limited or nonexistent. In addition, an effective programme in urban aquaculture would introduce new industries leading to economic development in poor countries as well as in the more established ones. Employment is also diversified in the many phases of urban aquaculture, from production to processing and to the purveying of products, thus helping to resolve issues of unemployment.

Finfish for consumption

The finfish aquaculture activities of AREAC have included the commercially important walleye (*Stizosledion vitreum*), summer flounder (*Paralichthys dentatus*), winter flounder (*Pseudopleuronectes americanus*) and tilapia (*Oreochromis niloticus*). The most intensive culturing programme, and one that is currently drawing major commercial attention, is growout of tilapia. Fingerlings are brought into the facility at approximately 1 g in weight. These fingerlings, reared in a 10 m^3 RAS, can be brought to market size (approximately 0.5 kg) after 6–6.5 months of culture (Figs 14.5 and 14.6). Tilapias are fast growers, resistant to disease, tolerant to overcrowding, highly compatible and, indeed, excel in activity and growth in these water reuse systems. It is our belief that tilapia, coupled with RAS, will be the keystone to the successful establishment of a major tilapia urban aquaculture industry in New York State. To this end, Brooklyn College's AREAC and Cornell University are moving to design a programme that will ultimately establish a 454,000 t industry in New York State (Timmons *et al.*, 2004). It is thus the intention of this effort to make tilapia the 'poster fish' for New York State Urban Aquaculture.

A New York tilapia industry can model itself after the successful catfish industry in the southern United States. The catfish industry has developed a 273,000 t per year production base over the last 20 years, adding nearly 45,000 t of production in just the last 2 or 3 years (Timmons *et al.*, 2004). In large measure, the catfish industry was patterned after the chicken broiler industry. The success of both these industries is attributed to their vertical integration of breeding, growing, processing and distribu-

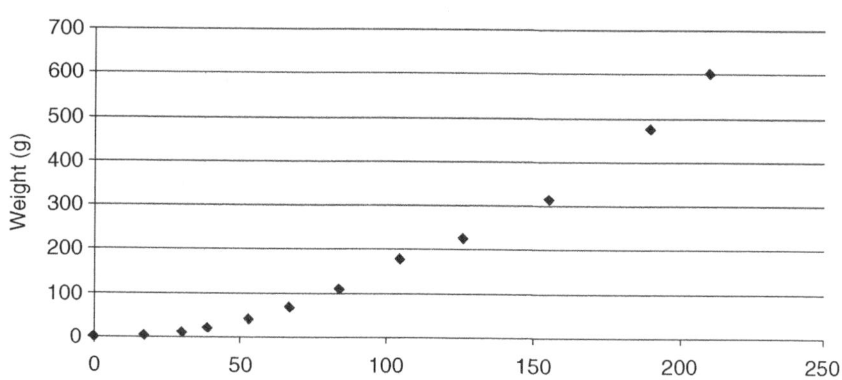

Fig. 14.5. Growth rate for tilapia cultured in a 10 m^3 RAS in a single experiment.

Fig. 14.6. AREAC staff culling market size tilapia.

tion operations under a single business structure. The success of this plan is augmented by the use of RAS. Contrary to outdoor pond and net pen systems, indoor fish production using RAS is sustainable, infinitely expandable, environmentally friendly and has the ability to guarantee both the safety and the quality of the fish produced throughout the year (Timmons *et al.*, 2001, 2004). Additionally, there is the spin off of creating a new job market. According to available figures (Fingerlakes Aquaculture, New York), each 453 t of production will require five people in the production facility – the processing facility will add an additional 10 people. Therefore, there are 15 jobs created for every 453 t of production, or 15,000 jobs will be generated by a 500,000 t per year industry in New York State.

Ornamental aquaculture

AREAC continues to do research and development in an attempt to ascertain the technical and marketing feasibility of bringing successful ornamental aquaculture ventures to non-traditional parts of the United States, most notably to New York and the Northeast. Over 800 aquaria, ranging in size between 9.5 and 284 l are used in the culture of ornamental organisms, both vertebrate and invertebrate. Some aquaria are RAS, while others are balanced, static systems containing snails and aquatic plants, in addition to fish. Currently these organisms include guppies (*Lebistes reticularis*), platyfish (*Xiphophorus maculatus*), swordtails (*Xiphophorus helleri*), clownfish (*Amphiprion ocellaris*), several species of soft corals and peppermint shrimp (*Lysmata wurdemanni*).

In our so-called 'freshwater facility', aquatic organisms are being propagated for ornamental as well as for research use. Prize-winning guppies raised in our laboratory are being studied to demonstrate the economic feasibility of ornamental fish production. For example, offspring of competition-winning guppies can command high prices on the wholesale and retail markets, sometimes as high as US$40 or US$50 for a mating pair. The fact that females can deliver 50–60 such young every 28 or 29 days (Axelrod, 1996), plus the ever-increasing demand of fish hobbyists, clearly indicates the feasibility of profitable business ventures (Fig. 14.7).

Soft corals also command high market prices, and are generally easier to propagate and less environmentally challenged than hard corals. Soft corals are attractive by virtue of size, shape and the colourful colonial aggregations that they form (Fig. 14.8). Cells of a soft coral colony can be dissociated by mechanical and chemical methods that we are developing. A small group of cells has the potential of developing into a new colony, suggesting the commercial attractiveness of these organisms (Ruppert and Barnes, 1994).

The well known, easily recognized, orange and white-banded saltwater clownfish commands high prices in a demanding market. A single fish has the potential to sell for US$30 or US$40 in retail shops. Considering that a female can produce 300 eggs per clutch twice a month (Ross, 1978), 7200 possible offspring per year, we have an exceptional candidate for profitable urban ornamental aquaculture ventures.

In order to realize the market potential of ornamental organisms, note that the US retail aquarium market is valued at approximately US$1 billion per year. In addition, over 15 million ornamental fish are imported into the US each month, worth about US$50 million each year. Although many species are still collected from tropical lakes and streams of South and Central America, Asia and Africa, imports now come from aquaculture farms in Thailand, Malaysia, Singapore, Hong Kong and Germany. In the USA, production of ornamental fish is valued at US$52 million per year, with most fish produced in Florida (Chapman *et al.*, 1994, 1997; Rowland and Cox, 2003).

Approximately 90% of freshwater ornamental fishes are bred in captivity (Dawes, 1998) and have an estimated price per half kilogram of US$35–60 (Hoff, 1996). By comparison, marine ornamentals draw a much higher price (US$400–600); however, their captive breeding and culture is much less advanced. The culture of ornamental fish and invertebrates is now recognized as a feasible alternative to the wild harvest of

Fig. 14.7. Guppies cultured at AREAC as part of our ornamental aquaculture programmes.

Fig. 14.8. Propagation of soft coral at AREAC.

specimens (Timmons *et al.*, 2004). In addition, many collecting localities around the world limit either the number of fish or the number of species taken, or both (Tlusty, 2002). The environmental challenges to aquatic ecosystems, coupled with poor animal husbandry after capture, and pressure from conservation groups and governments, have also challenged the success of the ornamental fisheries business. These problems can be addressed successfully and hopefully resolved by employing the technology and programmes proposed by RAS-driven urban aquaculture.

Aquaculture to produce research organisms

From its inception, AREAC has been involved in the use of aquatic animals in basic and applied research. This had its foundation in research activities in the early 1960s, when live-bearing teleosts – platyfish and swordtails – were used to study such diverse physiological topics as osmoregulation, cancer, maturation, development, reproductive system structure and function, aquatic toxicology and ageing. Our central research theme has been to understand the mechanisms underlying the genetic, environmental and neuroendocrine regulation of physiological processes. Species of the genus *Xiphophorus* have been our dominant research models. In recent years, much of our efforts have been to study the impact of space travel on the neuroendocrine regulation of physiological processes, especially reproduction. This programme reached a pinnacle when adult and juvenile swordtails (*X. helleri*) were flown on two recent space shuttle flights (STS-89 and STS-90) in a Closed Equilibrated Biological Aquatic System (CEBAS) developed at Ruhr University in Germany in a cooperative programme between NASA and DARA, the German space agency. This could truly be called the first orbiting RAS (Bluem *et al.*, 1995). The technology from space exploration studies can easily be extrapolated for the further development of urban aquaculture design, especially in the areas of remote monitoring that will come from the creation of improved sensing and control systems and in the realm of co-culturing of organisms in limited volumes.

Platyfish and swordtails are live-bearing aquatic vertebrates with well-studied genetic and physiological systems, and for these reasons they have become invaluable test organisms for understanding the impacts of human perturbations on the physiology of organisms in aquatic ecosystems (Magliulo-Cepriano and Schreibman, 1999). Saltwater animals have similarly been used for basic and applied research. Winter flounder (*Pseudopleuronectes americanus*) are currently being utilized to study the impact of endocrine-disrupting chemical pollutants, which pervade aquatic ecosystems and have a negative impact on the reproductive system structure and function of vertebrates. Combined field and laboratory studies now in progress in collaboration with SUNY-Stony Brook examine the impact of nonylphenols, a potent oestrogen-mimicking pollutant, to see the impact of this substance on reproductive system development in winter flounder in local estuaries. The need to collect and house adults, juveniles and young-of-the-year as well as to induce spawning for embryonic samples is resolved by the use of RAS technology.

The study of bivalves – oysters (*Crassostrea virginica*) and hard clams (*Mercenaria mercenaria*) – constitutes an important part of our research programme. The AREAC facility has been able to harness the RAS technology and the information gained from past endeavours in order to produce bivalves for research in closed systems. Although our practices follow standard hatchery procedures, the use of recirculating systems allows for complete control of all water parameters, thereby increasing our ability for controlled experimentation. In AREAC, bivalves spawned are cultured in conicals and then transferred to downwelling systems where they are held until they are 2 mm in size. The use of the RAS as a bivalve hatchery is attractive in that it offers many advantages to urban aquaculturists in comparison to the traditional hatcheries that use waterfront property and bay water. RAS hatcheries can eliminate the need to deal with fickle, but stringent, regulations imposed by city, state and federal agencies. Using RAS would also eliminate the need to acquire expensive waterfront property, meet issues of poor quality of urban waters, and bring the product closer to market.

Animals spawned in our centre are used to study bivalve physiology in order to address pressing concerns and problems of the shellfish industry. A major study focuses on causes of the overwintering mortality of juvenile clams that are planted in the autumn on the bay bottom for continued growth. These studies will detail mortality rates as well as physiological changes, especially in the dynamics of energy stores and their utilization by clams as the environment challenges them. This problem is of major importance to commercial hard-clam growers in the northeastern USA. We are also using the young bivalves generated at AREAC to re-examine the feasibility of growing these organisms in RAS for commercial markets (Sea Grant, 1982). Advances in RAS filtration and temperature control give hope that bivalves may be brought to market size under intensive culture conditions. However, research is needed in the mass production of microalgae, or a nutritionally equivalent alternative, in order for animal production in these systems to be cost effective. For now, however, they are useful particularly in the hatchery phases of the life cycle. In other field studies, oysters and oyster larvae, similarly generated, will be set on artificial reefs in New York and New Jersey waters in order to study life history successes in areas now devoid, but once among the most productive in the world. In addition, these organisms are being utilized to study the feasibility of harnessing their ability to filter large volumes of water in enriched marine ecosystems with potential restoration applications.

The 350 million year old invertebrate the horseshoe crab (*Limulus polyphemus*) has taken an important position in our research endeavours. AREAC is one of a very limited number of laboratories that is capable of successfully spawning and rearing horseshoe crabs in controlled laboratory conditions utilizing RAS systems. Currently, we are rearing and studying more than 2000 animals that range in age between several weeks to adults. Horseshoe crabs are of immense importance to the pharmaceutical industry for the harvesting of *Limulus* amoebocyte lysate (LAL), which occurs in their blood. LAL serves the important function of detecting gram-negative bacteria, but has never been synthesized and can only be extracted from living *Limulus*. Furthering our understanding of the behaviour and physiology of *Limulus* is essential to effective management strategies. The ability to culture them in the laboratory opens up important studies that heretofore could not be addressed. The depletion of horseshoe crabs by industries involved in fishing and fertilizer production is rapidly having negative impacts on their population size (Tanacredi, 2001). Our programme of captive breeding could play a significant role in restoring depleted natural populations.

Laboratory research on fishes and aquatic invertebrates is prevalent throughout the world. One problem that many researchers face is the acquisition of experimental animals. Collection can be time consuming, costly and often inconsistent (i.e. seasonal changes, declining fisheries stocks). For example, in order to collect winter flounder for a current research programme, AREAC sub-contracted a boat and crew that was costly. In the final analysis, the cost for collecting a barely adequate sample size was US$100 per fish. The production of research animals, including winter flounder, could meet the research demands of academia and industry. There are many species of fishes and invertebrates that have attributes which make them valuable to researchers and thus excellent potential candidate species for research aquaculture. AREAC's ability to generate large numbers of horseshoe crabs provides a new source of material for study on naturally depleted stocks.

Education

An assessment of the current public education status for New York clearly indicates that there is a blatant need for programmes to enhance science teacher skills, and AREAC has been working to fill this void by offering off-hour programmes, sabbaticals and workshops. Our first summer science institute intended for New York City science teachers, and held in 2002, was received with enthusiasm. Teachers learned, through laboratory and field exercises, to incorporate local urban marine and freshwater ecosystems into their curricula, thus making their students more aware of environmental issues as they focus on the natural and social sciences of aquatic communities. Bringing the parents of these students in for evening and weekend classes and workshops completes the loop for effective education.

The introduction of urban aquaculture and the technology of water reuse systems into the curriculum of the New York City school system generates a knowledgeable public base who could be marshalled to support our lagging development of urban aquaculture in the USA, and perhaps elsewhere as well. The opportunity to have young students learn by hands-on experiences will provide the next generation of urban aquaculturists. The development of vocational high schools based on aquaculture has already proven highly successful as measured by a population of young individuals with a sound education and a dedication to a new and needed industry (see Chapters 15 and 16 in this volume).

Community Outreach

The urban aquaculture programmes at AREAC have led to a number of exciting community outreach activities. In one of the more successful programmes, several hundreds of kilograms of tilapia, raised at Brooklyn College's AREAC, have been given to homeless shelters. Our primary recipient has been 'Project Renewal', a not-for-profit organization based in Manhattan that provides housing, sustenance and job training to the homeless and chemically challenged. Among their several successful job training programmes, culinary arts is the one that has been most closely associated with AREAC. Students in this programme receive whole AREAC-grown tilapia and are taught to prepare them in a number of ways, either whole or filleted. The final product is then served to the residents of Project Renewal's several shelters. Our desirable tilapia has also been used for community events, political rallies and staff appreciation rallies at Brooklyn College. Tours of AREAC and educational lectures are provided on an irregular basis to regional organizations in order to familiarize citizens with the technology, need and current status of urban aquaculture in their community.

Conclusions

Despite an untapped potential for enormous success, Brooklyn, indeed New York City, does not have a single successful major aquaculture venture. The inertia and obstacles in moving successful urban aquaculture programmes from discussion to reality must be overcome to face the ever-growing needs imposed by over-fishing, pollution and increasing demands for aquaculture products. These demands can be met locally and state-wide by way of effective, dynamic and significant initiatives. As a member of a university family, we firmly believe that educational collaborations within New York State can move towards these ends. The state, city and private universities are numerous and scattered throughout the state, with concentrations in the metropolitan area. They must partner and collaborate to make the case known for the importance of developing a state-wide urban aquaculture industry. They must trumpet the important reasons for economic development, job training, education and environmental stability to the policy makers, land users and to the individuals with capital who can support these essential ventures (Fig. 14.9). These three groups, in collaboration with

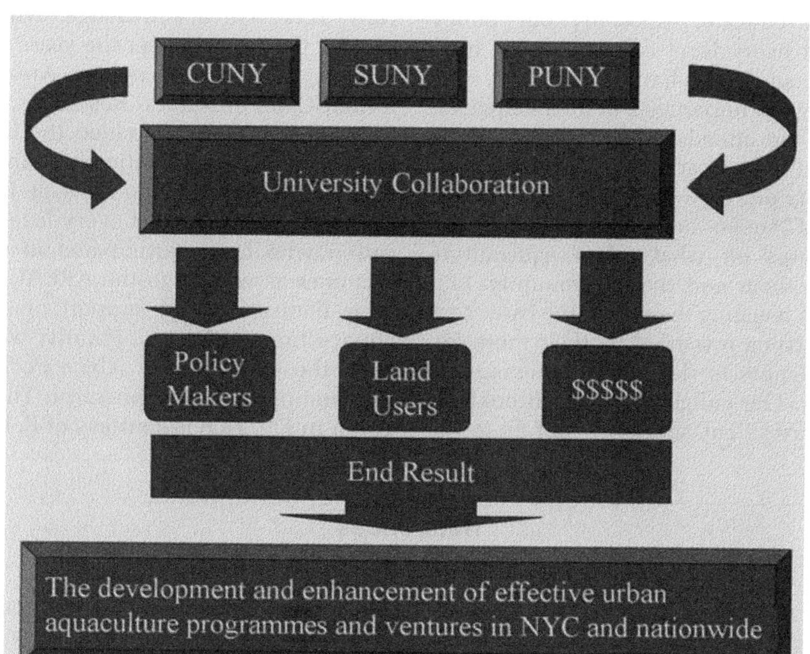

Fig. 14.9. A schematic suggesting the role of universities in developing a programme for urban aquaculture in NYC and NY State. CUNY is the city university system, SUNY the state university system and PUNY the network of private universities in the state of New York.

the universities, can and must lead to the development and enhancement of effective urban aquaculture initiatives and ventures in New York City, as well as nationwide.

We are literally missing the boat. Internationally recognized business management expert, Peter F. Drucker (1999), predicts that aquaculture, or fish farming, will be one of the three major economic opportunities in the new millennium. Aquaculture has been identified as a major economic opportunity due to the diminishing supplies of wild caught species in the oceans, concern with rampant accumulation of pollutant chemicals and the inability of this natural supply of seafood to meet the increasing demands of the consuming public for healthy, nutritious and tasteful products. The university and their collaborators need to sail the boat towards the productive, promising and rewarding goals promised by urban aquaculture development.

In New York, AREAC, in conjunction with Cornell University and its Extension Program, has been attempting, with a series of white papers, conferences and workshops, to educate, especially our policy makers at every level of government from local boroughs to the halls of Congress, as to the merits and importance of urban aquaculture and the attendant economic development and job creation. In addition, our community outreach activities include tours of AREAC, open houses and lectures to civic groups on what urban aquaculture means to them and their community. Our success in reaching the ears of the New York media is being rewarded by their more frequent attempts to deliver 'the message' to the public. Our enlightenment of Brooklyn's Borough President and their staff as to the virtues of fish farming in the city is reaching discussion and planning levels as to how to develop this industry in economic development zones in Brooklyn that were recently designated by the Mayor of New York City. Our experiences in New York show that the key to the development of an urban aquaculture industry, and to overcome the general inertia, is to lobby support from all levels of municipal and government agencies and educate them in the socio-economic impacts that the industry will have in New York City. We believe that this approach could also be effective on a national and international basis as well.

Acknowledgements

Our programmes have been funded by: United States National Park Service, United States Army Corps of Engineers, the Department of Environmental Conservation, Con Edison, NASA, New York Sea Grant, The New York City Board of Education and the City University of New York. We acknowledge, with many thanks, the support over the years from the Gateway National Recreation Area and its former Chief of Natural Resources, Dr John T. Tanacredi. They provided the funds for our first small recirculating system. We are also grateful to our students at Brooklyn College, and students at every level of education who have participated in our programmes as well as to the AREAC staff for their dedication and support of AREAC's aquaculture initiatives. Finally, we appreciate the assistance of Alyce Pech in the preparation of this manuscript. The photograph in Fig. 14.8 is courtesy of R. Dickie.

References

Axelrod, H. (1996) *Encyclopedia of Tropical Fishes* (5th edn). TFH Publications, New Jersey.

Bluem, V., Andriske, M., Eichhorn, H., Kreuzberg, K. and Schreibman, M.P. (1995) A controlled aquatic ecological life support system (CAELSS) for combined production of fish and higher plant biomass suitable for integration into a lunar or planetary base. *Acta Astronautica* October, 361–371.

Chapman, F.A., Fitz-Coy, S., Thunberg, E., Rodrick, J.T., Adams, C.M. and Andre, M. (1994) *An Analysis of the United States of America International Trade in Ornamental Fish*. University of Hawaii Sea Grant, Honolulu, Hawaii.

Chapman, F.A., Fitz-Coy, S., Thunberg, E., Rodrick, J.T. and Adams, C.M. (1997) United States of America trade on ornamental fish. *Journal of the World Aquaculture Society* 28, 1–10.

Dawes, J. (1998) International experience in ornamental marine species management. Part 1: perspectives. *Ornamental Fish International Journal* 26 February 1999. http://www.ornamental-fish-int.org/marinespecies1.htm

Drucker, P.F. (1999) Beyond the information revolution. *The Atlantic Monthly* 284(4), 47–57.

Hoff, F.H. (1996) Conditioning, spawning and rearing of fish with emphasis on marine clownfish. Aquaculture Consultant, Dade City, Florida.

McCall, H.C. (1999) *Brooklyn: an Economic Review.* Office of the State Deputy Comptroller for the City of New York, New York.

Magliulo-Cepriano, L. and Schreibman, M.P. (1999) Methods to assess the effects of environmental chemicals on the brain–pituitary–gonad axis of the reproductive system. In: Henshel, D.S., Black, M.C. and Harrass, M.C. (eds) *ASTM Eight Symposium on Environmental Toxicology and Risk Assessment: Standardization of Biomarkers for Endocrine Disruption and Environmental Assessment.* Standard Tech Pub., 1936 ASTM, West Conshohocken, PA, Vol. 8, pp. 207–225.

Ross, R.M. (1978) Reproductive behavior of the anemonefish, *Amphiprion melanopus*, on Guam. *Copeia*, 103–117.

Rowland, L.W. and Cox, L.J. (2003) *Opportunities in Ornamental Aquaculture.* National Coastal Resources Research and Development Institute, Portland, Oregon.

Ruppert, E.E. and Barnes, R.D. (1994) *Invertebrate Zoology* (6th edn). Saunders College Publishing, Philadelphia.

Sea Grant (1982) *Intensive Marine Bivalve Cultivation in a Controlled Recirculating Seawater Prototype System.* University of Delaware Sea Grant Program, publication number DEL-SG-07-82.

Sea Grant (2001) *The Economic Contribution of the Sport Fishing, Commercial Fishing and Seafood Industries to New York State.* New York Sea Grant publication number NYSGI-T-01-001, Stony Brook, New York.

Tanacredi, J.T. (2001) *Limulus in the Limelight.* Kluwer Academic/Plenum Publishers, New York.

Timmons, M.B., Ebeling, J.M., Wheaton, F.W., Summerfelt, S.T. and Vinci, B.J. (2001) *Recirculating Aquaculture Systems.* Northeastern Regional Aquaculture Center Publication 01–002. Cayuga Aquaculture Ventures Ithaca, NY.

Timmons, M.B., Rivara, G., Baker, D., Regenstein, J.M., Schreibman, M.P., Warner, P., Barnes, B. and Rivara, K. (2004) *New York Aquaculture Industry: Status, Constraints and Opportunities.* White Paper. Cornell University, Ithaca, NY.

Tlusty, M. (2002) The benefits and risks of aquacultural production for the aquarium trade. *Aquaculture* 205, 203–219.

US Census Bureau (2000) *Census SF 1.* Population Division – NYC Department of City Planning.

This page intentionally left blank

15 Growing a Future Crop of Aquaculturists: Creating an Urban Aquaculture Education Programme in New Haven, Connecticut, USA

John J. Roy
The Sound School Regional Aquaculture Center, New Haven, USA

Abstract

The Sound School Regional Aquaculture Center is a comprehensive high school in the New Haven Public School System, a Connecticut State Vocational Agricultural Educational facility. The Interdistrict Marine Educational Program (IMEP) provides students at the Sound School with hands-on experience through shellfish hatchery and instructional lobster hatchery programmes. Of the 320 students who attend the school, 50% are from New Haven, with the remaining students coming from 21 surrounding region school districts. The Sound School promotes the attitude of 'learning through doing', where students enrolled in the school are encouraged to become involved in areas of study common to all vocational agriculture education. These areas include career awareness, investigations of aquaculture and marine trades, enhancement of leadership techniques and the practical application of problem-solving strategies through application of learned skills in the field. Faculty and staff seek to provide students with experiences, at an early age, which may be transferred directly to the work place or used to facilitate post secondary education. Using an interdistrict cooperative grant from the Connecticut State Department of Education's Office of Urban and Priority School Districts, the Sound School has developed one of the first aquaculture-based hatchery curricula in the nation.

History of Aquaculture Education in the Connecticut Public School System

The development of aquaculture education at the high school level in the state of Connecticut began in 1980. When the United States Congress passed the National Aquaculture Act (PL-96-362) to support and expand aquaculture research, it prompted Roger Lawrence, the Vocational Agriculture Education Consultant for the State of Connecticut at that time, to contact the Sea Grant Marine Advisory Program at the University of Connecticut (UCONN). The Marine Advisory Program was able to provide pertinent information regarding aquatic science and technology. Two events occurred the following year that made the introduction of aquaculture into the Vocational Agriculture System (VAS) a viable concept.

(eds B. Costa-Pierce, A. Desbonnet, P. Edwards and D. Baker)

First, as levels of industry and the economics associated with the urbanization of Connecticut continued to grow, more and more people were 'moving away' from the farming profession. The number of students who lived in coastal areas of Connecticut that were interested in attending Vocational Agriculture Centers for traditional terrestrial agriculture coursework had been declining steadily for several years. In 1981 a survey was developed and given to students in several middle schools along the Connecticut shoreline. The survey was designed to assess the percentage of students interested in aquatic science and associated technologies if the course work were incorporated into the vocational agriculture curriculum. The results of the surveys showed interest ranging from 25% to 55% among the children who had been raised on or near the waters of Long Island Sound (T.C. Visel, Connecticut, 2002, personal communication). The development of an aquaculture curriculum was seen as a means to strengthen the interest in the Vocational Aquaculture Schools (VAS) that had started in Connecticut in 1981. Vocational aquaculture consists of applied coursework, supervised occupational experience and leadership activity with the National Future Farmers of America Organization (FFA).

Secondly, the harvesting of oysters had been a flourishing Connecticut business until the turn of the century. Typical yields approaching 2,000,000 bushels (7×10^7 l) were recorded annually from 1898 to 1910 (Galtsoff, 1964). The New Haven oyster beds had been described as the 'richest in the world', with recorded recruitment of over 1 million oyster spat set on a single bushel of cultch (shell left from the process of shucking oyster meats) in Morris Creek (Loosanoff, 1940). Pollution in New Haven harbour, particularly the discharge of copper from munitions factories, took a toll on oyster production. There was no recorded wild set of oysters for 22 years (1935–1957). By 1959, approximately 95% of the oyster production had been lost in New Haven harbour (George Mcneil, 1980, personal communication).

In 1967, the state of Connecticut requested, and received, relief from the federal government to rebuild the failed oyster industry. During the early 1970s several years of good oyster sets were recognized and the industry began to revive. By 1978, the oyster business was re-established in the state. In 1981, Connecticut became the first state 'to recognize aquaculture as a form of agriculture and farming' (Public Act, PA-81–269). The recognition of aquaculture as a form of farming, and the results of the surveys showing strong student interest in the marine environment, opened the door for the development of aquaculture curriculum materials for the VAS. In 1982, Erroll Terrell, the Bureau Chief of Vocational Services for the State of Connecticut, authorized the creation of the Stamford Vocational Aquaculture Center saying, 'Many are predicting that aquaculture is a new and emerging industry which will provide food for a hungry world. Stamford's work in these initial stages of developing a vocational educational programme in aquaculture for secondary students will serve as a model for others to follow.'

The first meeting to discuss the creation of these aquaculture centres was held in Waterford on 21 March 1983, where Richard E. Schneller, the senate Majority Leader of the Connecticut General Assembly, delivered the keynote address. A request for proposals (RFP) was solicited to establish Regional Marine Science Technical and Vocational high schools within the Connecticut VAS. As a result of the RFP, plans were developed to establish, or expand upon, five vocational training centres for high school students interested in the study of aquaculture, marine trades or related businesses. There was concern that the schools be located in different geographic regions along the Connecticut shoreline to ensure that no coastal towns would be excluded from participation. Five sites were therefore proposed that would maximize coverage of the Connecticut coastline: Stamford at the western end of the state; Bridgeport; New Haven; Old Lyme; and either Groton or New London in the eastern portion of the state. It was hoped that by establishing the schools in separate locations, equal access would be provided to all interested students along the coastline.

The development of each centre is financed completely by the State of Connecticut. The state pays 100% of construction costs, the cost of the land, any environmental clean-up of that land, marine improvements (i.e. docks, etc.), the complete cost to renovate any existing structures and the expense of purchasing the necessary equipment for these schools. Additionally, these centres have total funding for a vessel(s) related to sail training, commercial fishing, aquaculture or oceanographic research.

The pilot programme initiated at the Stamford Center lasted until 1984. The first expansion grant for aquaculture was approved at the Ledyard Center in 1986. In 1987, the State of Connecticut looked to build its first aquaculture centre at the mouth of the Connecticut River in the town of Old Lyme, but that effort was abandoned by Regional School District 18. The first new aquaculture facility was achieved in Bridgeport in 1993. The Sound School in New Haven became the second aquaculture facility to be established in 1994. The completion of a US$28 million facility in New Haven was achieved in the spring of 2003. Recently the Stamford Center broke ground for a US$10 million expansion, US$5 million of which is dedicated to aquaculture. Unfortunately, the municipalities selected to house the last school in the eastern end of the state did not approve the creation of the regional facility in their school district. It was supposed, at the time that these schools were conceptualized, that there would be a tremendous amount of interest from shoreline towns; however, this has not proven to be the case. The limited inclination for communities to act as a host district is perplexing, as levels of interest among eighth grade students in the eastern portion of the state ranged from 55% to 81% (*Old Saybrook Survey*, 1997).

Development of the Sound School Vocational Aquaculture Center in an Urban Environment

New Haven, Connecticut is a coastal city with a population of roughly 120,000. Located between New York and Boston, it is home to a population that is culturally diverse. The city has several renowned institutes of higher learning, most notably Yale University. The 2000 US Census reports a median family income of almost US$30,000, with 325 individuals employed in either agriculture or fisheries. In the New Haven area, the leading aquaculture crops are oysters and clams raised in and around New Haven harbour. The Sound School is located on the shore of Long Island Sound in the City Point area of New Haven. The site the school is built on was the original home of the Thomas and McNeil Oyster Companies. Although no longer present on City Point, the historical significance and influences of New Haven's turn-of-the-century oyster industry on the city of New Haven are well documented. As such, it seems fitting that this high school aquaculture facility is located on this portion of the city's waterfront. The Sound School was created in 1982 as a comprehensive high school of choice in the New Haven Public School System. In 1994, when the school became a vocational agriculture centre, it was determined that 50% of the students who attended would come from the city of New Haven with the remaining 50% coming from the surrounding communities. Presently the student body size is slightly over 300. Twenty-one school districts send students to the school, with students coming from a wide geographic area. Every student who attends is enrolled in vocational agriculture education; roughly 90% are involved with aquaculture, with the remaining 10% of the students enrolled in traditional terrestrial agriculture classes. At present, the Sound School is the largest vocational agriculture high school aquaculture facility in the United States.

The Sound School's Philosophy

The Sound School Regional Aquaculture Center is a unique public high school offering students a blend of academic and practical education necessary to succeed in today's ever-changing world. The Sound School enrolls students from New Haven and 21 surrounding towns creating a community of diversity that develops students' social and intellectual skills. Our mission is to help students become full participants in the global, multi-cultural society of the 21st century by involving them in a broad-based high school experience that focuses on aquaculture, marine trades and sciences (Fig. 15.1). We encourage all students to develop sophisticated, problem-solving and critical-thinking skills in order to apply scientific and ecological principles to everyday life. In this way, we believe that we will prepare Sound School graduates for employment in society, as well as for the pursuit of additional opportunities at the post-secondary level.

Fig. 15.1. Students conducting water quality analysis as part of classroom activities in one of the laboratories at the Sound School.

Vocational Aquaculture Center Purpose

The aquaculture courses are greatly influenced by both the Future Farmers of America (FFA) mission statement and agricultural education mission (National FFA Organization, 2002). Learning through doing, the development of leadership skills and preparation for post secondary school or the labour force are incorporated into the curriculum. Scientific research using marine life as a learning tool and the development of student problem-solving skills through hands-on, outcome-based learning are key concepts in helping to facilitate the students' learning experience (Fig. 15.2). Being a vocational agriculture centre, the Sound School is mandated to have an advisory committee made up of individuals who are professionals in the curricular areas that are taught. The school draws on a wide range of expertise and is influenced in three ways. One group of advisory committee members who had input into the curriculum is made up of persons who are employed in the industry. The relationship between the school and these individuals makes it possible for the students at the school to experience unique vocational educational programmes within the aquaculture industry. As a graduation requirement, each student is expected to spend 200 h each year, a total of 800 h prior to graduation, outside of school in employment or internship related to aquaculture or the marine trades. A second group of advisory committee members who are involved in the ecological aspects associated with Long Island Sound (i.e. water quality monitoring, wastewater assessment, etc.) encouraged the creation of an environmental science component of the curriculum that included instruction that provides the student with the concepts of wise natural resource management practices and an appreciation of the value of coastal waters for commerce, recreation and food production. The third group of advisory committee members is associated with the colleges and universities in the State of Connecticut. They focused their curriculum develop-

Fig. 15.2. Student checking growth and performing maintenance tasks as part of course work at the Sound School 'hands-on' shellfish hatchery.

ment on the creation of classes that help prepare young people for future educational opportunities beyond a high school education in aquaculture, marine technology and related marine science fields.

The Development of a Shellfish Hatchery Curriculum

Connecticut is now the third largest producer of oysters in the country, and the quality of Long Island Sound oysters is preferred worldwide. Connecticut has become the largest producer of seed oysters in the northeast USA with approximately 200 seed oystermen involved in a business worth US$300,000–700,000 per year (Simlick *et al.*, 2001). A consistent supply of high quality seed is vital to the industry. Three sources of seed in Connecticut are: shellfish hatcheries (<10%), prepared (cultched) private beds (leases and grants ~60%) and public beds – those areas harvested by the seed oystermen (~30–40%) (Simlick *et al.*, 2001). The development of a shellfish hatchery component of the curriculum was deemed necessary for a vocational aquaculture centre.

Among the first of its kind in the nation, the Interdistrict Marine Educational Program's Instructional Shellfish Hatchery (IMEP/ISH) was designed to provide an opportunity for the enhancement of knowledge in aquaculture and marine sciences through the development of curricula dedicated to assisting students in applying lessons in 'hands-on' situations. The programme is funded by the Office of Urban and Priority School Districts and the

Connecticut State Department of Education as a Continuing Interdistrict Cooperative Grant, and is administered by the science department at the Sound School. The IMEP/ISH focuses on the techniques and methods employed in the commercial shellfish industry along the New England coast, with particular emphasis being placed on propagation techniques currently being used in Long Island Sound.

The IMEP/ISH programme incorporates the vocational agriculture attitude of 'learning through doing' as the method of education that best enables young learners to conceptualize the abstract points in their investigations as well as providing them with the techniques necessary to achieve the definable outcomes being sought. The IMEP/ISH provides science teachers and their classes with an opportunity to learn together by exploring the science of shellfish aquaculture. The IMEP/ISH further provides the students who attend with a sense of understanding and a realization of the extreme importance of protecting Long Island Sound as a natural resource and that any use of the Sound, either commercial or recreational, should be done responsibly.

The IMEP/ISH programme goals are to increase the students' interest in science through hands-on applications as well as to promote interaction and develop relationships between high school students from various coastal Connecticut high schools. Students achieve the IMEP/ISH programme goals through: (i) facilitated awareness of the importance of aquaculture on the eastern seaboard of the USA, with particular emphasis placed on farming in Long Island Sound; (ii) increased familiarity with the various scientific methods employed in the aquaculture industry; and (iii) highlighting the cultural diversity that exists among the students and staff who participate.

The IMEP/ISH programme was begun in the spring of 2001. The programme was developed to introduce students to the various tasks and physical labour involved in shellfish culture, while working with experienced personnel in the aquaculture industry. Site selection for the first year of the programme was based on the need for access to a running seawater delivery system, and adequate space for construction of a small-scale shellfish hatchery. Space for the initial shellfish hatchery was made available by Purity Processed Seafood, in Noank, Connecticut, and technical assistance was provided by Aeros Cultured Oyster Inc., Long Island, New York. The owners of Aeros Cultured Oyster, Inc. also assisted with the development and testing of laboratory exercises, broodstock conditioning, and facilitated development of the shellfish hatchery to be used to instruct students in culture techniques. Spawning events were erratic in the first year, suggesting an association with a commercial hatchery operation is critical to development of a successful spawning protocol.

During the second year of the IMEP/ISH, the programme's operations were moved to the Noank Aquaculture Cooperative, in Noank, Connecticut, a commercial hatchery facility, allowing the instructional hatchery to be further supported by the full-scale technical staff associated with the Cooperative. A total of 78 students participated in the spawning and production of several hundred thousand oyster (*Crassostrea virginica*) and bay scallop (*Argopecten irradians*) seed, used in both restoration efforts by area shellfish commissions as well as for growout by students at the Sound School.

Commercial and recreational shellfisheries in Connecticut are regulated on a town-by-town basis by individual shellfish commissions. Shellfish commissions are generally composed of volunteers, and one or two paid wardens, responsible for setting and enforcing shellfish harvest regulations within municipal waters. Shellfish commissions grant commercial leases, set recreational use regulations and may be responsible for coordinating supplemental stocking efforts. Wardens are responsible for posting the status on conditional recreational shellfish beds, and enforcing recreational creel limits and commercial leases. The Sound School IMEP/ISH supports supplemental stocking efforts of Connecticut shellfish commissions by donating seed stock and breeders.

Students from three other school districts (Ella T. Grasso-Southeastern Vocational Technical School, and the Ledyard and New London High Schools) joined Sound School students for training at the IMEP/ISH hatchery facilities. The cooperating school districts do not offer applied shellfish aquaculture curriculum, providing the Sound School with a unique opportunity to facilitate interactive educational experiences for interested students with different technical and cultural backgrounds. Students from both the cooperating school districts and the Sound School were instructed in the scientific techniques utilized for oyster production in commercial hatcheries. Techniques included topics on water quality assessment, bivalve reproduction, and nursery and growout methods. Instruction during the third year resulted in the production of 400,000, 2–4 mm oysters. Oyster seed were returned to the Sound School facilities in New Haven, Connecticut for further growth in up-welling systems.

The 'traditional' vocational technical student studies a trade, for example, carpentry, plumbing or culinary arts. Ella T. Grasso-Southeastern Regional Vocational Technical School is a vocational technical high school in the Connecticut State Vocational Technical Education system. Students from Grasso represent more than a dozen districts, and represent diverse backgrounds. Ledyard High School, located in a rural community, is a public high school of the Ledyard School District and provides a traditional vocational agriculture education programme. The majority of the student body has not been exposed to a vocational agriculture programme, and those that have are trained in terrestrial agriculture. New London High School is a typical urban high school, located along the eastern Connecticut coastline. The educational programme does not provide a vocational agriculture component.

The IMEP/ISH programme is structured to provide an opportunity for cooperative learning to occur when students from the different high schools collaborate while working in the laboratory and while engaged in group work associated with problem-solving the aquatic culture methods being employed. The IMEP/ISH programme runs for 20 days each spring and includes instruction in: Water Quality, Shellfish Biology, Bivalve Reproduction and Spawning Techniques, Micro Algae Culture, Hatchery Technology, Nursery Systems, Growout Technology, Shellfish Health, Harvest and Handling, and Depuration Techniques.

The Interdistrict Marine Educational Program's Instructional Lobster Hatchery

A secondary result from the success of the IMEP/ISH programme has been the expansion to include construction of an Instructional Lobster Hatchery. In the late 1990s, the Western Long Island Sound population of *Homarus americanus* experienced a dramatic decline. Students of the Sound School who were actively involved in fishing lobster traps as part of their class and after-school activities became concerned with the health of the lobster population after pulling barren traps for many weeks during the winter of 1999. Several students began asking what they could do to help restore the lobster population. As a result of those conversations, the idea for the development of the Interdistrict Marine Educational Program's Instructional Lobster Hatchery (IMEP/ILH) began.

Researchers at the University of Rhode Island (URI) provided valuable support to the programme during the development and planning stages of the programme, including multiple tours of the URI lobster hatchery, reference materials and technical expertise on the husbandry techniques required to support juvenile lobsters. The IMEP/ILH was projected to be a multiyear undertaking. The initial housing for the IMEP/ILH programme was provided by Project Oceanology located on the Avery Point Campus of the University of Connecticut in Groton, Connecticut. Project goals during the pilot year were to involve students in the design and construction of a functional lobster hatchery in anticipation of juvenile lobster production during subsequent years of the programme.

Students participating during the first year designed and constructed:

- A recirculating system using a 375 l tank for maintaining lobster brood stock. System requirements included the regulation of temperature and the ability to invert the photoperiod, while maintaining a viable population of berried, female lobsters. Temperature was regulated using titanium heaters, and photoperiod was controlled by placing the tank in a dark room.
- A Kreisal system (Chang and Conklin, 1983) with eight tanks for holding Stage I–IV juvenile lobsters. System requirements included temperature regulation (titanium heaters), the ability to separate juveniles of differing developmental stages (by tank) and a forced air system for the reduction of cannibalism within individual tanks.
- A recirculating system to house the Stage V and older juveniles. Students explored various techniques and equipment designs primarily focused on reducing juvenile interaction and inhibiting aggression among individuals (K. Castro, personal communication).
- A system to produce *Artemia* in sufficient quantities (densities) to feed the planktonic and benthic juvenile lobsters (K. Castro, personal communication). System requirements included control of temperature and salinity, and an external air source. Students established protocols for culture enrichment and harvest through monitoring of stock viability.

Production of juveniles

The production of juveniles was not expected during the first year of the IMEP/ILH. A priority objective for the second year of the project was to produce viable young. Initially the young produced at the hatchery were to be released, as directed by the State of Connecticut Department of Environmental Protection, as part of a supplemental stocking effort. However, after reviewing the State of Connecticut's procedure for the introduction of farm-raised aquatic species into wild habitats (P. Howell, Connecticut, 2002, personal communication; Connecticut General Statue, 22–11g), it has become apparent that this could not occur. The juveniles that are produced would have to be made available to both education and aquaculture researchers rather than released. The hope is that lobsters grown as part of the IMEP/ILH programme will be used to help establish successful lobster culture ventures that may one day serve to lessen demands on the wild population of lobsters in Long Island Sound and the surrounding regions.

Conclusion

The interdistrict grant supplied by the Office of Urban and Priority School Districts and the Connecticut State Department of Education, coupled with the FFA attitude of learning through doing, has proven an effective means of introducing young people from a wide range of backgrounds to the science of aquaculture and to each other. With this grant, we have been able to achieve versatile and unique curriculum programming that can be adapted and applied by shoreline schools. Recently there have been requests for the shellfish curriculum from interested schools in New Jersey and Maryland. The IMEP programme will continue to work closely with those involved in the aquaculture industry to develop, strengthen and revise the curriculum for several years to come.

References

Chang, E.S. and Conklin, D.E. (1983) Lobster (*Homarus*) hatchery techniques. In: McVey, J.P. (ed.) *Handbook of Mariculture,* Vol. 1 *Crustacean Aquaculture.* CRC Press, Boca Raton, Florida, pp. 271–275.

Galtsoff, P.S. (1964) *The American Oyster (*Cassostrea virginica *Gmeln).* US Department of the Interior, Fishery Bulletin of the Fish and Wildlife Service.

Loosanoff, V.L. (1940) *Spawning and Setting of Oysters in Long Island Sound in 1937, and Discussion of the Method for Predicting the Intensity and Time of Oyster Setting.* US Department of the Interior, Bulletin of the Bureau of Fisheries, No. 33.

National FFA Organization (2002) *2002–2003 Official FFA Manual.* National Future Farmers of America Organization, Washington, DC.

Old Saybrook Survey (1997) *Old Saybrook Survey.* Old Saybrook Board of Education, Old Saybrook, Connecticut.

Simlick, T.L., Williams, L. and May, A. (2001) *Seed Oystering.* Connecticut Sea Grant Extension Publication No. CTGS-01–02. Groton, Connecticut.

This page intentionally left blank

16 Science in Action: Tools for Teaching Urban Aquaculture Concepts

J. Adam Frederick
Center of Marine Biotechnology, Baltimore, USA

Abstract

In the last 20 years K-12 aquaculture education has moved from its place as an extension of vocational agriculture in the United States to having solidified itself as an integral part of environmental, biological and research-related courses in classrooms across the USA. This attention to aquaculture in the classroom has been driven in large part through support by government agencies like the National Council for Agriculture Education, the US Department of Agriculture and the National Science Foundation, and other entities such as aquaculture industry, university outreach and education programmes, regional and state boards of education, and fishery management organizations. References give indications that aquaculture is an effective teaching tool (Caldwell, 1998) and students are highly motivated to learn in an environment that includes aquaculture (Wigenbach *et al.*, 1999). This chapter presents an overview of the role of aquaculture in the education of our youth, and how technology has increased the quality of such an educational experience for students involved in research projects.

Introduction

One cannot jump straight away into a discussion of aquaculture as a teaching tool without first looking at the history of environmental education and issues of efficacy surrounding the tools of environmental education as a learning strategy for all students. It is important to recognize the roots of environmental education, and use those as motivating guides for our teachers and students so that a greater understanding can be attained of the importance of the stewardship of the environment. However, it is important to note that environmental education programmes must be designed effectively to ensure their success and continuation by the user after training. Key points made by Athman and Monroe (2001) consider the planners and users of environmental education as a partnership. Aquaculture holds great promise in promoting effective environmental education, the understanding of science concepts and the management of natural resources.

A Brief History of the Development of Environmental Education

A collaborative paper by Judy Braus and John Disinger (1998) gives a comprehen-

(eds B. Costa-Pierce, A. Desbonnet, P. Edwards and D. Baker)

sive view of the history of environmental education in the USA. The key points summarized below are referenced from that document.

- 1891, *Nature Study for the Common Schools* (Jackman, 1891) gave the first definition to the nature study movement and aimed to educate urban dwellers that had lost touch with the natural world.
- 1911, *The Handbook of Nature Study* (Comstock, 1986), compiled from junior naturalist newsletters at Cornell University, was used to teach students natural history.
- 1930s, the conservation education movement develops an initiative by resource management agencies as a way to educate the public about vital natural resources in response to soil erosion, floods and dust storms.
- 1938, *Experience and Education* (Dewey, 1938) outlines a philosophy of experience and its relation to education, involving informal education and a philosophy of 'learning by doing'.
- 1950s, the rise of outdoor education developed from the concern that urban students were not having any contact or experience with the natural world. Teachers were encouraged to teach many subjects outdoors to increase these opportunities.
- 1962, *Silent Spring* by Rachel Carson challenged the practices of agricultural scientists and the government, and called for a change in the way humankind viewed the natural world.
- 1970, The National Environmental Education Act (P.L. 91–516) was signed into law, defining environmental education as 'The educational process dealing with man's relationship with his natural and man-made surroundings, and includes the relation of population, conservation, transportation, technology, and urban and regional planning to the total human environment.'
- 1975, *The Belgrade Charter* (UNESCO-UNEP, 1976) was developed, and states: 'The goal of environmental education is to develop a world population that is aware of, and concerned about, the environment and its associated problems, and which has the knowledge, skills, attitudes, motivations, and commitment to work individually and collectively toward solutions of current problems and the prevention of new ones.'
- 1977, The Tbilisi Report (UNESCO, 1978) built upon the Belgrade Charter and produced the following goals for environmental education: (i) to foster clear awareness of, and concern about economic, social, political and ecological interdependence in urban and rural areas; (ii) to provide every person with opportunities to acquire the knowledge, values, attitudes, commitment and skills needed to protect and improve the environment; and (iii) to create new patterns of behaviour of individuals, groups and society as a whole towards the environment.
- 1983, North American Association for Environmental Education (NAAEE) built upon the Belgrade Charter and the Tbilisi Report to develop the following definition for environmental education: environmental education is a process that promotes the analysis and understanding of environmental issues as the basis for effective education, problem solving, policy-making and management.
- 1990, National Environmental Education Act (P.L. 101–619) re-enacted the National Environmental Education Act, giving the US Environmental Protection Agency (EPA) a Congressional mandate to strengthen and expand environmental education as an integral part of its mission to protect the environment.
- 1998, *Closing the Achievement Gap: Using the Environment as an Integrated Context for Learning* (Lieberman and Hoody, 1998) defined the first data regarding the use of the environment as a context for learning in all curricula for all students, and is the most referenced document by environmental educators regarding the performance of students in

science, literature and other content areas when the environment is used as the focal point for learning.

The link between environmental education and using aquaculture as a means to teach science and environmental awareness is clear. Aquaculture has the ability to connect the student to the environment, while at the same time giving them 'hands-on' experience with the process of science, scientific concepts and problem-solving techniques. The key is to integrate science education and environmental education in a seamless way for all students regardless of their core interest. Aquaculture is a tool that can accomplish this in an extended community of educators and specialists in higher education.

Development of Aquaculture as an Educational Tool

Congress recognized the opportunity for making significant progress in aquaculture development in 1980 by passage of the National Aquaculture Act (P.L. 96–362). The 1990 Farm Bill, also known as the Food, Agriculture Conservation, and Trade Act of 1990 (P.L. 101–624), reauthorized the Regional Aquaculture Center (RAC) programme at US$7.5 million per annum. Five centres were established: one in each of the northeastern, north central, southern, western and tropical/subtropical Pacific regions of the country. These centres, or RACs, became the focal points of aquaculture and lent some support for educational efforts with teachers and students. However, a much larger boost of support came from The National Council for Agricultural Education (NCAE) by establishing aquaculture into the agricultural education curriculum as a priority in the late 1980s. The NCAE initiated the development of a national 'core curriculum' for aquaculture education in 1990 with funding appropriated by the US Congress. The National Council for Agricultural Education (1998) established the following long-term goals for its education initiative in aquaculture:

- Teach the principles of success in aquaculture.
- Impart scientific principles, and recognize importance of properly managing aquatic resources.
- Use aquaculture curriculum as a means to interest students in science, maths and other relevant subjects.
- Encourage the integration of aquaculture in other subjects and to augment greater cooperation among individuals.
- Inspire students to enter post-secondary study in aquaculture or environmental science, thereby providing new talent to the industry.
- Explore the potential of aquaculture as an alternative enterprise in rural communities.

Maryland Model of Aquaculture in Action

Aquaculture in the classroom was initially limited to traditional agriculture and vocational education programmes, but in many pockets of the country teachers were catching on to the attractiveness of aquaculture as a teaching tool for science. Two such schools – South Carroll and Westminster High Schools – located in Carroll County, Maryland, are model programmes for the use of aquaculture as a teaching tool in science and research classes and have been using aquaculture in the classroom for nearly 20 years. Each school has renovated a large shop/technical class area into an aquaculture research lab equipped with a number of recirculating systems, a separate classroom area for instruction, computers, and resources for teaching biology, chemistry and physics. Each has done this by developing an approved science research curriculum (Carroll County Public Schools, 1997), and partnering in many environmental education initiatives that engage their students in restoration projects on school grounds and beyond.

Maryland Sea Grant developed the education and outreach programme, Aquaculture in Action, in 1997 in partnership with Carroll County Public Schools.

The title Aquaculture in Action was chosen to emphasize that aquaculture itself is a dynamic tool for learning, and that participants in a focused training programme would be well equipped with both knowledge and materials to put into practice recirculating aquaculture in their classroom. The goal was to establish a network of Maryland teachers appropriately trained and equipped with the essential tools of an aquaculturist, so that they could raise and release native species of fish while sharing the experience with their students. In turn, the opportunity for students to learn the operation and management of fish in a recirculating aquaculture system lends itself to the integration of scientific concepts, their application and improved problem-solving skills. The Aquaculture in Action (AinA) model also employs the use of web technology so that teachers and students can ask questions related to their system operation, and share descriptions and data from their individual school projects via web-based interfaces.

Aquaculture projects extend beyond the classroom, with the improvement of environmental literacy as a major goal of science education today. These unique projects form life-long relationships between students and local species, and these relationships motivate students to investigate and perform stream restoration projects that can further improve conditions for local species. This is an essential part of the extension of aquaculture from the classroom to the field. Scientifically, aquaculture projects engage the process of experimental design and data collection, emulating the scientific process. The collection of data over the long term allows students to identify trends and make connections between the physical and biological factors within a particular controlled setting. The end result is greater student awareness of applied skills, scientific knowledge, environmental literacy and an appreciation for the effort required to apply concepts in the field. The following goals are adapted from the Carroll County Science Research curriculum design specifically for aquaculture in the classroom.

Goal 1: system design and construction

The AinA programme designed a prototype system which is the standard system used in the classroom. The system contains an 800 l central tank with a 5 cm bottom drain that flows into a 114 l sump/filter to collect solid waste. The water passes through filter medium in the sump into a 114 l upflow biological filter containing biofilm carrier media. Water flows up through the suspended biofilter medium and is returned to the tank by gravity. An air compressor is used to supply air to the tank and biological filter.

Objectives

- Learn about format, design and operation of recirculating systems.
- Inoculate the biological system and monitor physical and biological parameters.
- Collect and analyse physical data over an extended period.
- Monitor water quality after the introduction of fish.

Goal 2: species selection

The selected species of fish for use in the classroom results from collaboration with local, regional and/or state agencies, universities and/or other appropriate partners (e.g. Sea Grant). Assistance and direction from state fisheries agencies help determine the location of the release site for all species at the end of the school year.

Objectives

- Select a target species based on availability and where the fish is to be restocked.
- Conduct research and write reports on all aspects of the species, physical characteristics, habitat requirements, reproduction, behaviour and food requirements.

Goal 3: environmental analysis

Conduct water and environmental analysis of local streams, and make comparisons between a natural system and a recirculating aquaculture system in the classroom.

Objectives

- Select an appropriate fish species for restocking in cooperation with local and/or state agencies.
- Research physical, chemical and biological parameters of the water.
- Apply biotic indices to streams to determine their habitat suitability for the reintroduced species.
- Collect, maintain, evaluate and expand a computer database of the stream analyses performed.
- Work with appropriate agencies and other environmental groups in stream improvement efforts aimed at increasing the habitability of local streams for fish species.

Goal 4: monitoring growth

Monitor and record growth of fish before release at the end of the school year. Fish growth is carefully monitored in relation to density and food conversion. Results will be recorded in a database so that projects can be shared between schools.

Objectives

- Collect, analyse and graph data on pH, dissolved oxygen, temperature, ammonia, nitrites and nitrates.
- Examine relationships between physical parameters and composition/amount of food to fish density.
- Transport and release fish.
- Design and carry out experimental and monitoring procedures with reproducible results.

Goal 5: comparing artificial systems to nature

The comparison of the recirculating system environment to the natural environment where the fish are being restocked is essential to the project. Identifying specific improvements for the aquatic system to increase the success of the reintroduction is very important.

Objectives

- Compare the aquaculture environment to the natural environment of the target species by examining chemical, physical and biotic parameters.
- Identify specific projects to enhance or protect the environment of the aquatic system.
- Reduce turbidity by proposing the construction of stream bank stabilization devices, reducing people impacts, establishing buffer zones, etc.
- Develop protocols for above activities to make sure they comply with local and state regulations.
- Research similar problems and solutions at various levels and geographic regions.

Aquaculture in the classroom affords all the luxuries of bringing a part of the environment indoors, but in a way that is different from a typical classroom aquarium. An aquarium environment is used to simulate a habitat for the inhabitants and maintain their health for display and appreciation. The goal in aquaculture is to maintain, growout, release, collect data and study the requirements of a species (e.g. a local fish species), investigate the limits of the recirculating system, and work as teams with other students to problem solve. A primary objective is to develop an understanding of the process of scientific investigation and how to apply that to the environment. Furthermore, students may gain a greater feeling of ownership related to their behaviours concerning local water quality issues when the impacts are illustrated in such a direct manner.

While aquaculture is an excellent tool and 'hook' for students, the key is a dedicated teacher with an interest in professional development and the flexibility and support to take on unique projects within their school. Currently, there are many efforts underway to infuse aquaculture into K-12 education as a tool for science education and a legitimate career choice. It is clear from our diminishing returns from wild fisheries, and the US national deficit in aquaculture products, that future key decision-makers will need a more comprehensive view of aquaculture.

Developing an Aquaculture Programme for the Classroom

What do teachers need to prepare themselves for aquaculture in the classroom, and how can it be provided? These are a couple of key questions that need to be addressed on an individual level for each teacher that wants to use aquaculture as a teaching tool. From the teacher perspective, Fig. 16.1 illustrates the needs of the teacher, and the support required within the school system to ensure that there is success for such an endeavour. While using aquaculture has many benefits, it is an undertaking that needs to be clearly defined for supervisors and building maintenance personnel, and to gain their support.

Once the appropriate administrative support is in place, a teacher needs to find a programme that will give them the professional development and skill training related to aquaculture and how it can most effectively be used as an educational tool. Specific training is required to understand aquaculture and the maintenance of fish. This is where professional development becomes an essential component of having a successful start-up in aquaculture, whether the endeavour includes a small, large or even multiple systems.

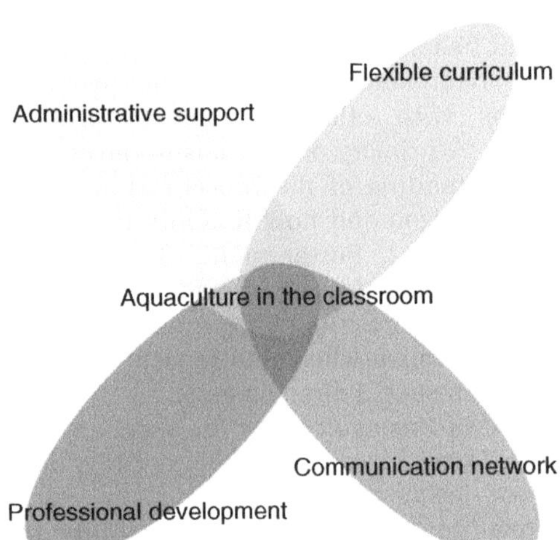

Fig. 16.1. Illustration of the needs of a teacher, school and school system to support a successful aquaculture programme in the classroom.

The professional development model used in the Maryland AinA programme contains specific features designed to assist a teacher in the successful start-up and maintenance of recirculating aquaculture as a tool for teaching science. The following components are based upon recommended guidelines from the National Research Council (1995) *National Science Education Standards.*

Component I. Learning from master teachers

In order for teachers to buy into a new method for enhancing science instruction, it is essential to have it endorsed and developed by well-respected peers. The AinA programme started with the input of two teachers in Carroll County, Maryland. These teachers had designed a course known as Science Research, in which students were required to choose a topic of research and design a semester-long research project, write a paper and develop a presentation. School principals and science curriculum supervisors were extremely supportive of the development of an aquaculture education programme. Over time a manual has been developed and used as a guideline by AinA for teachers to utilize as a resource. Key components of the manual that integrate science instruction and aquaculture are:

- Philosophy of the Aquaculture in Action model.
- Recirculating system design and maintenance.
- Aspects of biofiltration.
- Water quality and microbiology.
- Handling fish and collecting data.
- Aquaponics.

- Field assessment.
- Research project design.
- References, literature and the web.

The development of this guide has evolved along with the programme, with technological resources provided by the Maryland Sea Grant network of specialists. For example, the Sea Grant programme has developed a website where teachers and students around the state can enter information and data about their research projects, including fish and water quality parameters.

Component II. Partnerships with aquaculture specialists

The need for an increase in scientist–teacher interaction is well known as an effective content enhancement and enrichment experience for both teachers and scientists. The goal is to introduce aquaculture to the teachers in an applied manner by having them experience recirculating aquaculture by touring and working in an aquaculture hatchery (e.g. University of Maryland Center for Environmental Science Horn Point Lab) and/or state-of-the-art aquaculture research facility (e.g. University of Maryland Center of Marine Biotechnology) alongside scientists and specialists who could provide the fundamentals of research in aquaculture. This component of professional development has proved to be invaluable for teachers since very few have had exposure to aquaculture at this level, or even realized the importance of aquaculture to the economy and our daily lives.

Component III. Aquaculture in Action workshop

A 1-week workshop is the cornerstone of the professional development model. The selection of teachers for the workshop is based upon an application process that includes essays on how the teacher views the potential of aquaculture as a teaching tool, and how it would be used to enhance the curriculum in their school. The teacher must also have the appropriate administrative support and letters of recommendation for participation in the workshop. The objective is to work with teachers that can then become mentors for other teachers. The workshop brings together teachers from various regions and forms a community experience for teachers that are unfamiliar with one another, as well as from varying backgrounds, both educationally and culturally.

The first day and a half are full of activities and talks by scientists and specialists in aquaculture, with recirculating aquaculture fundamentals and water quality as the main topics of discussion. The final three and half days are spent at a local school that has integrated aquaculture into the classroom, and where participants undergo a 'learning by doing' philosophy (Fig. 16.2) as teachers perform techniques that combine science and aquaculture. In the process, participants build their own individual recirculating systems using a team approach that teaches them valuable skills related to construction and design. The team approach also builds bonds within the group of teachers that will last throughout their professional career, and gives them another resource to tap for ideas and advice. At the end of the week teachers transport the system back to their school along with the other materials of instruction that have been provided for them during the workshop (Fig. 16.3).

Component IV. Follow-up and support

To provide easy access to support after the workshop, a website was developed (http://www.mdsg.umd.edu/Education/AinA/index.html) that includes information about the workshop, participating schools and their location, a link to school projects, a link to aquaculture resources and a *Fish Talk* link, where questions can be asked of the whole network of participants and specialists. One of the most anticipated parts of follow-up and support is the acquisition of fish for the classroom after the system has been appropriately prepared.

Fig. 16.2. A teacher works on fitting a bulkhead for the outflow of the sump during the Aquaculture in Action workshop.

Fig. 16.3. Teacher-built recirculating system used in the Aquaculture in Action Program in Maryland.

Follow-up and support is further provided by the development of a communication network for teachers to use to help answer questions, as well as for student involvement. As was previously mentioned, a portion of the AinA website is dedicated to the display of research projects. Each teacher is given a name and password that gives them access to an internal portion of the web that provides an opportunity to enter information and data regarding their project. In this way students can quickly observe and display their own work to others. Teachers can evaluate this work and give feedback to students on how to improve their writing and data collection methods and, in turn, students and teachers can edit and change information that has been entered. The grand finale of the follow-up and support is the participation of students and teachers in the release of fish at the end of the year (Fig. 16.4). Some of the greatest learning experiences come when those students that have studied their fish and their requirements during the academic year start to ask questions about water quality and other issues on release day. Additionally, the release of the fish relieves the pressures of maintaining the system over the summer.

Other resources to access for further exploration of models for integrating aquaculture into the classroom are:

- New England Board of Higher Education (http://www.nebhe.org/aqua.html) – In August 1997, a 2-year grant from the National Science Foundation Advanced Technological Education (ATE) programme was received and used to form the *New England Aquaculture Educators Network, AQUA* (Stewart, 1997). The project was used to prepare middle and high school teachers, and 2- and 4-year college and university faculty, to introduce aquaculture curricula into existing programmes. This publication describes the steps necessary for introducing an aquaculture programme using a freshwater recirculating system into a classroom environment. A second product is the *Aquaculture Curricula Resource Guide: a Tool for the Aquaculture Educator* (Soares *et al.*, 1999). This guide reviews 13 educational resources classified as aquaculture curriculum, and advises on strengths and weaknesses of each based upon five categories: Coverage and Quality, Organization and Structure, Format and Readability, Assessment,

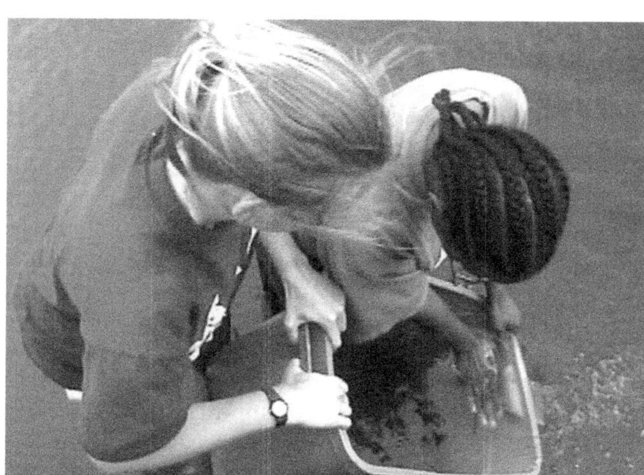

Fig. 16.4. Students from a Maryland high school release striped bass in Chesapeake Bay that they raised in the classroom.

and Teacher Resources. This NRAC publication is available at http://www.state.ma.us/dfa/aquaculture/education_curricula.pdf.

- The Freshwater Institute (http://www.freshwaterinstitute.org/), Shepardstown, West Virginia – The Freshwater Institute has developed a curriculum for an Aquaculture in the Classroom (Simmons, 2001) programme, and has worked with schools to offer students and teachers in traditional agriculture classes an alternative and integrated approach to teaching skills and scientific concepts.
- Massachusetts Institute of Technology Sea Grant Aquaculture Education Program (http://web.mit.edu/seagrant/k12/curricula/index.html) – At MIT Sea Grant a focus on Raising Saltwater Fish in the Classroom (Moran *et al.*, 2000) has been produced through collaborative efforts with MIT scientists. Resources are available on-line at the web address above, and workshops are held to educate teachers on the set-up and maintenance of recirculating aquaculture with an emphasis on fingerlings. The resources are comprehensive and detailed in terms of system set-up and are well organized into three sections: Introductions and Contact Information, Teacher Edition, Student Workbook.

Aquaculture Meets Standards Locally and Nationally

Many states have established goals for high-school educators to use as guidance in developing curriculum. The AinA programme can be directly aligned with at least some of these goals. The following are some links between specific areas of aquaculture and an integrated curriculum.

Skills for success

Learning skills

Students can plan, monitor and evaluate learning experiences throughout the year by setting goals for specific content learning related to the aquaculture project. Specific goals can be established that relate directly to the career interest of the student. For instance, a student may want to pursue a career in writing, and with the help of the teacher will set goals that focus on organizational writing skills like those involved in the preparation of reports. Throughout the year the student can build these skills by preparing some reports that are used as an assignment in English class. The feedback from the English teacher and content knowledge gained from the aquaculture project will help improve these skills.

Thinking skills

Students will generate creative ideas as they design and implement an aquaculture-based project that will last throughout the school year, and will require adaptation and modification as it progresses. A project on fish food and growth rates, for instance, will require that students analyse data and make modifications to their plan based upon feedback from water quality, fish growth and feed quality. Critical analysis will allow students the opportunity to develop new strategies to improve fish growth and improve their effectiveness in the decision-making process.

Communication skills

Presentations of the findings of aquaculture-based science projects can be shared with the class, the student body and the community. Students can develop presentations that describe the goals of the project, displays of the data and conclusions, and present these in a journal club format. This is a common practice in the scientific community and enhances the learning experience when your peers are allowed to analyse and critique your work. Local media can be encouraged to interview students about their projects, extending their communication skills further.

Technology skills

Aquaculture projects can facilitate the use of the latest computer technology and monitoring devices for data collection and analysis by students. Programs may range from spreadsheets to statistical packages, giving students access to a variety of computer applications.

Interpersonal skills

The team-based approach of aquaculture projects is an excellent way to promote the enhancement of interpersonal skills. The interactions of students in a research team, and amongst teams, will assist in the development of positive attitudes for group decision-making, the acceptance of constructive feedback and appropriate changes in behaviour.

Science-based skills

Process

Aquaculture-based projects engage students in the scientific process of experimental design, implementation and analysis. Students learn these important skills by participating in all phases of setting up and maintaining an aquaculture system and managing a controlled experiment. These projects are interdisciplinary as well, and enhance the application of content from other subjects.

Concepts for earth/space science

Students monitor natural cycles of water, carbon and nitrogen, by monitoring water quality of the aquaculture system. These cycles can be related to the natural world, and how they are altered by human activity such as development, agriculture and industry. These concepts can readily be applied to other science disciplines.

Concepts for biology

Students perform 'hands-on' work that reinforces the connection of matter and energy to cell processes and functions. The role of biological molecules, cell components and environmental conditions which impact those processes can be studied with the use of macro- and microbiological techniques applied to the aquaculture project. Concepts of genetics, reproduction and genetic engineering can be researched by students, enhancing their ability to understand how industrial aquaculture producers use these concepts to improve their market product, whether ornamental fish or game fish. The aquaculture system can also be used as a tool for the study of the interaction between the abiotic and biotic factors in a controlled environment, and how limiting factors play a role in the growth of their fish.

Concepts for chemistry

Integrating chemistry into an aquaculture project can be accomplished by giving students experience with the tools used to measure various properties of matter in a lab setting. Chemicals used to treat water or disease can be used as a vehicle to study properties of matter, and their reactions in water and with the biological components of the system. Water quality measurements provide another avenue for the integration of chemistry by investigating how the stability of water quality parameters are essential to aquaculture.

Concepts for physics

Investigating the laws of physics is an integral part of aquaculture-based education that emphasizes the importance of fundamental engineering and design in the set-up and maintenance of an aquaculture system. For instance, relating water pressure and friction in pipes to output from a pump is important in choosing the right water pump, and is a fundamental consideration in designing a system. In turn, understanding how system volume and flow rate from the selected pump determine the overall turnover time for the system is critical.

Conclusions

Aquaculture can play a role in secondary education as an integrator of many disciplines and areas of student interest. Furthermore, integration of aquaculture into secondary education systems provides students insight into another possible career track. If nothing else, it provides better comprehension of the processes that will increasingly come into play to provide food for an ever-growing global population. Aquaculture is in its infancy throughout much of the USA as a component of secondary education curricula, but it is beginning to spread as technology improves the capacity for profitability of aquaculture ventures, and as the probability of earning a good living through aquaculture grows. Furthermore, the continuing pressures put on wild stocks of aquatic organisms, and the dwindling ability of those stocks to meet demand, makes the acceptance of aquaculture as a way of life inevitable. As this inevitability approaches, the need and desire for integration of aquaculture into secondary education will become paramount. In the meantime, integrating aquaculture into mainstream secondary education curricula can help bolster lagging student interest in the sciences by pulling together many disciplines, and then focusing them on real life, practical issues and problems.

References

Athman, J.A. and Monroe, M.C. (2001) *Elements of Effective Environmental Education Programs.* School of Forest Resource and Conservation, University of Florida, Gainesville, Florida.

Braus, J. and Disinger, J. (1998) Educational roots of environmental education in the United States and their relationship to its current status. In: Archie, M. (ed.) *Environmental Education in the United States: Past, Present and Future.* North American Association of Environmental Education, Troy, Ohio, pp. 9–19.

Caldwell, J. (1998) *Why Use Aquaculture as an Educational Tool?* The Conservation Fund's Freshwater Institute, Shepherdstown, West Virginia.

Carroll County Public Schools (1997) *Science Research Curriculum Guide.* Carroll County Public Schools, Westminster, Maryland.

Carson, R.L. (1962) *Silent Spring.* Fawcett World Library, New York.

Comstock, A.B. (1986) *The Handbook of Nature Study.* Reissue by Cornell University Press, New York.

Dewey, J. (1938) *Experience and Education.* The Macmillan Company, New Jersey.

Jackman, W. (1891) *Nature Study for the Common Schools.* Henry Holt & Company, New York.

Lieberman, G.A. and Hoody, L.L. (1998) *Closing the Achievement Gap: Using the Environment as an Integrating Context for Learning.* State Environmental Education Roundtable. Science Wizards Poway, California.

Moran, B., Goudy, C. and Hankin, A.L. (2000) *Marine Aquaculture: Raising Saltwater Fish in the Classroom.* MIT Sea Grant College Program, Marine Advisory Services, Cambridge, Massachusetts.

National Council for Agricultural Education (1998) *Aquaculture Education – Making a Splash!* http://www.teamaged.org/aquaculture/aquaculture1.htm

National Research Council (1995) *National Science Education Standards.* National Academy Press, Washington, DC. http://www.nap.edu/html/nses/html/

Simmons, J. (2001) Learning happens. *Now & Then – the Appalachian Magazine* 18(1), 18–21. Eastern Tennessee State University Center for Appalachian Studies and Services, Johnson City, Tennessee.

Soares, S., Buttner, J.K. and Leavitt, D.F. (1999) *Aquaculture Curricula Resource Guide: A Resource Tool for the Aquaculture Educator.* NRAC Project No. 98–7 Subcontract No. 556507, NRAC Publication No. 01–001. Northeastern Regional Aquaculture Center, University of Massachusetts Dartmouth, North Dartmouth, Massachusetts.

Stewart, L. (1997) *New England Board of Higher Education AQUA Program.* National Science Foundation (Grant # 9752050) for the development of educational materials and training for secondary education teachers in aquaculture practices in the classroom. http://www.nebhe.org/aqua.html

UNESCO-UNEP (1976) The Belgrade Charter. *UNESCO-UNEP Environmental Education Newsletter* 1(I), 1–2.

UNESCO (1978) *Final Report: Intergovernmental Conference on Environmental Education,* Tbilisi, USSR, 14–26 October 1977. UNESCO, Paris.

Wingenbach, G.J., Gartin, S.A. and Lawrence, L.D. (1999) Students' perceptions of aquaculture education in the northeast region. *Journal of Agriculture Education* 40(1), 14–22.

This page intentionally left blank

17 Urban Aquaculture: a Necessary Reality

Joseph K. Buttner
Salem State College, Salem, USA

Abstract

Planet Earth is currently home to an estimated 6.3 billion people and approximately 80 million individuals are added annually. Increasingly, humankind congregates into urban centres, isolated and buffered from nature. Often urbanized humanity professes a sincere interest in the environment, 'thinking green', but this desire to protect and preserve our natural heritage is compromised by misinformation and limited hands-on exposure to the synergistic interdependence of biotic and abiotic elements that define our biosphere. Urban aquaculture provides a widely accessible and effective conduit to re-link people with nature by promoting public awareness, nurturing the 'ecoethos' characteristic of aboriginal cultures globally, and restoring an appreciation of agriculture's role in ensuring a secure food supply while generating jobs, income and product.

While the economic viability of commercial aquaculture in urban settings is viewed with trepidation and scepticism, the reality and significance of urban aquaculture cannot be disputed. Much of the apparent disparity between the perceived and realized potentials of urban aquaculture originates from a restrictive appreciation of 'what is aquaculture'. Production of aquatic organisms for human food represents only one of many ways that aquaculture can be and is pursued. Cultured organisms need not be limited to finfish and shellfish; organisms from the Kingdoms Monera, Protista, Plantae and Animalia can be successfully grown and marketed. Aquaculturists produce organisms for landscaping, aquaria, garden ponds, jewellery, leather, nutritional supplements, pharmaceuticals, enhancement and/or restoration efforts, preservation of endangered species, fee fishing, live-bait, bioassay, education and research activities. In urban settings, implementation of sustainable aquaculture practices that target niche markets, and the energetic integration of community education/involvement, will facilitate commercial success, while fostering a realistic appreciation of living within the confines of our biosphere.

Introduction

How can aquaculture in an urban setting be considered a necessary reality when its legitimacy and economic viability are frequently viewed with trepidation and scepticism? Much of the apparent disparity between the perceived and realized potentials of urban aquaculture originates from a restrictive appreciation of 'what is aquaculture', ecology's impact on human history and 21st century realities. When broadly defined and wisely pursued, urban aquaculture can expose large numbers of people to

(eds B. Costa-Pierce, A. Desbonnet, P. Edwards and D. Baker)

ecological concepts and nurture a realistic appreciation of living within the confines of our biosphere, while promoting social acceptance of aquaculture as a viable means to create jobs and generate revenue.

What is Aquaculture?

To embrace the diversity of species cultured and techniques employed, aquaculture has been variously defined. The Joint Subcommittee on Aquaculture (1993) defined aquaculture as 'the farming of aquatic animals and plants'. Acknowledging the increasing sophistication, complexity and importance of aquaculture, the US National Marine Fisheries Service (2002) recently refined the definition as 'the cultivation of aquatic animals and plants in controlled or selected environments for commercial, recreational, or public purposes'.

Organisms from at least four Kingdoms (Monera, Protista, Plantae and Animalia) are cultured commercially, and production of aquatic organisms for human food represents only one of many ways that aquaculture can be and is pursued. Aquaculturists also produce organisms for landscaping, aquaria, garden ponds, jewellery, leather, nutritional supplements, pharmaceuticals, enhancement and/or restoration efforts, preservation of endangered species, fee fishing, live-bait, bioassay, education and research activities (Avault, 1996; Guo and Kraeuter, 1999). Facilitating and sharing the explosive growth of aquaculture globally are a myriad of support providers that include feed manufacturers, suppliers of instrumentation and equipment, extension and research professionals, processors and distributors, regulators and elected officials as well as the consumer. Many components and supporters of the aquaculture industry are based in urban settings.

Evolution of Aquaculture

Aquaculture is pursued in a variety of open, semi-closed and closed systems that target a wealth of markets with a primary purpose that parallels terrestrial agriculture – to increase the amount of food available for human consumption (Avault, 1996; Open Ocean Aquaculture, 1996; Black, 2001; Bridger and Costa-Pierce, 2003). While aquaculture has been practised for centuries, its impact and importance have expanded dramatically over the last few decades as the quantity of seafood harvested by traditional capture fisheries stagnated and demand increased (United States Department of Agriculture, 2000; FAO, 2001; Tomasso, 2002). Aquaculture production has grown to mitigate the disparity between supply and demand. Pursuit of aquaculture, like its terrestrial equivalent, agriculture, has occurred out of necessity.

At the dawn of civilization, 4000–10,000 years ago, humankind on every habitable continent shifted from the hunting and gathering of nature's surplus to cultivation and domestication of desirable species (Cohen, 1977; Zvelebil, 1985; Smith, 1995). In prehistory, increased human population, extinction of terrestrial megafauna and advantages of agriculture catalysed this transition. Vastly increased food supplies and security were achieved through manipulation of natural processes that transformed light energy from the sun into chemical energy stored in cultured organisms. Plants with digestible leaves (e.g. spinach and lettuce), roots and tubers (e.g. potatoes and carrots), stems (e.g. asparagus and celery) and reproductive structures (e.g. fruits and grains) as well as relatively docile herbivores and omnivores (e.g. cattle, poultry, sheep and swine) were selected by our prehistoric ancestors. Further increases in productivity were achieved through selective breeding that enhanced desirable attributes in favoured organisms and by supplementing light energy from the sun with other forms of energy such as electricity, fertilizer, fossil fuels and labour (Smith, 1995; Brown *et al.*, 1996, 1998). Agriculture provided a local, abundant and safe food supply that supported much greater numbers of people than possible through hunting and gathering. Agriculture lead to the establishment of cities as people could remain in one spot, making specialization of effort, acquisition of materials and civilization possible (Cohen, 1977; Smith, 1995).

The inability to harvest sufficient seafood from the oceans through traditional capture fisheries necessitates cultivation of aquatic organisms much like shortfalls in food supply catalysed the prehistoric transition from hunting and gathering to agriculture in terrestrial environments. The 'agricultural revolution' occurred over millennia, dampening social disruptions and gradually allowing for general acceptance. The aquatic phase of this transition, the 'aquaculture revolution', will occur more quickly with exasperating societal impacts. Minimization of disruptions, maximization of benefits and expeditious implementation of sustainable aquaculture can be realized through education.

Why Urban Aquaculture?

Urban aquaculture, perhaps defined as *the* culture of aquatic organisms within the confines or shadows of a city, offers many benefits and opportunities that extend beyond food production, job creation and revenue generation. Earth is currently home to slightly less than 6.3 billion people and between 70 and 90 million individuals are added annually (Brown *et al.*, 1996; Population Information Program Center, 2001; World Population Calculator, 2002). Increasingly, humankind on all continents congregate into urban centres, isolated and buffered from nature (Population Information Program Center, 2001). Linkages between people inhabiting cities and local, as well as global, ecosystems have become frayed. An urgency, and a new receptivity, exists to rediscover the 'ecoethos' that characterized our species during most of its existence, and remains an important unifying trait of Native American and aboriginal cultures globally (North American Indian Travelling College, 1984; Buttner, 1997; Gray-Krueger, 1998; Beal, 2002). Native Americans view themselves as part of nature, illustrated by verses from an Iroquois Thanksgiving Address (Haudenosaunee Environmental Task Force, 1999):

> We who have gathered together are responsible that our cycle continues. We have been given the duty to live in harmony with one another and other living things. We give greetings that our people still share the knowledge of our culture and ceremonies and are able to pass it on. We have our elders here and also the new faces yet to be born, which is the cycle of our families. For this we give thanks and greetings. Now our minds are one.
>
> We give greetings and thanks to our Mother the Earth. She gives us that which makes us strong and healthy. We are grateful that she continues to perform her duties as she was instructed. The women and Mother Earth are one, givers of life. We are her color, her flesh and her roots. Now our minds are one.
>
> We give greetings and thanks to the plant life. Within the plants is the force of substance that sustains many life forms. From the time of the creation we have seen the various forms of plant life work many wonders. We hope that we will continue to see plant life for the generations to come. Now our minds are one.

Urban aquaculture provides a forum to not only generate significant amounts of food and other products for humankind as urban agriculture currently does (Population Information Program Center, 2001), but also to reconnect large numbers of people with ecological processes. Systems used to culture aquatic organisms are readily observable as miniature ecosystems. Impact of humankind and humankind's impact on ecological processes can be appreciated firsthand. Specifically, urban aquaculture can help people formulate a realistic 'ecoethos' based upon factual information:

- Urbanites often profess a sincere desire to learn about the environment and to act in environmentally sound manners. 'Thinking green' has gained popular acceptance (World Watch Institute, http://www.worldwatch.org/; Brown *et al.*, 1998). Unfortunately, people seeking to think and act in an environmentally friendly manner are often bombarded and swayed by incomplete information, misinformation and misconceptions. Wisely managed aquaculture operations provide a convenient conduit for environmentally conscientious individuals to observe an ecosystem and gain a realistic appreciation of its importance, resiliency and vulnerability.

- Urbanites have become isolated from their food supply and increasingly take that supply for granted. Circular slabs of meat displayed in grocery stores bear little semblance to cattle, swine or sheep. Canned goods, bakery products and microwaveable meals are rarely recognized as previously living organisms. A fundamental truism has been lost – that for us to eat, something must cease living. The role and value of agriculture in ensuring our food security is underappreciated; a very real potential exists to compromise humankind's recently achieved food security through sincere but misdirected initiatives. Urban aquaculture affords people an opportunity to observe the process of food production and better appreciate the challenges associated with feeding 6.3 billion individuals daily.
- Production of food in large amounts requires significant space and resources. Since most potentially arable land is being cultivated or has been lost through human impact, the oceans represent the largest underutilized resource for food production (Brown *et al.*, 1996). For aquaculture in the ocean to achieve its potential, private access and use of this historically common resource must occur, as has occurred on land. Urbanites exposed to sustainable aquaculture operations may become, through their experiences, more receptive to management and cultivation of the open ocean.
- Since the 1960s the USA has become increasingly a service-based economy; currently 80% of the workforce is employed as service providers (Atkinson and Court, 1998). Productivity has slipped from 3–4% annually to around 1% (Atkinson and Court, 1998). With diminished productivity, society has become increasingly polarized as low wage, low skill jobs (service providers) and high wage, high skill jobs (service recipients), while mid-wage, mid-skill jobs have become scarcer (Atkinson and Court, 1998). To secure and perpetuate a healthy middle class and democratic government, productivity must be supported and increased. Sustainable aquaculture pursued in an urban setting demonstrates that production of aquatic organisms can be practised successfully with benign impact on the environment and, perhaps, help to reverse the export of production jobs.

Urban aquaculture presents challenges and opportunities that extend dramatically beyond cultivation and marketing of product. To realize fully their potential, aquaculture operators in urban settings must assign education and public relations the same importance as sustainable and profitable production. Consideration of these activities should be incorporated into facility design and operation.

How Can Urban Aquaculture Educate People and Sway Public Opinion?

Aquaculture operations in urban settings must develop and follow stringent quality assurance measures (Nash and Green, 1999; Gall and Rivara, 2000; Alabama Aquaculture Best Management Practices, 2002; Tomasso 2002; United States Environmental Protection Agency, 2002) to ensure efficient, safe and sustainable operation. Once an operation is running smoothly, providing public access can prove a most effective method to educate people about the value of aquaculture and nurture their support for aquaculture. A variety of experiences may be considered appropriate:

- Tours, targeting youth who are typically more receptive to new ideas and approaches than their parents, represent a relatively easy and effective method to introduce significant numbers of people to aquaculture and ecological principles. At the Northeastern Massachusetts Aquaculture Center, we typically entertain multiple tours monthly at our Cat Cove Marine Laboratory (Table 17.1). Annually, several hundred K-16 students observe fish and shellfish culture, becoming excited and sharing their excitement with

Table 17.1. Between June 2001 and October 2003 inclusive, nearly 50 tours were scheduled in advance and conducted at the Cat Cove Marine Laboratory exposing slightly less than 1000 people to aquaculture in an urban setting. Unscheduled visits by students, administrators, elected officials and the general public, which typically number around two dozen annually, are not listed.

Date	Audience	Number of participants
2 Jun 01	New England Estuarine Research Society (scientists, students)	36
19 Jul 01	Bentley School (Salem, MA; K-2 youth)	100
3 Aug 01	Prospective students (Scanton, PA; college)	10
22 Aug 01	Prospective students (New Fairfield, CT; college)	3
12 Oct 01	Saltonstall School (Salem, MA; primary school youth)	40
19 Oct 01	Biology Faculty Alumni (Salem State College; general public)	20
24 Oct 01	GIS course (Salem State College; college)	15
2 Nov 01	Lynn Classical H.S. (Lynn, MA; high school youth)	40
9 Nov 01	Lynn Classical H.S. (Lynn, MA; high school youth)	40
16 Nov 01	Lynn Classical H.S. (Lynn, MA; high school youth)	40
20 Nov 01	Cub Scout Troop (North Shore communities, MA; youth)	25
8 Jan 02	URI faculty (Rhode Island)	2
9 Jan 02	Essex Agricultural School (Hathorne, MA; high school youth)	13
7 Feb 02	Ipswich Shellfish Advisory Board (Ipswich, MA)	8
21 Feb 02	Monserrat School of Art (Peabody, MA; college)	18
22 Feb 02	Biological Oceanography Class (Salem State College; college)	15
4 Mar 02	Wilmington H.S. (Wilmington, MA; high school youth)	7
5 Apr 02	Saltonstall School (Salem, MA; primary school youth)	30
19 Apr 02	Institutional Development (Salem State College; administrators)	5
7 Jun 02	O'Maley School (Gloucester, MA; middle school youth)	90
17 Jul 02	Project Splash (Lowell, MA; middle school youth)	60
25 Jul 02	Ocean Quest (Salem, MA; middle school youth)	8
22 Aug 02	Ocean Quest (Salem, MA; middle school youth)	10
9 Sept 02	Wilmington, H.S. (Wilmington, MA; high school youth)	10
7 Nov 02	Lynn Classical School (Lynn, MA; high school youth)	50
13 Nov 02	Massachusetts Watershed Initiative Representative	1
14 Nov 02	Lynn Classical School (Lynn, MA; high school youth)	50
30 Jan 03	Pingree School (Salem, MA; high school youth)	6
20 Feb 03	Montserrat School of Art (Peabody, MA; college)	41
21 Feb 03	Prospective students (college)	5
21 Feb 03	Revision House (Dorchester, MA; battered women)	4
23 Jun 03	Estuarine Ecology Class (Salem State College; college)	12
25 Jun 03	Massachusetts Department of Education	14
11 Jul 03	School for Field Studies (Salem, MA; faculty)	3
24 Jul 03	Ocean Quest (Salem, MA; middle school youth)	8
31 Jul 03	Ocean Quest (Salem, MA; middle school youth)	12
14 Aug 03	Ocean Quest (Salem, MA; middle school youth)	10
6 Sep 03	Salem Now (Salem, MA; local television station)	2
1 Oct 03	Wilmington H.S. (Wilmington, MA; high school youth)	5
4 Oct 03	Open House (local community, fishers, elected officials)	38
10 Oct 03	Ecology Class (Salem State College; college)	12
15 Oct 03	Marine Biology Class (Salem State College; college)	7
Total		925

siblings, friends, parents and relatives (Fig. 17.1). Through positive youth experiences a conduit to inform and educate adults has been forged that can be implemented elsewhere.

- Youth projects conducted at an urban aquaculture facility can provide young people with a personalized, educational experience. High school students, Eagle Scouts and undergraduate students from other institutions have conducted projects at the Cat Cove Marine Laboratory (Fig. 17.2). Their enthusiasm and experiences were shared with friends and family, while the participant gained personal experience in and about aquatic ecology. Similar opportunities could be made available to motivated young people elsewhere.
- Teaching teachers about aquaculture and how it can be employed as an instructional tool in the classroom is a particularly effective method to inform and impact large numbers of young people (Fig. 17.3). While some students may elect to pursue aquaculture as a career, many more are positively influenced by the hands-on aquaculture experience. Aquaculture breathes life and relevance into biology and laboratory experiences. Although instructional guides and information exist to assist teachers with integrating aquaculture into their classrooms (Hanes and Cookson Lewis, 2000; Hanes *et al.*, 2001, 2002; Soares *et al.*, 2001), urban operations can provide a nearby resource for tours and technical assistance.
- Commercial fishers in coastal areas can tour and become familiar with aquaculture facilities. Commercial fishers ply their vocation in open waters away from urban areas; however, they frequently reside in cities such as Bridgeport, CT; Gloucester and New Bedford, MA; Portsmouth, NH. Opening doors to commercial fishers will build upon commonalties that exist but are often ignored between culturist and fisher, facilitating the pooling of talents and resources to promote wise and innovative management of aquatic resources (Fig. 17.4).

Fig. 17.1. Feeding time at Cat Cove by visiting students on tour.

Fig. 17.2. High school student explores aquatic biology at the Cat Cove Marine Laboratory as part of a school-sponsored research experience.

Fig. 17.3. Teachers learn water chemistry analysis at Cat Cove during a summer course.

- Elected officials and the media (radio, television, newspapers) tend to be concentrated in urban areas. Inviting them to tour operations, meet with principals and become familiar with the many positive aspects of aquaculture can promote public awareness and receptivity (Fig. 17.5). Strong support from the public and elected officials will help shift approval for aquaculture from the all too often case-by-case litigation in courts to the more appropriate, less expensive and broadly applicable arena of legislation.

Fig. 17.4. Commercial fishermen integrate oyster aquaculture into existing lobster pot wild harvest fishery techniques.

Fig. 17.5. Elected town officials on a tour of the Cat Cove aquaculture facility.

- Hosting or sponsoring meetings for professional, trade or other associations at an aquaculture facility introduces the membership to aquaculture and its benefits. Recently the Cat Cove Marine Laboratory hosted a meeting of the New England Estuarine Research Society that incorporated a tour as well as a traditional clambake (Fig. 17.6). Over three dozen educators, students, researchers and regulators enjoyed a very positive exposure to aquaculture. Similar experiences have been afforded to conservation groups, town officials and a variety of organizations (e.g. retirees, educators) and could be widely duplicated.

While individual operations may generate local experiences and benefits, broader action is needed to universalize the impact. Trade associations (e.g. National Aquaculture Association, East Coast Shellfish Growers Association) provide a conduit to disseminate information within the aquaculture community and to targeted audiences. The Internet globalizes aquaculture's impact and acceptance by sharing it electronically (World Catch Report, http://www.worldcatch.com, Intrafish, http://www.intrafish.com). Advocacy groups (e.g. Global Aquaculture Advocate, Farm Fresh Salmon Organization, http://www.farmfreshsalmon.org) can extol the benefits of sustainable aquaculture.

Conclusions

Globalizing the positive attributes and multi-faceted benefits of urban aquaculture is the shared responsibility of producers, support industries, academics, researchers and all involved in aquaculture. Concerted, coordinated and well-orchestrated efforts will benefit individuals involved in aquaculture and all of society. In urban settings, implementation of sustainable aquaculture practices that target niche markets coupled with the energetic integration of community education and community involvement will facilitate commercial success, while further securing our food supply and kindling a realistic appreciation of living within the confines of our biosphere.

Fig. 17.6. New England Estuarine Research Society clambake at the Cat Cove aquaculture facility site.

Acknowledgements

David Arquette, Lloyd Benedict and Henry Likers (Mohawks of Akwesasne, New York, Ontario, Quebec) provided information and shared insight on ecological traditions of Native American culture. Dr Karl Wesolowski (Department of Economics, Salem State College, Salem, Massachusetts) assisted with interpretation and description of recent economic developments. Ms Bonnie McAneney reviewed earlier versions of this chapter.

References

Alabama Aquaculture Best Management Practices (2002) http://al.nrcs.usda.gov/Sosections/Engineering/BMPindex.html

Atkinson, R.O. and Court, H. (1998) *The New Economy Index, Understanding America's Economic Transformation.* Progressive Policy Institute, Washington, DC.

Avault, J.W. (1996) *Fundamentals of Aquaculture. A Step-by-Step Guide to Commercial Aquaculture.* AVA Publishing Company, Baton Rouge, LA.

Beal, K.L. (2002) TEK: Traditional ecological knowledge, diversity, and NW tribal practices. *Fisheries* 27(4), 4.

Black, K.D. (ed.) (2001) *Environmental Impacts of Aquaculture.* Sheffield Academic Press, Sheffield.

Bridger, C.J. and Costa-Pierce, B.A. (2003) *Open Ocean Aquaculture: From Research to Commercial Reality.* World Aquaculture Society, Baton Rouge, LA.

Brown, L.R., Abramovitz, J., Bright, C., Flavin, C., Gardner, G., Kane, H., Platt, A., Postel, S., Roodman, D., Sachs, A. and Starke, L. (1996) *State of the World.* W.W. Norton and Company, New York.

Brown, L.R., Flavin, C., French, H., Abramovitz, J., Bright, C., Dunn, S., Gardner, G., McGinn, A., Mitchell, J., Renner, M., Roodman, D., Starke, L. and Tuxill, J. (1998) *State of the World 1998.* W.W. Norton and Company, New York.

Buttner, J.K. (1997) First Nation people and Great Lakes aquaculture. *Aquaculture Magazine* 23(1), 27–40.

Cohen, M.N. (1977) *The Food Crisis in Prehistory, Overpopulation and the Origins of Agriculture.* Yale University Press, New Haven, Connecticut.

FAO (2001) *FAO Annual Report: The State of Food and Agriculture 2001.* http://www.fao.org/docrep/003/x8002e/x8002e04.htm

Gall, K. and Rivara, G. (2000) *HACCP Guide for the Aquaculture Industry.* NRAC Publication No. 00–005. University of Massachusetts, North Dartmouth, Massachusetts.

Gray-Krueger, D. (1998) Tribal perspectives in natural resource management and environmental technology. *Journal of Environmental Health* 61(3), 34–37.

Guo, X. and Kraeuter, J. (1999) Aquaculture and technology. *The Jersey Shoreline, Special Edition* 19(3).

Hanes, F. and Cookson Lewis, C. (eds) (2000) *Recirculating Aquaculture Set-up Chronological Assistance Letters RASCALs.* New England Board of Higher Education, Boston, Massachusetts.

Hanes, F., Audet, R., Brown, C., Buttner, J., Leavitt, D. and Soares, S. (2001) *Explorations in Aquaculture, Volume 1: Marine Resources for the Classroom.* Aquatic Ecosystems, Inc., Apopka, Florida.

Hanes, F., Audet, R., Brown, C., Buttner, J., Leavitt, D. and Soares, S. (2002) *Explorations in Aquaculture, Volume 2: Freshwater Resources for the Classroom.* Aquatic Ecosystems, Inc., Apopka, Florida.

Haudenosaunee Environmental Task Force. (1999) *Words That Come Before All Else; Environmental Philosophies of the Haudenosaunee.* Native North American Travelling College, Cornwall Island, Ontario.

Joint Subcommittee on Aquaculture (1993) *Aquaculture in the United States: Status, Opportunities, and Recommendations.* Joint Subcommittee on Aquaculture, Washington, DC.

Nash, B. and Green, D. (1999) *Hazard Analysis and Critical Control Point: Model Safety Plans for Small Seafood Dealers, Packers and Processors.* Publication UNC-SG-99-01. North Carolina Sea Grant, Raleigh, North Carolina.

National Marine Fisheries Service (2002) *A Code of Conduct for Responsible Aquaculture Development in the U.S. Exclusive Economic Zone.* http://www.nmfs.gov/aquaculture.htm

North American Indian Travelling College (1984) *Legend of Our Nations.* North American Indian Travelling Center, Cornwall Island, Ontario.

Open Ocean Aquaculture (1996) *Proceedings of an International Conference*, 8–10 May 1996, Portland, ME. Marie Polk, editor. New Hampshire/Maine Sea Grant College Program Report UNHMP-CP-SG-96–9.

Population Information Program Center (2001) *Population Growth and Urbanization: Cities at the Forefront.* http://www.jhuccp.org/pr/urbanpre.shtml

Smith, B.D. (1995) *The Emergence of Agriculture.* W.H. Freeman and Company, New York.

Soares, S.J., Buttner, J.K. and Leavitt, D.F. (2001) *Aquaculture Curricula Resource Guide: A Resource Tool for the Aquaculture Educator.* NRAC Project No. 98–7, NRAC Publication No. 01–001. University of Massachusetts, North Dartmouth, Massachusetts.

Tomasso, J.R. (ed.) (2002) *Aquaculture and the Environment in the United States.* US Aquaculture Society, Baton Rouge.

United States Department of Agriculture (2000) *National Agricultural Statistics Service, 1997 Census of Aquaculture (1998), Volume 3, Special Studies, Part 3.* USDA, Washington, DC.

United States Environmental Protection Agency (2002) *Effluent Limitations Guidelines and New Source Performance Standards for the Concentrated Aquatic Animal Production Point Source Category; Proposed Rule.* http://www.epa.gov/waterscience/guide/aquaculture/aquaFRN.html

World Population Calculator (2002) *Current World Population.* http://www.onlineconversion.com/world_population.htm

Zvelebil, M. (1985) *Hunters in Transition. Mesolithic Societies of Temperate Eurasia and their Transition to Farming.* Cambridge University Press, Cambridge.

This page intentionally left blank

18 Ecolabelling and Urban Aquaculture

Cathy A. Roheim
University of Rhode Island, Kingston, USA

Abstract

Ecolabelling has been identified as a potential means to create market-based incentives for environmentally friendly products and production processes by creating consumer demand for these products. The increased demand is expected to result in higher prices and/or an increase of market share, other economic returns and environmental benefits. The aquaculture industry is in the infant stages of creating environmental labelling programmes, beginning primarily with cultured shrimp from ecosystem-friendly production methods.

Much has been said about ecolabelling in the last few years, with several studies investigating consumer demand for ecolabelled seafood product and showing that there is a demand for these products. This chapter will report on a recent survey done in the USA in which consumers were asked to trade-off amongst seafood products when there were four different species, ecolabelled and not, and varying prices.

Introduction

We often find in aquaculture that the focus of producers is on production. This is where the 'cool' technology is found – the gadgets and critters that entice and occupy producers' time. Marketing? Well, there are no cool gadgets, and some thought has to go into the question of 'How much are consumers willing to pay for my product?' Surely, this is not as much fun as playing with the technology!

Producers who allow their firm to have such a lop-sided view of production versus marketing are often those which go out of business shortly after their first shipments of product are available for purchase – at prices high enough that no one wants to buy. The potential of urban aquaculture to fall into this trap is probably greater than for more traditional aquaculture, in large part because of the greater cost of water recirculation technology and perhaps higher property values in urban areas. There are two ways out of this trap – lower the costs of production, or affect an increase in prices that consumers are willing to pay for the product. The objective of this chapter is to focus on the latter, and to discuss potential niche markets urban aquaculture producers may be able to take advantage of, in part, *because* of their technology.

In particular, this chapter will explore ecolabelling of urban aquaculture products.

Urban aquaculture firms are in the position to exploit, perhaps better than most other producers of farm-raised seafood, the concept of sustainable production. Sustainable aquaculture production can be defined as producing fish or shellfish without compromising future production due to harmful environmental practices resulting from current production. Urban aquaculture producers have control over several aspects of production, including incoming and outgoing water purity, waste treatment, chemical treatments, medical applications such as vaccines and antibiotics, and feed composition. Sustainable aquaculture practices are those that do no harm to the surrounding ecosystem, which includes treatment of wastewater, prevention of chemical residues from affecting surrounding or downstream ecosystems, and the protection of the world's ecosystems in general (e.g. the resource from which the feed is produced is itself sustainable). Thus, with good environmental practices, the urban aquaculture industry could potentially differentiate itself from both sea-based aquaculture products and from wild-harvested fish using an ecolabel.

The remainder of the chapter will demonstrate how the urban aquaculture industry might benefit from ecolabelling. First, a pair of examples of competition in global aquaculture markets will be presented, with particular references to use of labelling as a means to create a market niche. Next, ecolabels are defined, followed by a presentation of empirical evidence that indicate consumers are indeed interested in how their fish is produced, and that they may have a preference for sustainable production, whether it is farm raised or harvested from the wild. The chapter concludes with recommendations for the urban aquaculture industry.

Competition in the Marketplace: the Role of Specific Attributes

Competition in global seafood markets is fierce. Aquaculture in particular has increased the level of competition. For the past several years, southern states were the primary sources of farm-raised catfish in the US market. Demand expanded as supply expanded due to aggressive marketing campaigns. The farm-raised catfish industry climbed to a value of US$590 million in 2001 (Johnson, 2002). However, that industry has recently found itself in the centre of a trade war with Vietnam. Imports of Vietnamese tra and basa (the appropriate names for Vietnamese catfish) into the US increased from 900,000 kg in 1999 to 7.7×10^6 kg in 2001 – which represents 12% of the US catfish market. Prices for US-farmed catfish have declined substantially as a result.

The approach of the US catfish industry to solve this problem has been to focus on trade restrictions. An alternative approach, country-of-origin labelling, may help US catfish growers as well. A change in policy has occurred in the US government that is not directly addressing the catfish case, but will influence the catfish market and any other segment of the seafood market that relies on imported product. As part of the 2002 Farm Bill, the US government instituted a 2-year voluntary country-of-origin labelling programme for fish, meats and other perishable agricultural commodities. Country-of-origin labelling will be mandatory for fish by 30 September 2004. In the case of catfish, country-of-origin labelling may be a policy beneficial to the US catfish industry. Consumer demand for Vietnamese basa and tra, when it is labelled as originating from Vietnam, may change due to consumers' perceptions of Vietnam. If those perceptions include a concern that there are less stringent environmental standards in the water from which the fish are harvested in Vietnam than in the USA, that would lead demand away from Vietnamese product toward US product. In that respect, country-of-origin labelling may be somewhat of a substitute for ecolabelling.

In another example, prior to salmon aquaculture, the primary sources of salmon were wild salmon in the Pacific, predominately from Alaska and Japan. In real terms, prices were relatively high and at least 3 of the 5 salmon fisheries (Chinook, coho, sockeye) were highly valued. In the 1980s and through the 1990s worldwide salmon aquaculture grew dramatically, to the point where farm-raised salmon are now more plentiful than wild salmon harvests (Johnson, 2002). Prices have correspondingly gone down, having a big effect in all 5 Pacific salmon fisheries (Chinook, coho, sockeye, chum and pink).

The question becomes, given an increase in supplies of some seafood products, how do we maintain or raise price levels? Producers need to affect consumer demand. In economics jargon, we want to shift the demand curve to the right, i.e. achieve a consumer willingness to pay an increase in price at any level of quantity demanded. How? Ecolabelling is one possible method to affect the perceptions and beliefs that consumers have regarding the product. In the case of wild salmon, salmon from Alaska is certified by the Marine Stewardship Council (MSC), reflecting sustainability of the fishery. Salmon from British Columbia may soon follow. This ecolabel has led to an increased market share of Alaskan salmon in speciality supermarkets, such as WholeFoods, the nation's largest natural-foods grocery chain, and reportedly is taking back some market share from farmed salmon in the European smoked salmon market (Roheim, 2003).

Ecolabelling as Product Differentiation

Environmental labelling, in general, has been in existence around the world for many years and is defined as 'making relevant environmental information available to appropriate consumers' (United States Environmental Protection Agency, 1993, 1998). A subset of environmental labelling is ecolabelling that relies on independent third-party verification that the products meet certain environmental criteria or standards (United States Environmental Protection Agency, 1998). Once the product meets those criteria or standards, a 'seal-of-approval' – an ecolabel – may be affixed to the product.

Ecolabelling has many potential societal benefits, including environmental improvement, accurate information dissemination to consumers, improved market share for producers and increased awareness and interest by the public about environmental issues. Accurate information dissemination is necessary for consumers to make informed decisions regarding their purchases. As consumers grow increasingly aware of environmental issues and the role their purchases may play in environmental degradation, market shares of products with ecolabelling may grow at the expense of products without ecolabelling. This may be true even if the labelled product is more expensive, because informed consumers may be willing to pay more for the product they feel has the least impact on the environment.

As MacMullen (1998) and Deere (1999) note, little environmental labelling has been applied to seafood products. However, that is changing. There are already several national, international, industry-sponsored, NGO-led and consumer–supplier partnership certification and standards schemes under development in the seafood sector. The range of possible labels is broad. The focus of claims regarding capture fisheries can range from not over-fished, to no marine mammal by-catch and not over-fished, to no by-catch of any sort and not over-fished, to ecosystem friendly where the entire ecosystem with its complicated food chain is sustainable. The focus of claims regarding farm-raised seafood can include environmentally benign production and no harm to surrounding ecosystems.

Environmental attributes related to a product's production are often impossible for the individual consumer to assess. Ecolabelling programmes offer an approach to provide consumers with that information, while at the same time creating a market-based approach to address environmental issues. In theory, ecolabelling programmes are meant to affect consumer behaviour as follows:

- An independent third party develops criteria for environmentally preferable products, and then evaluates products to determine if they meet those criteria.
- This complex information is presented in a product label.
- Consumers incorporate the information conveyed by the logo with the other attributes of the product, such as quality and price, to determine demand.
- If the labelled product is preferred by consumers, then producers will alter their behaviour such that their production processes conform with the environmental goal.

The most well-known aquaculture industries to suffer from attention to negative environmental impacts are shrimp and salmon aquaculture. Both are holding the attention of environmental groups due to the impact farming is having on coastal ecosystems. The National Audubon Society (http://www.audubon.org), Environmental Defense (http://www.environmentaldefense.org) and the Monterey Bay Aquarium (http://www.mbayaq.org) have all published guides for consumers on which seafood species are problematic, and encourage consumers not to buy them. For example, the Audubon Society evaluated aquacultured Atlantic salmon, stating that salmon farming pollutes, displaces wild fish, and prompts the shooting of predatory seals near farms. Consumers are urged not to eat farmed salmon. The Monterey Bay Aquarium notes that most farmed shrimp comes from Asia, where clearing of coastal land for shrimp farming has damaged habitat and displaced sustainable fisheries. Consumers are urged not to buy farmed shrimp.

To address these concerns, the Global Aquaculture Alliance (GAA) developed the Responsible Aquaculture Program (RAP), based in part on the FAO Code of Conduct for Responsible Fisheries. The initial focus of the GAA was on responsible shrimp farming, although it is expected that elements of the RAP will be applied to other species (http://www.gaalliance.org). The RAP for shrimp aquaculture includes nine codes that address the following topics: mangroves, site evaluation, design and construction, feeds and feed use, shrimp health management, therapeutic agents and other chemicals, general pond operations, effluents and solid wastes, and community and employee relations. The Aquaculture Certification Council, Inc. offers process certification for shrimp production based on GAA's Best Aquaculture Practices standards.

Given that ecolabelling is seeing a growing use in seafood marketing, the questions then become: If we gain certification of a well-managed, environmentally friendly facility and are able to ecolabel our products, will consumers buy it? And will we be able to shift the demand curve, thus potentially gaining higher prices for our products compared to others? To answer those questions, we must examine consumers' preferences for seafood products with varying environmental qualities.

Empirical Evidence of Consumer Preferences for Ecolabelled Seafood

To date, there has been no empirical evidence regarding consumer preferences specifically for ecolabelled aquacultured fish and seafood. Therefore, we cannot speak specifically toward ecolabelling of urban aquacultured products. Instead, we use the empirical evidence that exists for consumer preferences of ecolabelled seafood from capture fisheries as a barometer for ecolabelled aquacultured products.

Wessells *et al.* (1999) investigated the demand for ecolabelled seafood (cod, cocktail shrimp and salmon, specifically) in the US market. The methodology used involved gathering data with a survey administered to a random sample of 1640 US consumers by telephone. The survey was designed so that respondents compared certified (i.e. with an ecolabel) and uncertified (i.e. without an ecolabel) products, whose prices differed according to a premium paid for the certified product. With the exception of differences in certification and price, the two products were identical in all regards, including quality and freshness. Certification was described as a:

> programme ... that would label seafood in order to guarantee that it is caught under strict controls that prevent too much fishing. Certified seafood will have [a] new label that guarantees no overfishing. Uncertified seafood will not have this guarantee.

Consumers were presented with three paired comparisons, in random order, for salmon, cod and cocktail shrimp. The base price varied for each species, depending on the range of common retail prices for each product at the time of the survey. Prices ranged between US$4.00 and US$10.00/kg. The certifying agency alternated between the World Wide Fund for Nature (WWF), the National Marine Fisheries Service (NMFS) and the Marine Stewardship Council (MSC). It is important to note that certified salmon was compared to uncertified salmon, certified cod to uncertified cod, etc. The survey did not ask respondents to choose, for example, certified cod versus uncertified salmon.

Data were collected in the summer of 1998 on the household's geographic location, trust in specific agencies as providers of certification, seafood consumption habits, household seafood and grocery budgets, memberships in environmental organizations, perceptions of the status of Pacific salmon and Atlantic cod stocks, and a variety of other factors with potential impact on preferences for labelled seafood products. On average, about 70% of respondents chose ecolabelled shrimp, salmon or cod over non-ecolabelled.

Econometric analysis was performed to determine what factors influence the choice of ecolabelled over non-ecolabelled products. Using a logit analysis, results suggested that respondents' preferences for ecolabelled fish are most affected by the size of the premium. As the premium increases, the likelihood that the respondent would choose the ecolabelled product over the non-ecolabelled product declines. In addition, the likelihood of choosing ecolabelled fish differed by species, geographic location of the household, consumer group and was slightly affected by certifying agency. For example, the effect of the premium was negative for all species, but smaller in magnitude for salmon, and greatest for cod. Households on the west coast of the USA were more likely to choose certified salmon than those in other parts of the nation. Households that were members of environmental organizations were more likely to choose certified fish over uncertified. Other factors found to influence choice of ecolabelled fish were gender – females were more likely to choose ecolabelled products than men; and seafood budgets – those households with larger seafood budgets were more likely to choose uncertified products. These results also indicated that significant consumer education must take place, as fully two-thirds of respondents indicated that they were unsure of the status of Pacific salmon and Atlantic cod stocks.

Johnston *et al.* (2001) extend the paper by Wessells *et al.* (1999) by examining cross-country differences in preferences. In addition to the US data discussed above, the authors collected data from a virtually identical telephone survey administered to 2039 Norwegian residents during the autumn of 1999. The primary differences in information collected between Norway and the USA were that: (i) instead of cocktail shrimp, the Norwegian survey asked about a smaller coldwater shrimp; (ii) instead of using the NMFS as a governmental certifying agency, the Norwegian survey used the Norwegian National Fisheries Directorate; and (iii) the premiums were specified in Norwegian kroner, not US dollars. Norwegian consumers were less likely to choose certified seafood products, averaging closer to a 50% preference for certified. Approximately 34% preferred uncertified, while a fairly large percentage, 15%, gave no answer.

To test the hypothesis that consumer preferences for ecolabelled seafood differ across nations, econometric analysis of an equation similar to that discussed above was performed. In this case, the Norwegian and US data were combined, with appropriate variables specified to determine if there are differences in the two sets of respondents. In order to do the comparison with premiums that were in both kroner and dollars, the premiums were converted into a percentage. Results indicate that there are differences. Again, results indicate that as the premium grows, consumers will be less likely to choose ecolabelled seafood. This effect is even stronger in Norway, thus consumers in Norway may be more price sensitive. In addition, consumers in Norway are more likely than those in the US to be influenced by the certifying agency. Those Norwegians who belong to an environmental organization are less likely to choose certified compared to US respondents who are members of environmental groups.

There are several implications from the results of Johnston *et al.* (2001). Most importantly, respondents to these surveys were educated about what the product was being certified for, i.e. why it had an ecolabel. Once educated, they were then asked to make their choices between certified and non-certified. The results showed that the majority of respondents chose ecolabelled products, however, that was very dependent on the size of the premium. In addition, the sample of consumers who were surveyed in each country could be considered 'educated' consumers – i.e. educated about the meaning of the ecolabel. In reality, when these choices are no longer hypothetical and consumers may be more or less educated about the ecolabels, consumers may be more or less likely to choose products from certified fisheries. That choice will certainly depend on the premium paid for ecolabelled fish over non-ecolabelled, but will also depend on how aware the consumer is about the issue the ecolabel addresses. Furthermore, the consumer must understand the content of the label, i.e. the link between their purchasing decision and effective management of stocks. The analysis of the paper does not provide the authors with the means to compare choices with and without the information on what certification means.

Roheim and Donath (2003) addressed the question whether environmental convictions carry more weight in consumer purchasing decisions for seafood than their taste in fish. The study directly addresses the question of whether species (taste) is the overriding factor in the consumer's decision to buy seafood, and how ecolabelling affects that decision, in addition to the effect of prices.

The study is based on a survey designed and pre-tested in focus groups, and in personal interviews during August 2001. Surveys were mailed out to 1500 randomly drawn addresses in Connecticut in September 2001. The surveys were mailed out with a representative number of surveys mailed to each Connecticut county. The response rate was 29%. Compared to census data, the survey results indicate a bias towards older age groups, higher income and higher education among respondents. This result is consistent with results obtained in mail surveys.

The survey itself contains several sections; the first of which asks about the respondent's general seafood consumption, followed by a section on environmental purchasing behaviour. The next section contains information about ecolabelling, and two ranking questions that are the heart of this survey. A section on demographics is included at the end of the survey.

The survey describes the ecolabel as a programme that avoids overfishing. The label is included in the ranking question as one of three attributes and guarantees no overfishing of the specific species described. The survey also emphasizes that seafood with this label has the same colour, quality and freshness as the same species without the label.

The ranking question represents the basis for this analysis. In each ranking question, the respondent was asked to rank four different kinds of seafood with numbers from 1 to 4. Each ranking question contained the four species: salmon, swordfish, flounder and cod. Each species was assigned one of three different price levels using information from supermarkets in the New England area. The ranking questions also contained information on whether the product was awarded the ecolabel or not.

Results show that consumers' ranking of choices is sensitive to different price levels. The price sensitivity is strongest for swordfish, the most expensive species in absolute terms, followed by salmon and flounder. Respondents give a higher rank to those species they like best, according to taste. The presence of an ecolabel has a significant effect on respondents' rankings. In general, the presence of an ecolabel increases the probability of a high ranking for all choices. This increased probability is present in the case of respondents who consider the ecolabel to be important, respondents with high income and members of environmental organizations. Male respondents are significantly less likely to assign a high ranking to choices that include the ecolabel than their female counterparts.

When the relative importance of the various attributes, species, price and presence of an ecolabel are compared, this is the order of importance in the respondents' ranking of a product. For these respondents, the type of seafood is the most important factor in their ranking decision. Ecolabelling of seafood is important to consumers, though not differentiated across different species. Price levels do influence consumers' decisions, but are relatively less important in their ranking decision.

Thus, the answer to the question: Which is most important in a consumer's seafood purchasing decision? Given choices between price, ecolabel and species, it is species that is most important.

Conclusions

This chapter has shown that there are consumer preferences toward 'environmentally friendly' fish and seafood. What does this mean for urban aquaculture? Given the technology used in urban, closed-systems aquaculture, and caution on the part of the grower, the environmental impacts of production may be quite small. Taking advantage of that fact may create a niche market for these products. Depending on the product, they may be environmentally benign – even the Audubon Society's Seafood Watch programme gives farmed striped bass, scallops, mussels and clams a green light – assuring consumers that it is 'OK' to eat these products. Therefore, as production costs may be higher for urban aquaculture, utilizing ecolabelling, and perhaps other labelling such as locally grown, organic or socially responsible, should be explored as they may be a feasible means of achieving higher prices for the products.

References

Deere, C. (1999) *Ecolabelling and Sustainable Fisheries.* IUCN – The World Conservation Union and the Food and Agriculture Organization of the United Nations (FAO), Washington and Rome.

Johnson, H. (2002) *Annual Report on the United States Seafood Industry.* H.M. Johnson & Associates, Bellevue, Washington.

Johnston, R.J., Wessells, C.R., Donath, H. and Asche, F. (2001) A contingent choice analysis of ecolabeled seafood: comparing consumer preferences in the United States and Norway. *Journal of Agricultural and Resource Economics* 26(1), 20–39.

MacMullen, P.H. (1998) *A Report to the Fish Industry Forum on the Marine Stewardship Council and Related Topics.* Technology Division, Consultancy Report No. CR 152, November. Sea Fish Industry Authority, UK.

Roheim, C.A. (2003) Early indications of market impacts from the Marine Stewardship Council's ecolabeling of seafood. *Marine Resource Economics* 18, 95–104.

Roheim, C.A. and Donath, H. (2003) *The Battle of Taste Buds and Environmental Convictions: Which Drives Consumer Demand for Ecolabeled Seafood?* Working paper, Department of Environmental and Natural Resource Economics, University of Rhode Island, Kingston, Rhode Island.

United States Environmental Protection Agency (1993) *Status Report on the Use of Environmental Labels Worldwide.* Office of Pollution Prevention and Toxics, EPA 742-R-9–93–001. September. US EPA, Washington, DC.

United States Environmental Protection Agency (1998) *Environmental Labeling: Issues, Policies, and Practices Worldwide.* Office of Pollution Prevention and Toxics, EPA 742-R-9–98–009. December. US EPA, Washington, DC.

Wessells, C.R., Johnston, R.J. and Donath, H. (1999) Assessing consumer preferences for eco-labelled seafood: the influence of species, certifier and household attributes. *American Journal of Agricultural Economics* 81(5), 1084–1089.

19 Aquaculture in Future Urban Ecosystems

Alan Desbonnet and Barry Costa-Pierce
Rhode Island Sea Grant College Program, University of Rhode Island, Graduate School of Oceanography, Narragansett, USA

Throughout the course of this volume, authors describe the needs, as well as the opportunities, for aquaculture in future urban ecosystems. They point to broad opportunities for humankind to integrate the production of aquatic foods with the processing of wastes and/or the intensive reuse of water resources, mimicking naturally functioning ecosystems to a degree unseen and unpractised by most of the populated world. As resources, particularly water, become scarcer, and as people, particularly in rapidly developing countries, become more numerous, the need to shape societies as 'closed loop' ecosystems, particularly in urban areas, becomes paramount. This volume identifies the start of a path towards attaining sustainable urban ecosystems using aquaculture as an integrating mechanism.

In Chapter 1, Costa-Pierce and Desbonnet postulate that, given wastewater discharge increases in light of projected population growth, nutrient inputs to coastal oceans will increase, further stressing and degrading coastal ecosystems. The authors note that the incorporation of tertiary and reverse osmosis treatment technologies can reduce overall loading of nutrients to coastal waters, but they pose several critical questions: (i) Are the reductions enough to halt current trends of degradation? (ii) Would anticipated loadings reductions allow recuperation, and then sustenance of, healthy coastal ecosystems? (iii) Can and/or will society be able, or want, to foot the bill for such an expensive solution?

Costa-Pierce and Desbonnet suggest that societies worldwide will need to turn the 'wastewater pipes' inland away from the coasts and develop innovative solutions, one such being aquaculture wetland ecosystems (AWEs), to move towards a sustainable nutrient cycling loop in urban ecosystems. This technology has been employed successfully, and provides for habitat improvements along with the benefits of wastewater treatment and food production. The opportunity to further develop this model on a more widespread basis throughout a variety of cultures exists, and could be a significant benefactor to urban ecosystem health and sustainability into the future. The authors point out, in closing, that what needs to occur now and into the future is a greater emphasis on research, and a shift in urban planning processes to incorporate innovative thinking about the use of wastewater as an asset, rather than as a waste that is a debit of urban area function.

In Chapter 2, Lutz focuses on a major challenge to the development of sustainable urban aquaculture activities – planning. According to Lutz, urban planners must begin to recognize the opportunities and benefits of aquaculture as an urban-based agroindustry, and aquaculturists must become better planners for integrating operations as part of the urban environment and to take advantage of urban markets and other urban-based opportunities. Furthermore, Lutz suggests that new entries into urban aquaculture businesses must consider more than just the technologies needed to grow fish, despite its importance to success. They also need to understand marketing and consumer preferences.

Lutz notes that the cost of conducting urban aquaculture, even with good planning, will be high, at least at the onset as the fledgling industry is developed. In recirculating aquaculture, back-up systems, all intimately related to life support, must be developed in the planning phase, sometimes at considerable cost. Since all ecosystem functions in an urban, recirculating aquaculture system are bound to be nearly, if not entirely, of artificial construct, back-up systems for power, water flow, temperature, gas mixtures, etc., will be critical to success, profitability and sustainability. Lutz states that pioneers with innovative ideas are needed to blaze the trail for the development and implementation of these new types of systems. The innovators must look beyond the basics of physics, chemistry and technology, and acutely hone an emphasis on defining inputs and outputs, and particularly the market for outputs. Lutz notes there is currently too much emphasis on site costs, and not nearly enough consideration of the flow paths of energy and other needed materials into and out of urban aquaculture systems. Site costs are pro-rated over the life of the operation, and are therefore fairly simple to conceptualize and deal with. Input and output flows are everlasting, and success will hinge, as Lutz notes, on their careful and efficient management.

Lutz concludes, as Roheim later corroborates in Chapter 18, that in order for sustainable urban aquaculture to develop, time and effort must be put into researching the market, particularly regarding consumer preference for aquaculture products produced in an urban setting. This is particularly true in the more developed countries where per capita income is greater, and choice is often driven by more than just cost to the consumer. The opportunity for aquaculture to become integrated into a sustainable part of functional urban ecosystems exists, but innovative, market-oriented efforts and concerted planning will have to be deployed in order to achieve that end.

In Chapter 3, Little and Bunting give an overview of the history and development of urban aquaculture in Southeast Asia. The authors note that urban aquaculture has the opportunity to provide great benefit in alleviating at least some of the conditions that the poor must contend with in urban areas by providing both a means of economic livelihood and a source of inexpensive food. Urban aquaculture provides on-site jobs, and it also provides opportunities off-site for production of value-added fillets and processed, ready-to-eat produce. Certainly the development of urban aquaculture will not solve the problem of poverty in the urban areas of less developed, less wealthy countries, but it can be a significant part of a larger solution in moving towards sustainable urban ecosystems.

Little and Bunting go on to note that urban aquaculture, particularly that which utilizes wastewater, is an important component of waste treatment in rapidly urbanizing areas of less wealthy countries. Furthermore, aquaculture provides the services of wastewater treatment free of charge to municipalities, though Little and Bunting note this is hardly ever a consideration in urban planning venues. Unfortunately, as the authors state, as wealth accumulates the acceptability of aquatic products grown in wastewater diminishes, stymieing or reversing the benefits that accrue by the poor. As urbanization rapidly progresses, access to wastewater often becomes limited as wastes are treated, or more likely, shunted away from the urban ecosystem for disposal, thereby robbing aquaculture producers of an input vital to their sustainability. As if all this weren't bad enough, the authors show that as urban development progresses, industrialization ensues and wastewater often becomes contaminated with industrial wastes which produce poorly flavoured fish products that are unacceptable to consumers. Goudey and Moran, in Chapter 13, document some of the limitations that occur when aquaculture and heavy industry try to coexist.

Little and Bunting show that in Southeast Asia the demand for fish and fish products is high, and will continue to be high into the future, and that wastewater-based aquaculture is already an established, and accepted, method of food production in the region, and markets exist for the products produced. The authors conclude their essay by stating that education is needed to stop the exodus from wastewater-cultured product as affluence accrues, and that one proposed direction to move in is integrating urban aquaculture into the larger picture of urban sustainability, which would entail incorporating aquaculture into urban planning processes, and particularly into planning for wastewater collection, transport and/or treatment schemes. Urban aquaculture must also be linked to the larger cycle of economic planning, providing avenues for economic growth by poor members of urban society on a sustainable basis.

In Chapter 4, Edwards echoes many of the sentiments and insights brought forth in earlier chapters, though he delves deeper into the constraints and opportunities for wastewater-based aquaculture in urban areas, clearly stating that wastewater treatment is not going to be the norm in tropical, less wealthy nations in any foreseeable future. This is not only due to the high costs of construction, but also to the high costs – economical, technical and knowledge-based – of maintenance to these systems over the long term, even if initially constructed with foreign aid funds. Edwards then develops a typology that includes wastewater-fed aquaculture as an integral component of the functioning of the urban ecosystems of tropical, rapidly developing, rapidly urbanizing countries. If planned properly – and like in previous chapters, Edwards notes this has to become integrated into the larger processes of both economic and urban planning venues – aquaculture can serve the role of sustainable wastewater treatment for urban centres, and at the same time build economic opportunities.

Edwards, like Little and Bunting, acknowledges that the produce from wastewater-fed aquaculture will provide more benefit to the poor because the affluent will tend to shun the product. Edwards provides an example of a functional system which has already been developed, in Lima, Peru, and is ready for application, and that integrates high-tech wastewater treatment using aquaculture as an integral part of the treatment process. Unfortunately, this example of aquaculture as part of a sustainable urban ecosystem has not been readily accepted for implementation elsewhere. Why? Edwards concludes, as have Little and Bunting and Costa-Pierce and Desbonnet previously, that there is a basic lack of research on the safety of consuming aquaculture products grown in wastewaters. Some work has been done, as pointed out by Edwards, and by Costa-Pierce and Desbonnet, but it is clear that considerably more research needs to be done. Once completed, outreach and extension must follow to make it readily known that wastewater-fed aquaculture produces foods safe for human consumption, and that it does so in a way that benefits society at large by producing not only jobs, but in processing the stream of wastes produced by society.

Bunting *et al.*, in Chapter 5, further elaborate on urban aquaculture with their description of the past, present and future of wastewater-fed aquaculture in Kolkata, India. The authors note that while wastewater-fed aquaculture in urban Kolkata has been well developed, it has declined significantly over the past half century, mainly due to the encroachment of rapid urban development and urban renewal schemes, a theme noted previously by Bunting and Little, and Edwards. What remains at present in Kolkata is a series of smaller 'homestead' aquaculture operations, some of which are run as farmer cooperatives. But the authors note that these operations are at risk, mainly, as echoed in previous chapters, due to poor urban planning and a lack of maintenance to existing wastewater collection systems, which over time, forces the fish farmers to relocate out of urban areas, or change occupations.

The authors note that in Kolkata, as did Bunting and Little previously, aquaculture operations are subsidizing the costs of the wastewater collection system – they move water through a network of ponds that remove silt and nutrients, returning the treated water to major rivers in a more refined and clarified state, and at no cost to the municipalities. Of course, this benefit of urban aquaculture to society is neither well known nor well publicized, and it is certainly not taken into consideration by government officials or by the environmental community, which often views aquaculture as an unfavourable and undesirable activity. Making these benefits more widely known, and integrating these services into existing urban planning and management, as well as into environmental protection and restoration schemes, would go a long way in bringing urban aquaculture into the fold of sustainability as a component of functioning urban ecosystems. The authors also note the 'multiplier' effect of urban aquaculture on job creation through the interconnected web of relationships between input and output functions. The creation of jobs is not just in on-site working in the farm, but also via inputs – fertilizers, fry and fingerlings, netting and other mechanical devices, and outputs – transport mechanisms for product, skinning and fillet operations, display and sales at market, etc. Ultimately, thousands of jobs are created in this vast network of support for urban aquaculture operations, but again, this benefit is not generally recognized.

In closing, Bunting *et al.* note that most jobs created benefit the poor and the very poor, and that these jobs provide a sustainable source of income and benefits. The authors show that urban aquaculture provides for alleviation of poverty, for at least some portion of the urban poor population of Kolkata, and therefore its societal benefit is significant. Perhaps, the authors conclude, if aquaculture operations could be integrated into the municipal wastewater treatment system, and culturists were compensated for the services they provide to municipalities, this might improve the viability of these farms, leading to growth of existing operations – with increased societal benefit – as well as providing an avenue for expansion into the growth of new species and/or expansion into new markets. In all cases, society benefits by the integration of urban aquaculture as a functioning component of urban ecosystems.

Phan Van and De Pauw, in Chapter 6, further elaborate on the urban experience of aquaculture in rapidly developing countries, noting that massive water pollution in the mid to late 1980s led to the collapse of many aquaculture ventures in Ho Chi Minh City, Vietnam. Phan Van and De Pauw show how the ingenuity of farmers to adapt to circumstances beyond their control revitalized urban aquaculture via trial and error experimentation. From this experimentation has arisen a fairly robust co-culture system. Phan Van and De Pauw echo the findings of previous chapters – that one of the major limitations and challenges to wastewater-fed aquaculture is the lack of control over changes in the quality of the wastewater received. Integration of aquaculture into urban planning processes would go far in improving sustainability of wastewater-based aquaculture in Ho Chi Minh City, as well as other places.

In Ho Chi Minh City, like other rapidly developing country urban areas, the opportunities presented by aquaculture for social and economic improvement are significant. Poverty alleviation is a possibility for many, and an improved diet and improved access to protein is nearly assured in those areas where wastewater-fed aquaculture is practised. What is different, and encouraging regarding Ho Chi Minh City, is the openness of urban planners to begin considering the benefits, and therefore the needs, of urban, wastewater-fed aquaculture in their planning processes. Studies are being funded by the city to better comprehend and understand the interplay of aquaculture with urban growth, and this will be a significant step forward in acquiring sustainability of a practice that is not only traditional in nature, but fully accepted as a method of food production. This may prove to be a useful model for other urban areas to consider as a starting point for incorporating aquaculture into urban planning.

In Chapter 7, Vo and Edwards describe urban aquaculture in Hanoi, where a long history of wastewater-fed aquaculture exists. Wastewater is used as a source of nutrients to the aquaculture farm, and the techniques for using wastewaters, as was seen for Kolkata, are trial and error, based on a farmer's experiential knowledge. While the authors note this works well under current circumstances, the need exists to standardize and formalize this knowledge so that aquaculture can expand into the future, and best practices can be transferred between growers. With good planning and careful reuse of wastewater, a series of farms could be planned and developed to utilize wastewater to its maximum potential, thereby gaining the greatest benefit to farmers while simultaneously providing the greatest social good via the processing of wastewater for the municipality.

Some progressive efforts to integrate wastewater treatment and aquaculture appear to be occurring as city planners and leaders, through planning efforts already underway, seek ways to improve the integration of wastewater-fed aquaculture into urban functions. The authors, as has been previously stated by earlier authors, agree that one of the major limitations for urban, wastewater-fed aquaculture in Hanoi is the lack of control over the quality, as well as the quantity, of wastewater available to aquaculture farms. But in Hanoi the future seems brighter for a more rapid integration of aquaculture as a sustainable, functional element of urban ecosystem processes. The benefits of aquaculture appear to be clear to urban planners and municipal officials, and the limitations are being defined, and in doing so, sustainability gets closer to reality as urban planning processes move forward.

In closing, Vo and Edwards suggest that expanding aquaculture will also improve the lot of women and children, particularly those experiencing life at poverty level. This would be true not only in Hanoi, but other Asian urban areas. This theme has recurred throughout the chapters focusing on urban aquaculture in rapidly developing nations.

In Chapter 8, Bunting and Little move us from urban aquaculture in rapidly developing countries to Europe, where urban aquaculture practices developed early on in history. The authors note that there are many similarities between European and Asian aquaculture development. For instance, while urban aquaculture is not currently synonymous with wastewater aquaculture in Europe, though it is nearly so today in much of urban Asia, the authors note that this was true historically when the use of wastewater for the culture of fish was a very common practice. So while differences may be rather stark today, over a broader historical perspective, similarities abound.

An interesting point is that during the development of urban aquaculture in Europe, the practice was mainly one that was associated with royalty and the wealthy. Earlier chapters of this volume have not shown this to be the case in the rapidly developing nations of Asia and, in fact, have all pointed out consistently the relationship between poverty and self-subsistence and urban aquaculture. Furthermore, urban aquaculture today in Asia is driven by personal necessity, which was not the case historically in Europe, where religious bans on the consumption of meat made aquaculture a profitable, business-like endeavour rather than one of personal subsistence.

The authors note that urban aquaculture declined in Europe due to rapid urban expansion gobbling up lands, a trend also brought forth in previous chapters focusing on Asian countries. Interestingly enough, as population continued to expand throughout Europe, driving up the demand for fish products, capture fisheries became over-exploited, opening the door for aquaculture to develop profitably. Bunting and Little note, however, that aquaculture grew in rural areas mainly because of the very high prices for urban area land. As aquaculture product infiltrated the urban ecosystem as an import product, rural production became synonymous with high quality, cleanliness and a pollutant-free product. Urban-cultured product has not been able to gain a hold in the market to date, as it is looked upon as 'dirty' or polluted, or at the very least, not as 'pure' as rural-grown product, a perception also applicable to urban aquaculture in the USA.

The authors conclude by suggesting trading permits for pollution abatement provided by aquaculture in urban areas might be a way to improve profitability, as well as to improve public acceptance for such products. They also, like Vo and Edwards, suggest horizontal integration to more effectively and efficiently use wastewater, and water in general, while at the same time improving profitability and sustainability through product diversification. And, like Lutz in Chapter 2, the authors suggest that a systems approach is necessary for success – the technology is mostly straightforward, it is the economics and the marketing that are most important to success, and are the aspects typically ignored during an aquaculture start-up. A systems-based approach will force new ventures, particularly those in urban areas, to more fully assess options, opportunities, threats and limitations so that they can best plan for success.

In Chapter 9, Timmons brings forth the realities of the present by formulating a solid case for application of recirculating aquaculture systems (RAS) as the wave of the future for urban-based aquaculture. The author states the many benefits of RAS – excellent disease and disaster control, zero environmental impacts, less resource consumptive than pond or net pen culture, site independence and providing a non-seasonal product supply. RAS technology has everything going for it, except, as Timmons points out, adequate scale.

To date, according to Timmons, RAS technology has not been applied at a scale large enough to bring down the price of RAS-cultured product where it can compete with pond- and/or pen-reared product. Timmons states that the costs of labour are high in pond and pen culture, and that RAS could readily compete with, and even out compete, the price of product raised from pond and pen technologies – if it could achieve the appropriate scale. Timmons provides solid examples of other successful scaled-up technologies, such as that seen in the US chicken (broiler) industry and the very successful catfish production industry. In following the successes of other industries, Timmons sees a strong need for improved genetic selection to improve yield of high end/value added fillets, particularly from tilapia, which is a very promising candidate for RAS application. More effort in genetic selection will improve efficiency and boost the value added product line from RAS, improving profitability, marketability and sustainability.

In concluding, Timmons notes, as do other authors in this volume, a need for more effort into marketing, particularly for RAS-produced product as clean and pollution-free, environmentally friendly food stuffs. Timmons agrees with the findings of Roheim in Chapter 18, that the bottom line to consumers, at least in the USA and probably also in Europe and other more affluent regions, is value. Consumers might be drawn to 'environmentally friendly' product, but they will only pay so much extra for that product. Ultimately, as both Timmons and Roheim point out, it is unlikely that profitability can be achieved by looking solely towards charging premium prices for RAS-produced food stuffs. Premium pricing certainly should be a part of the overall strategy of improving market share, but it should not be relied upon as the primary factor.

In Chapter 10, Zohar *et al.* investigate using RAS technologies for the production of marine species, and detail the benefits, opportunities and current limitations of the technology. All the benefits and opportunities posited by Timmons in Chapter 9 extend to marine-oriented RAS, as do many if not most of the limitations and concerns. Zohar *et al.*, however, note that a major opportunity for marine-oriented RAS is the production of non-native species, at least in the USA and other nations where there are regulatory restrictions on the rearing of non-native species in ambient environments. The authors clearly set out that in order for marine-oriented RAS to be economically competitive, it cannot do so in direct competition against mass-produced pond or net pen-reared species, such as salmon or hybrid striped bass. The opportunity, according to Zohar *et al.*, lies in avoiding competition and raising high-value non-native species, for instance gilthead sea bream.

In a test case growing sea bream, the authors found that salt to make brine was a major barrier to cost effectiveness, and hence profitability and ability to produce a competitively priced product. They complete their test case by describing new research they are conducting that presents possibilities to reduce those costs. The authors present a very plausible methodology for the development of a marine-oriented RAS that can fit nicely into an urban setting, and that can fit well into urban markets for marine aquatic species. The technology presented, particularly in light of other ideas for RAS culture presented by Schreibman and Zarnoch in Chapter 14, provides a useful model for moving urban aquaculture forward.

In Chapter 11, Browdy and Moss take an in-depth look at the possibilities for pond-based, raceway shrimp production. The authors show that technologically, the intensive rearing of shrimp is feasible, but that economic feasibility does not necessarily follow. This finding mimics those of other authors of this volume. Browdy and Moss show that the advantages of such a system are many, including a relatively small footprint, making it more economical, but mainly in that their proposed layout uses considerably less water, making it site independent. This feature in particular makes intensive raceway-based shrimp culture possible in urban settings closer to established markets, as well as being applicable to water-scarce regions. The authors note that total recycling and reuse of culture water, which makes this form of aquaculture very sustainable, does not appear to limit or hamper shrimp growth, provided certain methodologies are employed, such as removal of solids and increased aeration. Despite this, the authors note certain limitations to economic viability and sustainability, as we heard in previous, and again in later chapters, that viability is marginal and more work needs to be done to get beyond this threshold and into an area where prospective new ventures (and expanding established ones scaling up) experience reduced economic risk.

In concluding, Bowdy and Moss state, as did Timmons for tilapia, that selective breeding is a key factor in crossing the threshold to sustainable production of shrimp in an intensive raceway system. Selection for maximal growth in high-density conditions, and selection for disease resistance, must be accomplished to increase profitability. Without this, hypothetical scale-up from their very successful pilots do not cross the threshold into economic sustainability. The authors also note that even when the threshold is crossed, the product will still be more expensive than frozen imported shrimp product, and therefore effort, as noted earlier by Lutz, and others, must be expended in creating a market that will bear a premium price for the cultured product.

In Chapter 12, Blake considers the use of restored/recovering urban ecosystems for the culture of shellfish in sub-tropical waters, using the bay scallop as a model. Growth and survival is good, despite the fact that natural populations of scallop have not returned to waters that they once inhabited. The author found that biofouling was a major detriment to success, particularly during months when water temperatures and organism growth were maximal. The author notes significantly reduced growth of scallops once biofouling began, and which intensified to the point where economic viability was degraded by having growth rates reduced to the point where a harvestable product could not be attained over the typical lifespan of the cultured organism. Blake states that much more practical research needs to be directed at the reduction and inhibition of biofouling.

A second major downfall described by Blake, and which may be unique to the US regulatory system under which the author works, is that even if successfully cultured, the harvested scallops could not be marketed in the one sector where it would have a competitive chance for success. The author clearly notes that cultured scallops, due to costs, will never be able to compete in the adductor muscle only market and would have to be marketed as a 'niche' product for whole animal consumption. By US regulations, however, any shellfish, scallops included, not grown in certified waters (e.g. of very high quality), cannot be marketed for direct consumption without being put through a depuration process. In either case, the added time and expense moves the cost to a point where consumers will not be apt to buy the product. These types of regulations, while existing in the USA, may also exist in other places where concerns over the risk to public health from the consumption of contaminated shellfish product is high. The bottom line is that while the culture of scallops in recovered, recovering and/or rehabilitated coastal urban environments is technologically feasible, assuming biofouling can be reduced and controlled, a competitive market may not exist. This is certainly a major limitation to 'field grown' product, particularly in light of regulatory control as regards public health increases over time. The solution may indeed be as iterated previously by Timmons, Zohar *et al.*, and later by Schreibman and Zarnoch, in their respective chapters – RAS is the way of the future for aquaculture conducted in urban ecosystems.

Goudey and Moran, in Chapter 13, describe their findings from employing aquaculture at a heavily industrialized site. Located along the shoreline of Boston Harbor, the authors developed a test site on an estuary with an historical background of heavy industry, which included significant and long-term municipal and industrial discharges. Despite this history, water quality in Boston Harbor has improved considerably in the past decades. Along with improved water quality, there is an abundance of waterfront sites suitable to adaptation for aquaculture purposes, and there is excellent proximity to the region's largest seafood market, also located on Boston Harbor.

While the site appears to have good credentials for the development of an urban aquaculture venture, the experiences of the authors suggest caution in developing aquaculture in heavily industrialized sites. The systems they put in place, which were designed to mimic commercial operations, experienced continuing problems with disease and mortality, nearly all of which could be accredited to conditions and/or circumstances in the industrial harbour. Heavily contaminated sediments still yield heavy metals concentrations in the water that exceed desirable levels, and which could account for at least some of the problems encountered. Various diseases, associated with ambient harbour water quality, also accounted for significant mortality in their operations. Episodic events – an oil spill and an unusually heavy rainfall event – both provided for significant mortality events in the system operated at the Boston Harbor site.

Goudey and Moran conclude by noting that while they experienced significant losses of product, they suggest that the addition of carbon filtration units, and high exposure UV to treat incoming water, might improve operations by significantly lowering mortality. The authors also note that attempting to use heavily industrialized sites as growout for adults rather than as a hatchery system also might improve results. Despite the authors ending on somewhat of a positive note, it is clear that the development of aquaculture ventures in areas that have been, or are, heavily industrialized, needs to be approached with caution.

Schreibman and Zarnoch, in Chapter 14, explore opportunities for urban aquaculture in New York City, one of the largest metropolitan areas in existence today. New York City, like many urban areas worldwide, possesses a mix of affluent and poor populations, and a very diverse array of ethnic communities. The authors note that each of these communities has various needs and desires for fish and seafood products, some of them very specific. Opportunity abounds for niche marketing of aquaculture products, and the authors note that these needs are not currently being filled, a finding which is most likely applicable to many urban cores worldwide. The authors also note great opportunity for niche marketing of high end, value-added aquaculture product to the vast array of restaurants that exist in all major urban centres.

Schreibman and Zarnoch state that RAS will be the most likely way to proceed in developing aquaculture in urban areas which are typically wastewater rich and potable water starved. RAS has great opportunity to utilize treated wastewater, thus providing a public service to municipalities they inhabit. Furthermore, aquaculture production can mean jobs which typically are sorely needed in urban areas. Schreibman and Zarnoch estimate that for every 45,000 t of product (from RAS tilapia culture) there would be 15 jobs created. With a significant push in the production of consumable fish products from RAS in urban regions, a significant job market could be created.

Outside of production of fish for consumption, Schreibman and Zarnoch describe several other promising areas for future urban aquaculture development. Ornamental species, particularly marine species, hold great promise as they command high prices, are in great demand and wild stocks are either dwindling or becoming less accessible over time due to restrictions on their harvest. The same holds true for species that are used for research and experimentation – laboratories and researchers need a steady, reliable supply of research organisms for the projects, and RAS culture could readily provide for this market, with the added benefit of being able to select for desirable traits and provide a 'pedigree' for laboratory stocks.

In closing, the authors, like others before them, note a continuing need for the engagement of universities in the future development of urban aquaculture. Limitations to the success of urban ventures can be researched at universities, and findings applied via traditional extension activities. For instance, a major current limitation to the culture of bivalve species in RAS systems is the ability to grow enough food to provide for the rapid growth of these organisms. The benefits of RAS-cultured shellfish cannot be achieved as yet due to this limitation, but directed research at the university level could develop solutions.

In Chapter 15, Roy provides an example of the successful integration of aquaculture science in a high school curriculum setting using ambient, on-site water resources. The author notes that the development of the high school vocational programme was largely dependent upon the assistance of various universities, as well as Sea Grant Programs located both in Connecticut and neighbouring Rhode Island, which provided the intellectual horsepower and guidance to initiate a hands-on, secondary education aquaculture endeavour. Roy also notes that the aquaculture programme was developed based upon an historical model successfully used in agriculture (e.g. Future Farmers of America), and that using the FFA parent organization as a larger umbrella for the initiative provided greater credence and acceptability at its onset, as well as a greater level of comfort to school systems considering adoption of the aquaculture curriculum model.

In moving towards a sustainable future for aquaculture in urban areas, development of a high school curriculum is critical according to Roy, as it is during this period that students begin exploring various career paths and options, whether for a professional career, or for planning a move into post-secondary education. Students not exposed to aquaculture during that period of career exploration do not consider it in the array of options open to them.

Roy brings out an interesting point regarding the blending of practical, applied science and more academic concepts and theories – by developing programmes that revolve around hatchery systems, first for shellfish and more recently for lobster, students track the life histories of organisms from start to finish, learning the basic biology and ecology of the organisms and the ecosystems in which they live. They also do this from a practical perspective, as they are applying ecosystem concepts to the rearing and harvesting of a crop. The students moving through this curriculum more fluently understand how aquaculture fits into urban ecosystems, which will be key to integrating aquaculture as a part of sustainable future urban ecosystems.

In Chapter 16, Frederick brings out another facet of integrating aquaculture into high school curricula, this time doing so using RAS. The author provides useful detail for those considering the development of an aquaculture curriculum, giving descriptions of multiple justifications and benefits. The author's focus on RAS makes it highly transportable to other areas, not just directly along the coast.

Frederick takes the approach of using aquaculture as a mechanism for restoration of natural ecosystems, rather than towards commercial application. Use of RAS also brings into play, as noted by the author, the opportunity to do comparative work between natural systems and the developed RAS, providing many opportunities to better exemplify the complexities of sustaining an ecosystem, and the even more horrendous complexities of creating them artificially. This style of education has the potential to show students, in a much more blatant fashion, the need for developing urban cores that function in a sustainable fashion. Using RAS technologies as a teaching tool in this regard can be a formidable benefit.

In concluding, Frederick lays out the need for a cadre of trained teachers who then can train other teachers and begin to form the core of a broad network of educators that teach science using aquaculture as a teaching tool. The Sea Grant Program, as it did for Roy, played a significant role in Maryland in developing the resource base for the aquaculture education initiative. Sea Grant facilitated the coming together of scientists and educators, and then provided the resources for follow-up support once the teachers completed their training course. Similar programmes should be sought out where Sea Grant is not present, or a Sea Grant-like model could be applied in international programmes to provide a similar role as facilitator to aquaculture education initiatives in developing nations.

In Chapter 17, Buttner expands upon the education model, showing that aquaculture is more than just growing fish and/or shellfish for food – it includes the rearing of many different kinds of organisms for many different purposes. The author suggests the opportunities for the application of aquaculture, particularly in urban areas, be articulated more vocally so that its initiation can be considered for more than just food production purposes. Indeed, it may be easier for urban managers and planners to more readily comprehend and accept the application of aquaculture for biotech purposes, for instance, than for food.

Buttner expounds upon the need to link people to their ecosystems, particularly in urban areas where the link to natural systems is distant, and often completely forgotten. Aquaculture in urban settings can begin to re-establish linkages to ecosystems, and can begin to illuminate ecosystem functions and functioning. The author notes the need for people to be connected to the urban ecosystem, and that they must understand that despite the ecosystem being urban, there are still material and energy flows that correspond to, and rely upon, natural ecosystem functions. Urban aquaculture can be a positive tool in forming and emphasizing those links.

For that to happen, however, Buttner suggests that the doors to urban aquaculture facilities be 'flung open' for public viewing. Urban schools should be encouraged to visit, and perhaps even participate in various aspects of operations. Teachers should be encouraged to earn 'experiential credits' for engaging with aquaculture operations, particularly if they link it back to classroom learning. Buttner suggests hosting public events at aquaculture facilities, and doing almost anything to make aquaculture in urban environments more visible. By being visible, it will become tangible and real. By becoming real, connections between urban ecosystem functioning and those of the natural world will begin to be built, resulting in a more sustainable future.

Roheim, in Chapter 18, brings the dialogue full circle into the realities of aquaculture from the marketing perspective. The author, like several previous authors, notes that it is easy to get caught up in, and focus on, the 'fun' technology end of aquaculture, and very easy to ignore or forget the 'boring' aspects of marketing. Roheim is quick to note that those endeavours that ignore the market generally enter it with a too high-priced product, and are shortly out of business.

Aquaculture, particularly when it is conducted in an urban setting, has the potential to be sustainable. Roheim suggests that in order to achieve sustainability, urban aquaculture must distinguish itself from similar products, much as US raised catfish is attempting to do in the face of tremendous pressure from a glut of inexpensive Vietnamese export products. The author suggests that one way to achieve this is through ecolabelling, through which a product is distinguished as being sustainably produced, organically produced or otherwise produced in a way that makes it superior to similar products, and thereby commanding a premium price.

The author's research, however, shows that there is a limit to how much a consumer will spend for ecolabelled products. Generally, cost becomes a guiding factor, particularly as the margin between sustainably produced and more generic products widens – consumers are willing to pay a premium, but only within limits.

Looking back across the suite of chapters presented, implementing cost-effective RAS in urban ecosystems, combined with clever and strategic ecolabelling, and a very public, informative outreach effort, could create the synergism needed to make aquaculture in urban regions highly profitable, particularly in more affluent, highly developed nations. In those nations that are already working with wastewater-fed aquaculture systems, a different approach will need to be taken, though not broadly different. The same parallels and links between aquaculture and the roles it plays in urban ecosystem functioning, as well as its broad-ranging societal benefits, must be the hub of sustainability over the long term. In both cases, emphasis should be placed on the potential for job creation, particularly with its benefits to the urban poor for both jobs and a source of protein.

In conclusion, urban aquaculture presents unlimited opportunities and potential in the development of future sustainable urban ecosystems. More research needs to occur to better determine impacts, if any, to human health resulting from wastewater reuse in urban aquaculture, and more research needs to focus on consumer desires and perceptions, with marketing moving in tandem with research so that urban-cultured products are not only found to be acceptable, but desirable by society at large. Efforts in education and outreach will also need to occur in a strategic and consistent fashion for the latter to occur, as well as to groom a next generation of aquaculturists that can better conceptualize 'urban aquatic farming' opportunities. And finally, urban planning needs to shift to a new paradigm – one that embraces urban aquaculture as a vital, functional and necessary component of urban ecosystems for waste processing, water resource recycling, job creation, poverty alleviation and food production. Urban aquaculture must begin to be viewed by urban planners and managers, and by all cultures and all societies, as an integrator of ecological and economic principles that create sustainable urban ecosystems.

Index

Printed and bound by CPI Group (UK) Ltd, Croydon, CR0 4YY
16/09/2026
14963436-0002